Artificial Intelligence for Smart and Sustainable Energy Systems and Applications

Artificial Intelligence for Smart and Sustainable Energy Systems and Applications

Special Issue Editors

Miltiadis D. Lytras
Kwok Tai Chui

MDPI • Basel • Beijing • Wuhan • Barcelona • Belgrade

Special Issue Editors
Miltiadis D. Lytras
Deree College—The American College of Greece
Greece
Effat University
Saudi Arabia

Kwok Tai Chui
The Open University of Hong Kong
Hong Kong

Editorial Office
MDPI
St. Alban-Anlage 66
4052 Basel, Switzerland

This is a reprint of articles from the Special Issue published online in the open access journal *Energies* (ISSN 1996-1073) from 2018 to 2019 (available at: https://www.mdpi.com/journal/energies/special_issues/Artificial_Intelligence_Smart_Sustainable_Energy_Systems_Applications).

For citation purposes, cite each article independently as indicated on the article page online and as indicated below:

LastName, A.A.; LastName, B.B.; LastName, C.C. Article Title. *Journal Name* **Year**, *Article Number*, Page Range.

ISBN 978-3-03928-889-2 (Pbk)
ISBN 978-3-03928-890-8 (PDF)

Contents

About the Special Issue Editors

Miltiadis D. Lytras Ph.D., is an expert in advanced computer science and management, and an editor, lecturer, and research consultant, with extensive experience in academia and the business sector in Europe and Asia. Prof. Lytras is a Research Professor at Deree College—The American College of Greece—and a Distinguished Scientist at the King Abdulaziz University, Jeddah, Kingdom of Saudi Arabia. Prof. Lytras is a world-class expert in the fields of cognitive computing, information systems, technology-enabled innovation, social networks, computers in human behavior, and knowledge management. In his work, Prof. Lytras seeks to bring together and exploit synergies among scholars and experts. He is committed to enhancing the quality of education for all. He has co-edited more than 80 special issues in international journals and is currently the Editor-in-Chief of the *International Journal on Semantic Web and Information Systems*.

Kwok Tai Chui Ph.D., received a B.Eng. in electronic and communication engineering, a business intelligence minor, and a Ph.D. from the City University of Hong Kong. He has industry experience as Senior Data Scientist for the Internet of Things (IoT) company. He joined the Department of Technology, School of Science and Technology, of the Open University of Hong Kong as a Research Assistant Professor. He was the recipient of the 2nd Prize Award (Postgraduate Category) of the 2014 IEEE (International Conference on Consumer Electronics) Region 10 Student Paper Contest. He also received the Best Paper Award from the IEEE, in China, in both 2014 and 2015. He has served in various editorial positions in SCI-listed journals, including as Managing Editor of the *International Journal on Semantic Web and Information Systems* and as the Topic Editor of *Sensors*.

Preface to "Artificial Intelligence for Smart and Sustainable Energy Systems and Applications"

The world has begun to focus on migrating energy systems and applications in a smart and sustainable manner. Environmental experts warn that the average temperature increase of the sea by 2 degree Celsius may lead to irreversible effects on ice melting and global warming. Government officials may face economic growth and carbon reduction dilemmas. Whilst continuing to utilize fossil fuel as the main energy source, artificial intelligence was introduced to improve existing energy systems and applications. Oweing to the advancement of big data infrastructure and high-performance computing tools, data collection via energy network sensing devices provides ground truth and valuable information for further processing and analysis.

This edition aims to share the latest research and results on artificial intelligence in energy systems and applications for smart and sustainable systems. Special attention is drawn to non-intrusive load monitoring, electric grids, wireless sensor networks, insulator detection, rheological property prediction and static Young's modulus estimation.

Many countries have already replaced traditional electric grids with smart grids by installing smart meters. In general, advanced metering infrastructure is set up to support two-way communication between end users and smart meters. With the total electricity consumption of rooms and buildings, the benefits of smart metering are limited. Therefore, the research problem of non-intrusive load monitoring (also named electricity disaggregation) from total electricity consumption to the electricity breakdown of individual appliances has been a key research direction for smart metering. Recently, granular, or high frequency electricity, data, and deep learning have become feasible approaches to escalate the performance of electricity disaggregation.

Attention has been drawn to the application of artificial intelligence techniques for electric grid and wireless sensor networks. Electric grid models for the prediction of peak-demand and tariff have been proposed. Wireless sensor networks are often linked with traditional wired networks to allow for mobility and quick installation. However, energy efficient schemes should be deployed to increase the lifespan of wireless sensors.

Artificial intelligence has been applied and its effectiveness evaluated as a way to improve the existing software, hardware, algorithms, tools, systems, and applications. The results may drive new insights that have not been realized by human beings. This collection has gathered works on insulator detection in inspection imaging, rheological property prediction for $CaCl_2$ brine-based drill-in fluid, and static Young's modulus estimation for sandstone formation. In the last decade, we have witnessed the success and effectiveness of applying artificial intelligence to energy systems and applications. We expect that energy research will continue to grow worldwide and will contribute to the vision of a smart city. In particular, edge computing, fog computing, and cloud computing will mature and the cost will decrease. Thus, the research and industry sectors may begin sharing computational work on edge, fog, and cloud devices. This is the key to carrying out data analysis globally and locally.

We would like to thank the professional staff at MDPI for their qualitative work and valuable support as well as all the contributors and reviewers that made this edition possible. We invite you to contribute your research to our next edition, Towards Optimization: Energy Monitoring, Modelling and Management, to which both reviews and technical articles are welcome.

Miltiadis D. Lytras, Kwok Tai Chui
Special Issue Editors

Editorial

The Recent Development of Artificial Intelligence for Smart and Sustainable Energy Systems and Applications

Miltiadis D. Lytras [1,2] and Kwok Tai Chui [3,*]

1 School of Business, Deree College—The American College of Greece, 153-42 Athens, Greece
2 Effat College of Engineering, Effat University, P.O. Box 34689, Jeddah 21478, Saudi Arabia
3 Department of Technology, School of Science and Technology, The Open University of Hong Kong, Ho Man Tin, Kowloon, Hong Kong, China
* Correspondence: jktchui@ouhk.edu.hk; Tel.: +852-2768-6883

Received: 5 July 2019; Accepted: 7 August 2019; Published: 13 August 2019

Abstract: Human beings share the same community in which the usage of energy by fossil fuels leads to deterioration in the environment, typically global warming. When the temperature rises to the critical point and triggers the continual melting of permafrost, it can wreak havoc on the life of animals and humans. Solutions could include optimizing existing devices, systems, and platforms, as well as utilizing green energy as a replacement of non-renewable energy. In this special issue "Artificial Intelligence for Smart and Sustainable Energy Systems and Applications", eleven (11) papers, including one review article, have been published as examples of recent developments. Guest editors also highlight other hot topics beyond the coverage of the published articles.

Keywords: artificial intelligence; computational intelligence; energy management; machine learning; optimization algorithms; sensor network; smart city; smart grid; sustainable development

---

1. Introduction

The world mission is not only improving energy systems to become smart but also progressing sustainable development. It can be further divided into environmental [1], economic [2], and socio-cultural [3] perspectives. However, there does not exist a perfect energy source as the solution for sustainable energy. It requires multidisciplinary techniques and systems to achieve the targets.

The rapid advanced development of computer science and engineering enables the implementation and adoption of artificial intelligence (AI) into various energy systems and applications. This special issue aims at consolidating recent developments of AI for smart and sustainable energy systems and applications. Pilot studies are especially welcome. Topics of interest for the special issue include (but are not limited to):

- New theories and applications of machine learning algorithms in smart grid;
- Design, development, and application of deep learning in smart grid;
- Artificial intelligence in advanced metering infrastructure;
- Multiobjective optimization algorithms in smart grid;
- Disaggregation techniques in non-intrusive load monitoring;
- Modelling and simulation (or co-simulation) in smart grid;
- Internet of Things and smart grid;
- Data driven analytics (descriptive, diagnostic, predictive, and prescriptive) in smart grid;
- Artificial intelligence techniques for security;
- Fraud detection and predictive maintenance;

- Demand response in smart grid;
- Peak load management approach in smart grid;
- Interoperability in smart grid;
- Cloud computing based smart grid;
- Vehicle-to-grid design, development, and application.

This Editorial is organized as follows. Section 2 summarizes the published articles in this special issue. In addition, editors discuss several hot topics beyond the coverage of the special issue articles in Section 3. Finally, the conclusion is drawn in Section 4.

2. Special Issue Articles

This section not only discusses each of the published articles in the main text but also summarizes these in the form of a table (Table 1) as a quick review. It is worth mentioning that the topic of non-intrusive load monitoring (NILM) has contributed five published papers.

Table 1. Summary of the application and methodology of the special issue articles.

Work	Application	Methodology
[4] #	Home energy management and ambient assisted living	Non-intrusive load monitoring techniques
[5]	Non-intrusive load monitoring for energy disaggregation	Genetic algorithm; support vector machine; multiple kernel learning
[6]	Optimizing residential energy consumption	Bacterial foraging optimization; flower pollination
[7]	Non-intrusive load monitoring for energy disaggregation	Long short-time memory and decision tree
[8]	Energy efficient coverage in wireless sensor network	Distributed genetic algorithm
[9]	Estimation of load and price of electric grid	Enhanced logistic regression; enhanced recurrent extreme learning machine; classification and regression tree; relief-F and recursive feature elimination
[10]	Detection of the insulators in power transmission and transformation inspection images	Improved faster region-convolutional neural network
[11]	Non-intrusive load monitoring for energy disaggregation	Concatenate convolutional neural network
[12]	Non-intrusive load monitoring for energy disaggregation	Linear-chain conditional random fields
[13]	Prediction of the rheological properties of calcium chloride brine-based mud	Artificial neural network
[14]	Estimation of Static Young's Modulus for sandstone formation	Artificial neural network; self-adaptive differential evolution

Review article.

The review article "NILM techniques for intelligent home energy management and ambient assisted living: A review" authored by A. Ruano, A. Hernandez, J. Ureña, M. Ruano, and J. Garcia provided a constructive review on home energy management and ambient assisted living [4]. The focus was on NILM, which aimed at producing a breakdown of the energy profile in equipment level based on the total energy profile of the apartment. Here, the energy profile could be information related to current, voltage, power, energy, power factor, harmonic distortion, etc. This review article divided NILM techniques into four parts: data collection, event detection, feature extraction, and load identification. The authors highlighted that NILM was a highly scalable but less accurate homecare monitoring system for ambient assisted living, compared to direct and indirect methods via biosensors and sensors for activity monitoring.

There are ten technical papers that proposed various AI techniques for energy systems and applications, towards the goal of smart and sustainable development. The first article "Energy sustainability in smart cities: artificial intelligence, smart monitoring, and optimization of energy consumption" [5] written by K. T. Chui, M. D. Lytras, and A. Visvizi formulated the NILM algorithm as a multiobjective optimization problem. Multiple kernel learning was introduced to the support vector machine classifier to enhance the classification accuracy by integrating valuable characteristics of kernels. Weighting factors between kernels were added and solved by a genetic algorithm to optimize the performance.

M. Awais, N. Javaid, K. Aurangzeb, S. Haider, Z. Khan, and D. Mahmood published an extended article "Towards effective and efficient energy management of single home and a smart community exploiting heuristic optimization algorithms with critical peak and real-time pricing tariffs in smart grids" [6] from the conference paper in the 2018 IEEE 32nd International Conference on Advanced Information Networking and Applications (AINA) [15]. The residential energy consumption problem was considered as a heuristic multiobjective optimization problem. The bacterial foraging optimization algorithm and flower pollination algorithm were utilized to minimize the electricity cost and peak-to-average ratio while maximizing the user comfort. The trade-off solution was recommended attributable to conflicting objectives.

In [7], T. T. H. Le and H. Kim presented an article "Non-intrusive load monitoring based on novel transient signal in household appliances with low sampling rate". Differing from [4], the authors divided the NILM framework into three parts: data acquisition, feature extraction, and classification model. The feature vector was constructed by transient signal. Long short-term memory and decision tree models were applied for the energy disaggregation problem. The performance evaluation of the proposed method was tested through five appliances: Samsung monitor, LG monitor, hairdryer, fan, and air purifier, where the accuracy was up to 97%.

Z. J. Wang, Z. H. Zhan, and J. Zhang in their article "Solving the energy efficient coverage problem in wireless sensor networks: A distributed genetic algorithm approach with hierarchical fitness evaluation" [8] proposed a distributed genetic algorithm to address the issue of energy efficient coverage (EEC) in wireless sensor networks. The authors evaluated the fitness using a hierarchical approach and constructed a two-level fitness function to determine the number of disjoint sets and its coverage performance. It was demonstrated to be effective in the maximization of the number of disjoint sets.

Predicting the load and price of the electric grid have been important tasks for utility in lowering the electricity cost and improving the service quality for end customers. A. Naz, M. U. Javed, N. Javaid, T. Saba, M. Alhussein, and K. Aurangzeb have addressed these topics via an article "Short-term electric load and price forecasting using enhanced extreme learning machine optimization in smart grids" [9]. Two-feature extraction approaches were adopted, namely classification and regression tree, relief-F, and recursive feature elimination. The extracted features were passed to enhanced logistic regression and enhanced recurrent extreme learning machines for load and price estimation. Results showed that the proposed two algorithms outperformed existing methods by 5%.

Z. Zhao, Z. Zhen, L. Zhang, Y. Qi, Y. Kong, and K. Zhang have published an article "Insulator detection method in inspection image based on improved faster R-CNN" [10]. This paper proposed an improved faster region-convolutional neural network to detect the insulators in power transmission and transformation inspection images. It yielded a precision of 81.8% with an enhancement of 28%. It was noted by the authors that the proposed method was customized to the specific type of insulator. For other insulators, it required further fine-tuning to achieve an optimal performance.

The article "Concatenate convolutional neural networks for non-intrusive load monitoring across complex background" was presented by Q. Wu and F. Wang [11]. The concatenate convolutional neural network was newly proposed as the NILM technique, which was evaluated based on key performance indicators: accuracy, robustness, and generalization of load recognition. A key observation was concluded regarding the background load, which is almost stationary in a given short period of time. This method improved the F1-score, precision, and recall by 12%, 19%, and 4% respectively.

In [12], H. He, Z. Liu, R. Jiao, and G. Yan proposed a novel algorithm named linear-chain conditional random fields for energy disaggregation in the article "A novel nonintrusive load monitoring approach based on linear-chain conditional random fields". This approach has eliminated some obstacles for performance improvement: (i) they relax the independent assumption of the hidden Markov model, and (ii) current and real power are included as representative features. As a result, it achieved an outstanding accuracy of 96–100%.

A. Gowida, S. Elkatatny, E. Ramadan, and A. Abdulraheem wrote an article "Data-driven framework to predict the rheological properties of CaCl2 brine-based drill-in fluid using artificial neural network" [13]. Artificial neural network was adopted to forecast the rheological properties of brine-based drill-in fluid so that it could avoid the loss of circulation, pipe sticking, and hole cleaning. The correlation coefficient and average absolute percentage error were 0.97 and <6.1%, respectively.

Finally, A. A. Mahmoud, S. Elkatatny, A. Ali, and T. Moussa contributed an article "Estimation of static Young's modulus for sandstone formation using artificial neural networks" [14]. The authors introduced self-adaptive differential evolution on top of artificial neural network to further enhance the performance of the estimation of static Young's modulus for sandstone formation. This approach reduced the average absolute percentage error significantly from 36% to 1%, as well as the perfect correlation coefficient.

3. Trends and Future Development

Besides the applications in Table 1, there are numerous applications that apply AI for smart and sustainable energy systems. The guest editors would like to summarize the key topics—renewable energy, cloud platform, edge computing, fog computing, as well as electric and plug-in hybrid electric vehicles—in this section. Also, we have attached a few related works as recommended readings in each field.

Renewable energy (specifically energy harvesting), acting as an alternative of fossil fuels, is one of the directions to fight against global warming. Typical sources are solar [16,17] and wind [18,19]. Focuses are basically on the reliability and efficiency of energy harvesting. Other emergent topics include vibration energy [20], water wave energy [21], acoustic energy [22], and waste-to-energy [23].

The cloud platform is not unfamiliar in today's era, as many smartphone users have linked their personal information to it. A more advanced technique of the cloud platform possesses not only big data storage but also massive computation power, which allows complex data analytics for smart grid data streaming, processing, analyzing, and storage. Readers are encouraged to read the following articles [24–27].

Edge computing and fog computing are alternatives to cloud computing that offer relatively local (lower latency) computation. In other words, data processing and loading can be distributed to edge, fog, and cloud services. For edge computing, it is recommended for readers to read [28,29], and the readings for fog computing are [30,31].

There is an increasing portion of new buyers purchasing electric or plug-in hybrid electric vehicles, which can reduce the usage of fuel. Various AI techniques have been applied to plug-in hybrid electric vehicles, for instance, artificial neural network [32] and integrated model predictive controller [33]. When it comes to electric vehicles, biased coupling, torque estimation, and cognitive heuristic techniques were adopted in [34] and deep neural networks in [35].

4. Conclusions

This special issue is composed of eleven papers (one review article) with various topics and methodologies in AI for smart and sustainable energy systems and applications. Contributors have shared many valuable insights on recent developments and beyond. The guest editors have briefly summarized the details of each work, as well as highlighted four groups of emergent topics in the energy industry. The guest editors would like to thank the contributions of all colleagues and reviewers. We hope to witness a lot of real implementation and adoption of AI techniques in the energy industry in the near future.

Author Contributions: K.T.C. and M.D.L. contributed equally to the design, implementation, and the delivery of the special issue. All co-editors contributed equally in all the phases of this intellectual outcome.

Funding: The authors would like to thank Effat University in Jeddah, Saudi Arabia, for funding the research reported in this paper through the Research and Consultancy Institute.

Conflicts of Interest: The authors declare no conflict of interest.

References

1. Cuce, E.; Harjunowibowo, D.; Cuce, P.M. Renewable and sustainable energy saving strategies for greenhouse systems: A comprehensive review. *Renew. Sustain. Energy Rev.* **2016**, *64*, 34–59. [CrossRef]
2. Chen, G.Q.; Wu, X.F. Energy overview for globalized world economy: Source, supply chain and sink. *Renew. Sustain. Energy Rev.* **2017**, *69*, 735–749. [CrossRef]
3. Radovanović, M.; Filipović, S.; Pavlović, D. Energy security measurement—A sustainable approach. *Renew. Sustain. Energy Rev.* **2017**, *68*, 1020–1032. [CrossRef]
4. Ruano, A.; Hernandez, A.; Ureña, J.; Ruano, M.; Garcia, J. NILM Techniques for intelligent home energy management and ambient assisted living: A review. *Energies* **2019**, *12*, 2203. [CrossRef]
5. Chui, K.T.; Lytras, M.D.; Visvizi, A. Energy sustainability in smart cities: Artificial intelligence, smart monitoring, and optimization of energy consumption. *Energies* **2018**, *11*, 2869. [CrossRef]
6. Awais, M.; Javaid, N.; Aurangzeb, K.; Haider, S.; Khan, Z.; Mahmood, D. Towards effective and efficient energy management of single home and a smart community exploiting heuristic optimization algorithms with critical peak and real-time pricing tariffs in smart grids. *Energies* **2018**, *11*, 3125. [CrossRef]
7. Le, T.T.H.; Kim, H. Non-intrusive load monitoring based on novel transient signal in household appliances with low sampling rate. *Energies* **2018**, *11*, 3409. [CrossRef]
8. Wang, Z.J.; Zhan, Z.H.; Zhang, J. Solving the energy efficient coverage problem in wireless sensor networks: A distributed genetic algorithm approach with hierarchical fitness evaluation. *Energies* **2018**, *11*, 3526. [CrossRef]
9. Naz, A.; Javed, M.U.; Javaid, N.; Saba, T.; Alhussein, M.; Aurangzeb, K. Short-term electric load and price forecasting using enhanced extreme learning machine optimization in smart grids. *Energies* **2019**, *12*, 866. [CrossRef]
10. Zhao, Z.; Zhen, Z.; Zhang, L.; Qi, Y.; Kong, Y.; Zhang, K. Insulator detection method in inspection image based on improved faster R-CNN. *Energies* **2019**, *12*, 1204. [CrossRef]
11. Wu, Q.; Wang, F. Concatenate convolutional neural networks for non-intrusive load monitoring across complex background. *Energies* **2019**, *12*, 1572. [CrossRef]
12. He, H.; Liu, Z.; Jiao, R.; Yan, G. A novel nonintrusive load monitoring approach based on linear-chain conditional random fields. *Energies* **2019**, *12*, 1797. [CrossRef]
13. Gowida, A.; Elkatatny, S.; Ramadan, E.; Abdulraheem, A. Data-driven framework to predict the rheological properties of CaCl2 brine-based drill-in fluid using artificial neural network. *Energies* **2019**, *12*, 1880. [CrossRef]
14. Mahmoud, A.A.; Elkatatny, S.; Ali, A.; Moussa, T. Estimation of static young's modulus for sandstone formation using artificial neural networks. *Energies* **2019**, *12*, 2125. [CrossRef]

15. Awais, M.; Javaid, N.; Mateen, A.; Khan, N.; Mohiuddin, A.; Rehman, M.H.A. Meta heuristic and nature inspired hybrid approach for home energy management using flower pollination algorithm and bacterial foraging optimization technique. In Proceedings of the 2018 IEEE 32nd International Conference on Advanced Information Networking and Applications (AINA), Krakow, Poland, 16–18 May 2018; pp. 882–891.
16. Zhang, J.; Lou, M.; Xiang, L.; Hu, L. Power cognition: Enabling intelligent energy harvesting and resource allocation for solar-powered UAVs. *Future Gener. Comput. Syst.* **2019**. [CrossRef]
17. Chander, A.H.; Kumar, L. MIC for reliable and efficient harvesting of solar energy. *IET Power Electron.* **2018**, *12*, 267–275. [CrossRef]
18. Mitiku, T.; Manshahia, M.S. Modeling of wind energy harvesting system: A systematic review. *Int. J. Eng. Sci. Math.* **2018**, *7*, 444–467.
19. Mahmoud, T.; Dong, Z.Y.; Ma, J. An advanced approach for optimal wind power generation prediction intervals by using self-adaptive evolutionary extreme learning machine. *Renew. Energy* **2018**, *126*, 254–269. [CrossRef]
20. Nabavi, S.; Zhang, L. Design and optimization of a low-resonant-frequency piezoelectric MEMS energy harvester based on artificial intelligence. *Proceedings* **2018**, *2*, 930. [CrossRef]
21. Liu, W.; Xu, L.; Bu, T.; Yang, H.; Liu, G.; Li, W.; Cheng, T. Torus structured triboelectric nanogenerator array for water wave energy harvesting. *Nano Energy* **2019**, *58*, 499–507. [CrossRef]
22. Chen, F.; Wu, Y.; Ding, Z.; Xia, X.; Li, S.; Zheng, H.; Zi, Y. A novel triboelectric nanogenerator based on electrospun polyvinylidene fluoride nanofibers for effective acoustic energy harvesting and self-powered multifunctional sensing. *Nano Energy* **2019**, *56*, 241–251. [CrossRef]
23. Nicoletti, J.; Ning, C.; You, F. Incorporating agricultural waste-to-energy pathways into biomass product and process network through data-driven nonlinear adaptive robust optimization. *Energy* **2019**, *180*, 556–571. [CrossRef]
24. Chekired, D.A.; Khoukhi, L. Smart grid solution for charging and discharging services based on cloud computing scheduling. *IEEE Trans. Ind. Inform.* **2017**, *13*, 3312–3321. [CrossRef]
25. Mehmi, S.; Verma, H.K.; Sangal, A.L. Simulation modeling of cloud computing for smart grid using CloudSim. *J. Electr. Syst. Inform. Technol.* **2017**, *4*, 159–172. [CrossRef]
26. Demir, K.; Ismail, H.; Vateva-Gurova, T.; Suri, N. Securing the cloud-assisted smart grid. *Int. J. Crit. Infrastruct. Prot.* **2018**, *23*, 100–111. [CrossRef]
27. Jegadeesan, S.; Azees, M.; Kumar, P.M.; Manogaran, G.; Chilamkurti, N.; Varatharajan, R.; Hsu, C.H. An efficient anonymous mutual authentication technique for providing secure communication in mobile cloud computing for smart city applications. *Sustain. Cities Soc.* **2019**, *49*, 101522. [CrossRef]
28. Liu, Y.; Yang, C.; Jiang, L.; Xie, S.; Zhang, Y. Intelligent edge computing for IoT-based energy management in smart cities. *IEEE Netw.* **2019**, *33*, 111–117. [CrossRef]
29. Chen, S.; Wen, H.; Wu, J.; Lei, W.; Hou, W.; Liu, W.; Jiang, Y. Internet of things based smart grids supported by intelligent edge computing. *IEEE Access* **2019**, *7*, 74089–74102. [CrossRef]
30. Barros, E.B.C.; Dionísio Machado Filho, L.; Batista, B.G.; Kuehne, B.T.; Peixoto, M.L.M. Fog computing model to orchestrate the consumption and production of energy in microgrids. *Sensors* **2019**, *19*, 2642. [CrossRef]
31. Maatoug, A.; Belalem, G.; Mahmoudi, S. Fog computing framework for location-based energy management in smart buildings. *Multiagent Grid Syst.* **2019**, *15*, 39–56. [CrossRef]
32. Jahangir, H.; Tayarani, H.; Ahmadian, A.; Golkar, M.A.; Miret, J.; Tayarani, M.; Gao, H.O. Charging demand of plug-in electric vehicles: Forecasting travel behavior based on a novel Rough Artificial Neural Network approach. *J. Clean. Prod.* **2019**, *229*, 1029–1044. [CrossRef]
33. Xie, S.; Hu, X.; Liu, T.; Qi, S.; Lang, K.; Li, H. Predictive vehicle-following power management for plug-in hybrid electric vehicles. *Energy* **2019**, *166*, 701–714. [CrossRef]

34. Xiong, H.; Zhang, M.; Zhang, R.; Zhu, X.; Yang, L.; Guo, X.; Cai, B. A new synchronous control method for dual motor electric vehicle based on cognitive-inspired and intelligent interaction. *Future Gener. Comput. Syst.* **2019**, *94*, 536–548. [CrossRef]
35. Huang, H.B.; Wu, J.H.; Huang, X.R.; Yang, M.L.; Ding, W.P. The development of a deep neural network and its application to evaluating the interior sound quality of pure electric vehicles. *Mech. Syst. Signal Process.* **2019**, *120*, 98–116. [CrossRef]

Review

NILM Techniques for Intelligent Home Energy Management and Ambient Assisted Living: A Review

Antonio Ruano [1,2,*], Alvaro Hernandez [3], Jesus Ureña [3], Maria Ruano [1,4] and Juan Garcia [3]

1 Faculty of Science & Technology, University of Algarve, 8005-294 Faro, Portugal; mruano@ualg.pt
2 IDMEC, Instituto Superior Técnico, Universidade de Lisboa, 1049-001 Lisboa, Portugal
3 Department of Electronics, University of Alcalá, 28805 Madrid, Spain; alvaro.hernandez@uah.es (A.H.); jesus.urena@uah.es (J.U.); jjesus.garcia@uah.es (J.G.)
4 CISUC, University of Coimbra, 3030-290 Coimbra, Portugal
* Correspondence: aruano@ualg.pt

Received: 30 March 2019; Accepted: 6 June 2019; Published: 10 June 2019

Abstract: The ongoing deployment of smart meters and different commercial devices has made electricity disaggregation feasible in buildings and households, based on a single measure of the current and, sometimes, of the voltage. Energy disaggregation is intended to separate the total power consumption into specific appliance loads, which can be achieved by applying Non-Intrusive Load Monitoring (NILM) techniques with a minimum invasion of privacy. NILM techniques are becoming more and more widespread in recent years, as a consequence of the interest companies and consumers have in efficient energy consumption and management. This work presents a detailed review of NILM methods, focusing particularly on recent proposals and their applications, particularly in the areas of Home Energy Management Systems (HEMS) and Ambient Assisted Living (AAL), where the ability to determine the on/off status of certain devices can provide key information for making further decisions. As well as complementing previous reviews on the NILM field and providing a discussion of the applications of NILM in HEMS and AAL, this paper provides guidelines for future research in these topics.

Keywords: non-intrusive load monitoring; home energy management systems; ambient assisted living; demand response; machine learning; internet of things; smart grids

---

1. Introduction

Non-Intrusive Load Monitoring (NILM) techniques have become one of the most relevant alternatives for energy disaggregation, since they provide a method to separate the individual consumption for certain appliances, respecting consumers' privacy and often using already-deployed smart meters. The rise of these NILM techniques has also been fostered by the recent importance of some emerging domains, such as Internet of Things (IoT), Smart Grids (SG) or Demand Response (DR) energy programs, where the information provided by NILM can be useful for deciding on further developments or services.

Most applications that use NILM techniques pursue energy efficiency, using itemised energy information to give feedback to tenants, who can consequently take actions to reduce their consumption through "energy awareness". One of the major advantages of the NILM approach is its non-intrusive nature; it is also easily deployed if smart meters are already installed.

On the other hand, with the increasing age of the population and medical advances, there is increasing demand for technology that supports the elderly with leading independent lives. Many digital solutions have been investigated to achieve personalized care, also taking into account other aspects such as acceptance and cost. Among them, NILM not only can provide information about activities within the home, but also has become an emerging alternative to be used in health and care applications. In this case, again, non-intrusiveness is the main and crucial advantage for NILM.

Consequently, as a new contribution and a complement to previous reviews in this field, this work will be focused on Home Energy Management Systems (HEMS) and Ambient Assisted Living (AAL), which are two domains where NILM has clearly contributed to the proposal of new solutions and services, with significant ongoing research, oriented to the achievement of a more efficient energy management, and to the enhancement of AAL systems in response to daily needs of an increasingly ageing population. The review has been conducted to include recent NILM proposals and work using NILM techniques, with a particular emphasis on the requirements that these two types of applications (HEMS and AAL) imply. The analysis includes aspects in the low-level processing (e.g., sampling rate and signal features) as well as in the high-level (e.g., algorithm considered for load identification). Additionally, it deals with the involved data sources and highlights the main contributions of each work.

The rest of the manuscript is organized as follows: NILM techniques are reviewed in detail in Section 2; Section 3 illustrates the application of NILM to Intelligent Home Energy Management; Section 4 deals with the use of NILM in the AAL domain; Section 5 points out current issues and presents guidelines for future research; and, finally, conclusions are drawn in Section 6. A summary of the most important characteristics of the works referenced in this review is presented in the Appendix.

2. NILM Review

A few reviews are already available in the literature about NILM techniques [1–4], which the reader is encouraged to read. This section briefly introduces NILM techniques and presents significant references, focusing on the most recent ones, not covered in previous reviews. For that purpose, the main stages in NILM are:

1. Data collection: electrical data, including current, voltage, and power data, are obtained from smart meters, acquisition boards or by using specific hardware;
2. Event detection: an event is any change in the state of an appliance over time. An event implies variations in power and current, which can be detected in the electrical data previously collected by means of thresholds;
3. Feature extraction: appliances provide load signature information or features that can be used to distinguish one from another;
4. Load identification: using the features previously identified, a classification procedure takes place to determine which appliances are operating at a specified time or period, and/or their states.

2.1. Data Collection

The first stage of energy monitoring system is dedicated to data acquisition or collection. This is an aspect frequently considered as less relevant, but it has major consequences in terms of the types of application that can later be tackled by NILM algorithms, as well as the performance, granularity, etc. This data acquisition is commonly related to a device or system, very close to the existing electrical facilities, where different approaches can be deployed in order to measure certain parameters, such as currents or voltages, in a certain household or building. Sometimes other parameters, actually coming from these voltage and current signals, can be determined, such as the real power, the apparent power, the power factor, or the I-V trajectory [5], and used as features. Not only these parameters, but also their variation over time, are clues to guide our approach to any further energy disaggregation and appliance identification. Taking these considerations into account, this section has basically considered two main criteria when analysing previous works: the sampling rate employed in the data collection and the type of hardware architecture implemented.

For simplicity's sake, maybe the most straightforward solution for data collecting is to think about available commercial plug-in devices. These provide off-the-shelf platforms, normally with the basic functionality ready to be used, but also with some significant drawbacks, especially in terms of sampling rates and flexibility. This trend was already stated in [6], where, after studying different commercially available smart meters and/or energy monitoring, it was concluded that these provide the required computational capacity to cope with advanced techniques, such as NILM. Neurio Technology Inc [7] and Smappee N.V. [8] provide similar energy monitoring solutions, both based on a current clamp, together with a set of utilities and applications intended to display and process the collected information as easily as possible. Furthermore, they provide different communication protocols to report data to other points; Ethernet or Wi-Fi links are the most popular, but this also includes other protocols such as ZigBee or RS-485. Other companies, such as ONZO Ltd. or Bidgely, Inc., propose similar approaches, most of them based on a smart meter/sensor and machine learning for energy disaggregation.

With regard to the drawbacks presented by the commercial solutions, it is worth noting that most of them are constrained to low sampling rates, 1 Hz maximum [9,10], thus limiting the achieved performance and the chance to use them in some demanding types of applications. Even worse, sometimes this sampling frequency is not consistent over time, thus adding a new challenge. In any case, it is widely accepted that systems providing higher sampling frequencies support deeper analysis of the measured features in order to achieve better energy disaggregation [11]. In some previous works, such as [12], the influence of the sampling frequency on the final performance was analysed, concluding that to implement more feasible and reliable appliance classifiers than those already proposed in the field, sampling frequencies should be higher than 4 kHz. As a counterpart, the use of high sampling frequencies is costly, both in terms of software and hardware complexity, and also requires larger communications bandwidth to transmit data to any monitoring or centralized station. Overcoming these difficulties is technically feasible nowadays, but the integration of these enhancements into commercial smart meters will definitely increase the final cost.

Although some smart meters are capable of acquiring signals in the range of kHz [13], their deployment is not actually so extended among electrical companies, likely due to their higher cost. This is the reason why those efforts focused on high sampling rates have been particularized in the design and development of ad-hoc acquisition systems, most of them based on a current clamp and a voltage sensor, together with fast enough analogue-digital converter. This trend is followed in [14,15], where an oscilloscope or a power analyser was used as the acquisition module. Furthermore, in order to employ less expensive and more specific and portable hardware, commercial or ad hoc dedicated data acquisition modules have been applied to measure voltages and currents [16–20]. A direct example of this approach is the BLUED database, acquired by a specific hardware design based on a commercial NI 16-bits acquisition board, which samples current and voltage [21]. Figure 1 gathers the different aforementioned alternatives for data collection in NILM applications, according to the sampling rate.

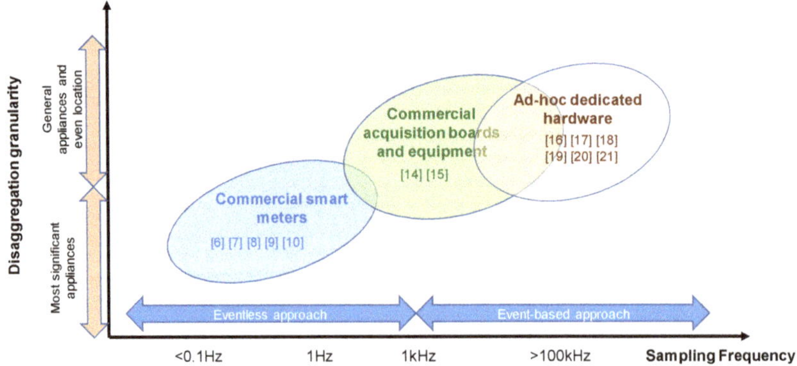

Figure 1. Data collection systems for NILM applications versus sampling frequency.

As has already been mentioned, the main drawbacks of the high sampling frequencies required by NILM algorithms to boost their disaggregation capabilities and identification performance are the increase in computational complexity and the real-time constraints associated with any implementation of these proposals, particularly when commercial smart meters or energy monitors are considered. For that purpose, different techniques have been proposed, aiming at reducing the algorithms' load. One of them is compressed sensing, which achieves a trade-off between the sampling frequency and the degree and accuracy in the disaggregation [22].

Figure 2 summarizes the above-stated aspects concerning data collection in systems oriented to NILM applications. It is also used to introduce the concept of locally and remotely computed tasks in the context of data collection for NILM applications.

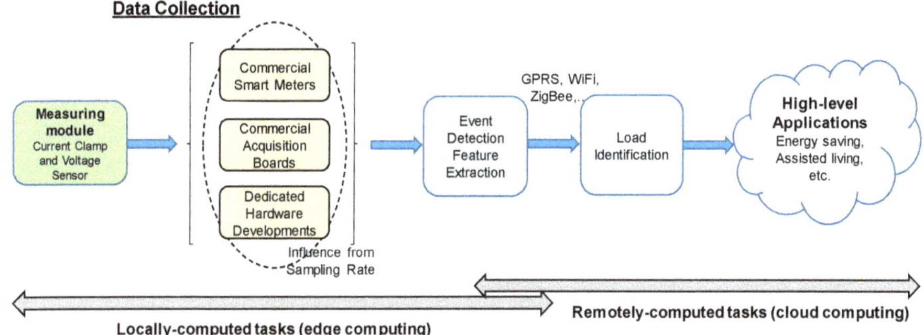

Figure 2. General view of the different aspects involved in the process of data collection for NILM applications.

After acquiring raw data, event detection (typically the on/off switching of electrical devices) should be tackled, as well as some parameters (often classified as steady-state features and transient ones) determined. These procedures, associated with data acquisition and first processing, can, depending on the computing capacity of local devices, be computed locally, thus reducing the amount of data to be transferred to a remote monitoring system. In this way, feature extraction tasks are sometimes implemented in local devices, especially when high sampling rates are available. In these cases, the hardware architecture should present a minimum computing capacity, and be designed keeping in mind that, as they will be finally installed in buildings and households, they should be portable, plug-in and easy to handle [23]. On the other hand, these features are often reported to remote centres where they are processed for further applications, such as load identification or even higher-level tasks, such as energy saving, assisted living, etc.

Keeping in mind the communication needs represented in Figure 2 between the local devices and the remote computing centres, a last relevant point must be considered: how to transmit event detection data as well as the features, locally determined. This data link can be tackled by means of a wide range of technologies and protocols, such as GPRS, PLC, Wi-Fi, Internet and so on, including in-home networks (ZigBee, Bluetooth, etc.) [24–26]. Another approach consists of subcontracting any telecommunication supplier or company, as shown in [27] with Orange. In [28] a gateway based on the OSGi framework is designed to collect information from sensors and smart meters via a ZigBee link.

It is also worth noting that many works in the literature avoid facing the issues that arise from practical and experimental implementations by verifying their proposals using existing databases composed of samples measured from real scenarios under different conditions [29]. Some of these popular databases are REDD [30], BLUED [21], PLAID [31], REFIT [32], TRACEBASE [33], WHITED [34], UK-DALE [35], DRED [36] or PECAN street (https://www.pecanstreet.org/). This approach allows researchers to deal in advance with the challenges and problems otherwise found in later stages, such as event detection or feature extraction, at the expense of limiting their results to the data collection system implemented during the creation of the database, with a particular key influence from the aforementioned sampling rate.

Summing up, the performance and type of hardware setup in the data collection determine the options available in later stages, enabling in some cases the detection of events in the signals of interest, and, specially, profiling a feature set that can be used for load identification. Both aspects are tackled in upcoming sections.

2.2. Event Detection

In NILM, any switch in a signal from a certain steady state to a new one is considered an event. It is often associated with high sampling rates, as this condition is necessary during the corresponding signal processing to achieve a suitable performance in the detection of events. Due to the fact that events are more clearly identified in current signals, compared to voltage ones, it is worth noting that most previous event detectors have dealt with this type of signal. Furthermore, event detectors typically use three different approaches, according to previous work [37]: expert heuristics, probabilistic models and matched filters.

Expert heuristics consist of the creation of a set of rules for each appliance. They commonly require the initialization of certain variables, such as the total power demand and power variation. Most previous works based on this approach were published in the 1990s and 2000s, focused on the detection of main appliances with significant power consumption. On the other hand, probabilistic models provide a probability, used to make a decision about the occurrence of events. For that purpose, they require a training process to fix certain variables and learn some statistical models for appliances and environments. A particularly well-known case is the Generalized Likelihood Ratio (GLR) method [37,38]. Finally, matched filters are characterized by extracting the signal waveforms and correlating them with known patterns. Although in this case no previous training or knowledge is needed about appliances or environments, this approach often implies high sampling rates. Techniques such as envelope extraction, advanced filtering, Kalman filter and Hilbert transform are usually involved here in a post-processing stage to achieve suitable event detection and even energy disaggregation [39–41]. Clustering and bucketing techniques have also been used in event detection [42].

Event detection is often evaluated in terms of certain metrics [37]. The most relevant ones are the true positive rate, the true positive percentage, the total power change and the average power change. The false positive rate and the false positive percentage are less frequently used metrics. In many cases, all the above metrics are combined into one, usually called a score function, where the different parameters can be weighted according to their desired influence on the final performance of the event detector.

In [43] a probabilistic method, based on a Goodness-of-Fit (GOF) methodology, is compared with an expert heuristic method on the REDD database; the authors found that the GOF event detection methodology achieves the smallest number of false positives. In [44] an event-based algorithm is proposed to identify load signatures, according to trajectories of real, reactive and distortion power. In [45] a simple and fast event detection algorithm is proposed for the variations of the current signal. Its main advantage is the higher determination accuracy of the beginning of the events. On the other hand, the detected events are used in [46] to drive a finite state machine based on fuzzy transitions that disaggregates different appliances on signal sampled at 2 Hz.

More recently, Decision Trees (DT) and Long Short-Time Memory (LSTM) models are used for event detection [47], obtaining 98.6% and 92.6% detection accuracy, respectively. Furthermore, [48] presents a very simple detection algorithm used in a low-complexity NILM proposal, achieving suitable performance in six houses from the REDD dataset.

Another novel approach is presented in [44], where, after pre-processing the voltage/current signals to enhance the event change detection, that event is classified into certain categories of appliances by applying principal component analysis (PCA) to the PQD (active, reactive, and distortion powers) trajectories captured during the event change. This approach follows the trend of considering event transients as an additional feature in later appliance identification [2].

In general terms, if event detection is applied, it leads to the determination and selection of the most representative features for a certain appliance, so they can be used in a later identification. These features are particularly significant around the change of state (event), thus justifying the importance of successful event detection when necessary. Figure 3 presents a general overview of blocks involving such event-based NILM algorithms. The next subsection is dedicated to introducing these sets of features and how they are employed in appliance identification.

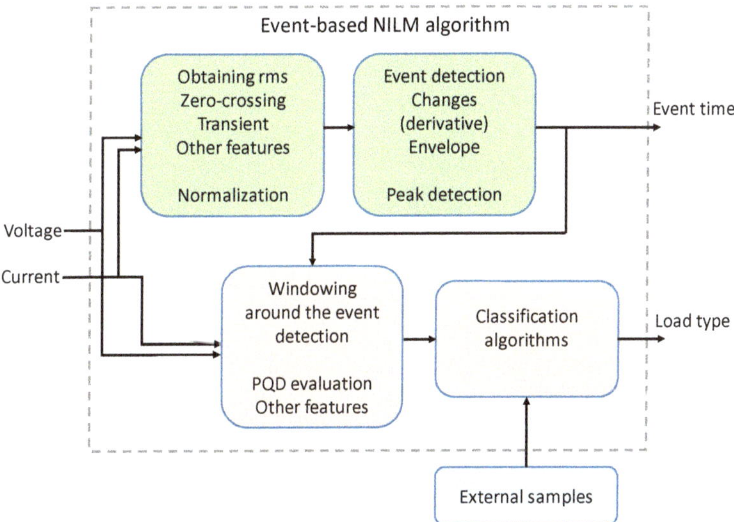

Figure 3. Block diagram of an event-based NILM algorithm to obtain event times and types of loads involved in each event.

2.3. *Feature Sets*

Energy disaggregation is achieved by identifying active appliances using a classification procedure. This way, a set of features must be available that should be closely related, on one hand, to the data collection and, on the other hand, to the methods that will be used for appliance identification. NILM features are highly dependent on the sampling rate used, which must be understood as the rate of the data output by the measurement device and that will be used for disaggregation, not the sampling rate of the current and voltage that constitute the device's input. A coarse division, using the threshold of 1 s for the sampling period, enables separating features between macroscopic or low-frequency and microscopic or high-frequency. A finer division proposed in [22] and used here divides the range of sampling rate into six classes: *very slow*, slower than 1 min; *slow*, between 1 min and 1 s; *medium*, faster than 1 Hz but slower than the fundamental frequency, *high*, from the fundamental frequency up to 2 kHz; *very high*, sampling frequency between 2 and 40 kHz; and *extremely high*, faster than 40 kHz. In this section we shall use this division, introducing the features used in representative NILM works and focusing on the most recent ones.

Most applications using very slow or slow sampling employ features obtained from the time series of power variables: voltage and current, apparent, active and or reactive power, power phase angle and power factor, etc. We shall assume in the following that the instantaneous values of the current voltage and power are denoted as i, v and p, respectively; their RMS values as I_{RMS} and V_{RMS}; the Active, Apparent and Reactive Powers as P, S and Q, respectively; the Total Harmonic Distortion as THD; and the Power Factor as PF.

The most employed feature is S, exclusively used in [26,49–51]; P and Q were employed in [52,53]. In a recent work [54] P and V_{RMS} measurements were used, sampled at 1 Hz, obtaining a high level of accuracy, even with varying supply voltages.

The various time series can be used in several ways. In a number of applications they are used directly, as it is the case of [55], where the P time series of the Individual Household Electric Power Consumption Dataset (IHEPCD) [56] is employed; in [57], where P and I_{RMS} are used, taken from the Almanac of Minutely Power dataset (AMPds); and [58], which employed the S and I_{RMS} series, from the AMPds and REED datasets. On the other hand, in [59] the power time series is segmented into sub-sequences that are used to compute the statistical moments of the load consumption, and in [60] the high-frequency current signal is subject to time-domain transformations.

Time-based features are typically used in eventless NILM algorithms, and the ones belonging to the low-sampling are typical steady-state ones. Within this sampling category, several other approaches have additionally been proposed. For instance, the authors of [61] split a power signal into "powerlets," which are the minimal group of short sequences (that represent the signal), obtained from Auto-Regressive models with eXogeneous inputs (ARX), characterizing each appliance. "Shapelets" [62] are similar, since every shapelet is a small subgroup of a time series.

Moving on to the next sampling category, the medium-rate range allows the characterization of transient electrical behaviour as appliances change state. While some transients may be visible from low-rate sampling, medium-rate sampling allows for much more detailed information on the transient shapes to be acquired. The authors of [63] proposed the use of seven features, extracted from the current waveform: number of spikes; number of semi-steady states (permanence in the state between 1 and 5 s); number of steady states (permanence longer than 5 s); total time in semi-steady states/length of the operating waveform; total time in steady states/length of the operating waveform; number of states per time window; and existence or nonexistence of repeating patterns.

As a time series often provides a high level of redundancy, increasing the model complexity and possibly leading to a low accuracy, it can be transformed into a frequency domain. This requires high sampling rates, however. Several features can be extracted from frequency information, such as harmonics [64] obtained with Fourier transform and multiple frequency bands using information entropy [65]. Due to its multi-resolution and time-frequency localization property, Discrete Wavelet Transform (DWT), is also employed [4,66]. Other transforms were also employed, such as the Stockwell Transform [67], and combinations of different techniques, such as DWT and harmonics [68], have also been proposed.

Very high rate data allows us to obtain much more detail about each appliance's waveform, either from the higher harmonics or from the shape of the raw current and voltage waveforms themselves. Two-dimensional voltage-current (V-I) trajectories, corresponding to the normalized steady-state voltage and current signals during one cycle, have already been considered as a likely method to identify load signatures in terms of features [69]. Generally speaking, the V-I trajectory presents unique characteristics for appliances with different working principles (resistive or inductive), which night be collected by wave-shape (WS) features, where it is possible to extract certain features, such as the looping direction, the enclosed area and the number of self-intersections. More recent applications [13] used additional features extracted from the V-I trajectory. Other features involving the shape of the waveforms [70] can be obtained from $p(t)$ and from the Instantaneous Admittance Waveform (IAW).

Higher-order harmonics can be obtained an using extremely high sampling rate, also enabling the capture of electric noise. In fact, the authors of [71] showed that the use of high frequency ElectroMagnetic Interference (EMI) signals enables the differentiation of similar switching mode power supplies in a home, which cannot be obtained with other techniques. Higher-order harmonics are employed in [18–20]. The first work is an extension of [65] for the simultaneous operation of various appliances, whereas the third one proposes to use, instead of the amplitudes of the current harmonics, the harmonic current phasors. The results show important improvements in performance when several combinations of appliances are considered. The second work uses the same type of features, although employing a different identification procedure. It achieves excellent performance for different combinations of small nonlinear loads. Unfortunately, as the data used are different and private, the performance of approaches [19,20] cannot be compared.

The features identified above can be computed from the main power feeder of the house. However, other information can be used. Variables such as time and duration of usage for a given event can be inferred just from the main power sensor [72]. In [73–75] the frequency of usage of an appliance, as well as the correlation of usage of multiple appliances, have been applied. This information can be extended with users' behaviour to express the uncertainty for each state of each appliance [76]. Occupancy, which can be measured or inferred in several ways, has been used to reduce the complexity of NILM algorithms [36]. For HVAC systems, external weather information has also been used [77].

It is not uncommon to use combinations of the features described above, leading in this way to hybrid approaches. For instance, [70,78] employ P, Q, IAW, p, eigenvalues and switching transient waveform, as features applied to a "Committee Decision Mechanism." More recently, feature selection algorithms have been employed to reduce an original dataset of 55 steady-state and 23 transient features to the 20 most relevant features [79].

2.4. Load Identification

Using the features described above, computed from the aggregate load, the objective here is to identify the appliances that are operating at a given time. This can be formulated as a not so simple optimization or classification problem, as four appliance models are usually considered:

- Type I—On/off devices: most appliances in households, such as bulbs and toasters;
- Type II—Finite-State-Machines (FSM): the appliances in this category present states, typically in a periodical fashion. Examples are washer/dryers, refrigerators, and so on;
- Type III—Continuously Varying Devices: the power of these appliances varies over time, but not in a periodic fashion. Examples are dimmers and tools.
- Type IV—Permanent Consumer Devices: these are devices with constant power but that operate 24 h, such as alarms and external power supplies.

This way, for the case of type II appliances, identification is not only translated into which appliances are active, but also their states. Additionally, some appliances can be replicated (for instance, two fridges might be available in a household), and it might be necessary to identify the operation/state of each replicated device using similar load signatures.

As a myriad of approaches has been proposed for this last step of NILM, the aim of this section is not to provide a deep review of the existing alternatives, but rather to point out important works on optimization and machine learning (supervised and unsupervised) algorithms used for load classification. Before introducing them, it should be noted that the performance of the different algorithms must be compared, using common datasets (please see Section 2.1) and similar performance criteria (please see [29,80] for a comprehensive list of performance metrics employed).

Optimization approaches use different methods to perform a combinatorial search. Examples are hybrid programming [81], genetic algorithm [82] and segmented integer quadratic constrained programming [83]. The main problem with this type of method, however, is their heavy computational burden. For this reason, most approaches belong to the so-called machine learning algorithms, involving both supervised and unsupervised methods.

Supervised techniques use offline training to achieve a database of information used to design the classifier (s). Some common supervised learning techniques that have been applied in NILM are (shallow) Artificial Neural Networks, mainly Multilayer Perceptron (MLP) [66,84], concatenated Convolutional Neural Networks (CNNs) [85], Deep Neural Networks [53,86–91], Support Vector Machines (SVM) [66,92], K-Nearest Neighbours (k-NN) [92–94], naïve Bayes classifiers [64,94,95] and, recently, linear-chain Conditional random fields (CRFs), which takes into account how previous states influence the current state and can deal with multi-state loads [96]. In [97] the performance of three classifiers, MLPs, Radial Basis Function (RBF) networks and SVM, with different kernels, is compared by employing odd harmonics (up to the 15th) from the current waveform, measured in a proprietary experimental setup. It has been concluded that all models provide excellent classification performance and correctly identified the existing devices, establishing the applicability of the proposed approach.

Unsupervised methods do not require any training prior to classification. This is an important advantage since, in this way, minimum effort is required from the user and the intrusiveness involved in building a database is reduced. Feature clustering, and the later labelling of each cluster with meaningful appliance names has been applied in [98,99]. A fusion of a supervised training process over available labelled datasets with an unsupervised training method over unlabelled aggregate data is proposed in [50].

The most recent unsupervised techniques applied to NILM belong to a family of methods that assume that the electrical signal is the output of a stochastic system, maintaining a representation of the whole system state, instead of dealing with individual events [100]. Examples are Hidden Markov Methods (HMM) and variants [14,26,83,100–105].

Another powerful option for solving data mining and signal processing problems is Graph Signal Processing (GSP). GSP applied to NILM [41,106,107] showed that this approach had remarkable performance related to the HMM approaches, offering additional advantages compared with conventional NILM methods, not requiring a training phase and obtaining good performance in low-sampling environments.

Tables A1 and A2, in the Appendix A, summarize the features employed, the load identification technique, the main contributions, the data source used, as well as the main application of the most important works referenced here. Notice that only two applications (HEMS and AAL) have been considered, identifying the context in which the referenced work was developed. All the other unlabelled references did not have a specific application in mind. No indication of performance was incorporated in the tables, as different data sources were used and, even in works using the same datasets, different houses/frequencies/number of appliances/performance criteria were involved, making a performance comparison not meaningful. For the sake of readability, references were ordered according to the sampling frequency employed and divided into two tables. The former considers approaches requiring data acquired up to medium sampling rates, and the latter proposals requiring higher sampling frequencies.

Having reviewed the steps comprising NILM methods and the most relevant and recent proposals in this topic, we will in the next two sections address their use in two important applications, HEMS and AAL.

3. Home Energy Management Systems

3.1. General Overview of HEMS

Buildings are actually the most demanding sector in terms of consumption, representing 40% of the total primary energy and accounting for 74% of the electricity sold in the USA [108]. For this reason, Home Energy Management Systems (HEMS) are becoming increasingly important to invert the continuously increasing trend in (electrical) energy consumption. Reviews on HEMS can be found in [109–114], as well as the works included in [115].

HEMS offer advantages to both residential occupants and electricity suppliers. For the former, HEMS are a means to reduce energy consumption in a household (or, perhaps more important, the electricity bill) while maintaining occupant's comfort. Notice that HEMS should not only perform real-time monitoring and scheduling of various home appliances, based on the user's preferences, but are also employed for the management of home renewable energy systems and energy storage systems, if available [115].

For the suppliers, the two-way communication enabled by smart grids allows much better management of the whole electricity network and the implementation of several mechanisms known as Demand Response. DR are those modifications in the electric usage of costumers, compared to other previous consumption patterns, as a consequence of the variations in the electricity cost over time, or incentives payments designed to ease a reduced electricity usage during those intervals with high prices, or suspected system reliability. Currently, DR are often grouped into two categories: price-driven and incentive or event-driven. The former can be sub-divided into several forms—time-of-use pricing, critical peak pricing, real-time pricing and peak-time pricing; while in the latter category we can find direct load control, emergency demand response programs, capacity market programs, interruptible/curtailable services, demand bidding/buyback programs and ancillary service market programs [112,116].

The first step of any HEMS is to monitor the electricity consumption of the several devices existing in a household. This can be achieved intrusively or using NILM techniques. In general terms, the non-intrusive approach is more popular both in academia and industry [3], mainly due to the fact that sub-metering installation is often expensive, difficult to upgrade, and involves certain privacy issues, thus avoiding any intrusive approach.

By reviewing previous literature [117], the availability of a disaggregated energy bill might be related to the reduction of domestic electricity consumption by 0.7–4.5% on average. This, as we know, is obtained with NILM techniques, by estimating the active appliances consumption. The availability of load disaggregation data via NILM can also enhance some other aspects, such as the load demand forecasting accuracy, and provide better criteria for companies to decide. For the grid operators, NILM additionally allows flexible resources management for demand response and tackling with uncertainty derived from renewable sources [118].

3.2. Use of NILM in HEMS

As mentioned before, a HEMS should schedule conveniently the electrical appliance's usage, as well as the electric energy flow, if renewable energy sources and/or storage are available at home. NILM techniques can also improve this overall goal, but some factors should be taken into consideration.

Firstly, it is important to classify appliances as non-deferrable (or non-schedulable) and deferrable (schedulable). The former comprises devices such as lighting, cooking or refrigerators, whose operation cannot be delayed. The latter includes washers and dryers, water pumps, and so on, whose period of operation can change according to the price of energy. Of special importance are HVAC systems, such as electric water heaters, and space heating/cooling systems, which sometimes are denoted as Thermostatically Controlled Loads (TCL). As NILM identifies the appliances that are active at any one time, it allows us to know in real time which schedulable and non-schedulable appliances are active.

Secondly, in previous sections we have essentially used NILM to identify appliances. For HEMS, electric consumption should also be estimated, and higher scheduling priority should be given to the appliances requiring high energy consumption. The level of consumption should also be estimated by the NILM module, and consumption can be predicted using forecasting methods. It is well known that HVAC systems actually are the largest part of energy consumption in buildings, and therefore correct HVAC control is important. Considering again the case of the USA [108], HVAC systems account for 35% of the primary energy and 45% of electricity consumed in buildings.

Thirdly, appliances' turn-on and turn-off times and time duration are important parameters for appliance scheduling. Note, however, that for Type II devices, these parameters should be available for all states of operation. The frequency of usage for each class of appliances can be obtained by means of these variables.

Finally, appliance flexibility is important for HEMS applications: This is a concept that is not universally accepted, with different forms proposed for its calculation. One definition, introduced in [119], is the possibility of the appliance getting involved in DR programs, taking into account not only the appliance characteristics but also the usage preferences from the user. Note that HVAC and power heaters are highly flexible loads, thanks to the inertia of an associated thermal storage and the need to fulfil some quality constraints [120].

The use of NILM techniques in HEMS has been increasing over the years. Perhaps the first proposal of using NILM in DR programs was in [121]. The authors analysed the requirements of DR and proposed a new NILM system with an enhanced load space and measurement approach.

Evolutionary multi-objective power scheduling using NILM techniques has been proposed for DR in [122]. Based on a real-home assessment of their proposal, the authors conclude that the automated mechanism is workable and feasible. They pointed out, however, that the power of each household appliance should be adaptively updated to improve the estimates of the daily power consumption. As their application did not include renewables, they proposed to include them, together with a forecasting mechanism for the electricity produced, in future work.

The same authors subsequently proposed a model of a residential consumer-centric Demand-Side Management [123], employing NILM, achieving, in simulations, a significant reduction (14%) of the Peak-to-Average Ratio (PAR). For future implementations, the authors proposed employing edge/IoT-based computing, in order to improve cloud computing technologies [124]. In a more recent work [125], the same group focused on the improvement of NILM classification, employing for that Particle-Swarm Optimization to the design of the ANN classifier.

Edge-computing is also advocated in [126]. The authors implemented a load-shifting mechanism, which allows non-time-constraint applications to be moved from rush hours to off-peak hours. This implies a reduction in the peak demand of the household, while maintaining the householders' comfort. Employing day-ahead pricing information, their system is composed of five modules: *energy production*, which consists of solar radiation and air temperature predictors, used to forecast the PhotoVoltaic (PV) energy generation; *solar energy management*, which manages the flow of energy between the grid, PV and battery storage; *NILM module*, which not only disaggregates the energy and estimates consumptions, but also computes usage patterns and features of each appliance; *classifier*, which labels the appliances as schedulable or not and, in the former case, passes this information, together with adjustable ranking, to the next module; and *appliances scheduling*, which, based on the information received from the previous module for deferrable appliances, proposes a dynamic algorithm to determine which state sequences in a certain appliance provide a lowest electricity cost over time. Using two test scenarios in a real testbed, they concluded that the use of the proposed HEMS achieves reductions of electricity consumption and cost of 73% and 82%, respectively. They pointed out that a better usage of solar energy could be obtained by merging solar energy forecast and appliances scheduling schemes.

The authors of [76] have addressed appliance-level dispatch with smart plugs for HEMS, employing in their application the D'hulst concept of appliance flexibility. Assuming that each appliance operation can be divided into states, these are estimated from the appliance power consumption using a combination of the minibatch k-means method [127] and the X-means technique [128], followed by an agglomerative clustering approach. User behaviour is characterized by different variables, such as state turn-on and off times, state on-duration, state energy consumption, state power value, etc.; to consider uncertainty, the features are often modelled as Gaussian distributions.

In operation, each state is assigned to an appliance type according to a k-nearest neighbours' classifier, where the likelihood is determined by means of the Hellinger distance. The appliance type is derived from weighted voting, where the weight is defined by the state's power consumption over a certain time T.

Appliance flexibility depends on the DR application and thus is a function of the start time, duration, controllability, user behaviour and power. Based on the desired DR event, the DR program is chosen. Then the HEMS searches for and selects suitable appliances. Finally, the flexibility of the selected appliances is calculated and inserted into a priority list and the appliances are dispatched according to that list. This approach has been evaluated on a REDD dataset, obtaining an excellent classification performance.

Finally, it is worth noting that new NILM methods have been proposed with application in HEMS in mind. This is the case, for instance, with [129], where, using only a single active power sample acquired at the general entry point with a rate of 1 Hz, it is feasible to distinguish turned ON appliances, their operating modes, as well as power consumption, together with the amount of solar power. In a more recent work [130], the authors extended their previous solution and were able to properly forecast the active power demand of a set of five households.

4. NILM in Ambient Assisted Living

4.1. AAL General Overview

Ambient assisted living (AAL) includes products and services for the physical independence of elderly people. In fact, the current increase of life expectancy has become a public health priority, mainly in developed countries, and most of the recent technological advances are used for constituting smart environments to assist the elderly. There are three important aspects or actuation levels to consider in AAL:

1. Using specific sensors (e.g., wearables, ambient sensors or even smart meters) to measure ambient (environmental) or physiological (person-related) parameters.
2. Monitoring a particular parameter of activity (e.g., physiological signals, movements or Activities of Daily Livings - ADL)
3. Taking appropriate decisions or recommendations (e.g., monitoring health deterioration in the long term or producing alerts for short-term intervention).

Figure 4 shows a general overview of current home monitoring systems in terms of accuracy and scalability. In general, accuracy is inversely proportional to scalability and to intrusiveness (and consequently to the grade of acceptance of the systems).

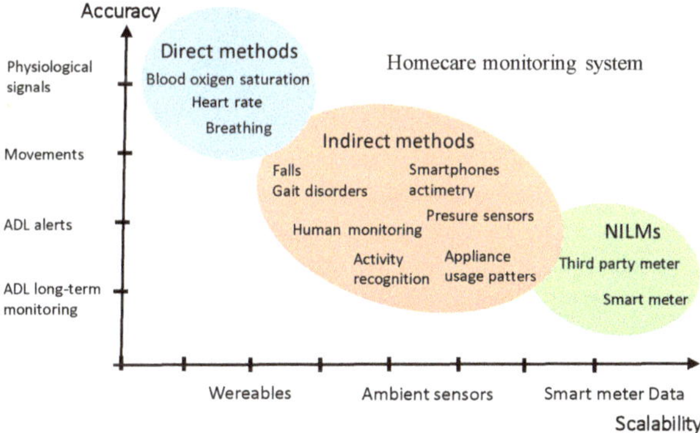

Figure 4. General overview of current homecare monitoring systems for AAL depending on accuracy and scalability (adapted from [131]).

As can be observed in Figure 3, four levels of accuracy have been considered depending on the outputs of the home monitoring system: ADL long-term monitoring, ADL alerts, movements and physiological signals. Scalability, strongly related to intrusiveness, depends on the kind of sensors needed (wearables, ambient sensors and smart meters). Direct methods may diagnose the health or monitor the activity directly by evaluating some physiological parameters; on the other hand, indirect ones can derive it from a parameter that may involve the health status or activity.

Physiological signals related to direct monitoring methods are often blood oxygen saturation, heart rate and breathing [132]. The acquisition of these signals is normally very accurate, but difficult to scale since the corresponding transducers required to be attached to the body. Accelerometers and gyroscopes in wearables and smartphones allow movement to be estimated, and can detect falls and gait disorders [133], but their acceptance is still reduced as users must carry them for proper operation.

Most of the approaches for AAL are based on ambient sensor on heterogeneous high-density sensor networks to perform Activity Recognition. These systems have to deal with overlapped activities; heterogeneous activity duration; the deployment of a complex and sometimes intrusive WSN; and with the need of a supervised training process for each individual household [134].

With a very low intrusiveness, a new approach to monitor ADLs by means of electrical signatures of appliances coming from plug-meters was proposed in [27]. Human activity can be inferred from the usage pattern of appliances, as they are strongly connected to daily activities. In this case, the activities monitored were food preparation and eating, hygiene and elimination. It should be noted that any labelling task, such as the weight of appliances on the activity and finding the activity duration, depends on the particular person monitored. Electrical events are mapped over daily activities using a k-means neighbours' classifier.

In a similar way, other works also proved the correlation between the appliance usage patterns with ADLs [135]. Here, the authors used the Latent Dirichlet Allocation (LDA) method to map appliance events with ADLs. The sensor density could be minimised, and the hardware cost and complexity reduced (of particular importance in large deployments). A major issue to be solved was again related to the overlapping of tasks and their heterogeneous duration.

The authors of [136] also proposed an approach to monitor the behaviour of the elderly based on detection of the usage of certain home appliances. In this case, the system is based on a smart meter that periodically acquires the global energy consumption in the house, associated with some smart plugs for punctually monitoring specific electrical devices. Although the system is simple and low-cost, it can detect unusual behaviour in the elderly.

All these methods are intended to measure health deterioration and are deployed for long-term monitoring. There are other methods that produce alerts during short-term monitoring of a particular health aspect. These only use the appliance usage pattern instead of inferring ADLs. For example, a relevant variable to detect changes in routines could be monitoring the kettle or the TV set [137]. Another example in [138] is the usage patterns of the kettle and fridge during the night to detect sleep disorders.

4.2. Use of NILM in AAL

There are several relevant health features that can be inferred from data obtained with smart meters or third-party devices installed as unique sensors at home (after applying energy disaggregation algorithms). These features can be inactivity, sleep disorders, memory issues, variations in activity patterns, low activity routines, occupancy and unhealthy living [139]. The main advantages of using smart meters are their flexibility, low cost, ubiquity and ability to generate data over time.

Figure 5 shows a general diagram that most systems follow when using NILM for AAL.

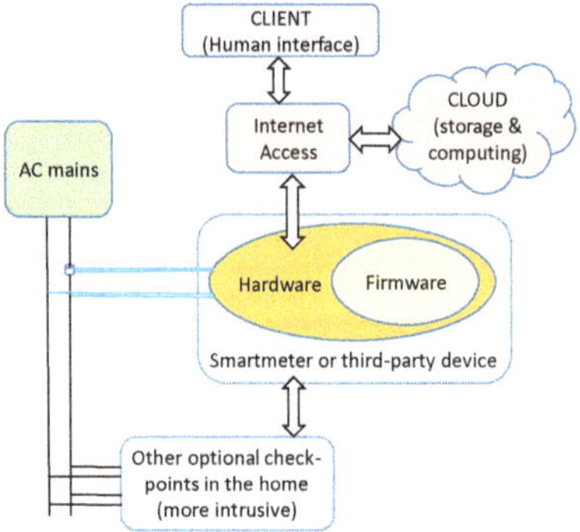

Figure 5. General diagram of systems that use NILM in AAL.

Applications can include launching alerts to caregivers or relatives whether unusual activity patterns are recognised [140], or even in a further extension, monitoring the progress of some treatments or living conditions (such as the use of specific devices). For instance, in [131] a unique power usage profile is derived for every appliance. The usage was categorized into usual and unusual patterns. Such appliance training methods are common in NILM, and the major challenge is to detect a wide range of devices with enough accuracy.

One of the first works on this topic, using only disaggregated data from a single home sensor, can be found in [141]. The system was developed with the main goal of determining load signatures of appliances to detect daily activities in a smart home. It was based on steady-state operations and signatures of appliances extracted with a single power analyser. Afterwards, in [101] the authors proposed the disaggregation of data coming from smart meters in order to monitor health. They employed an iterative time-dependent HMM to disaggregate appliances, according to a priori knowledge of the activities of people at home. After the disaggregation, every appliance was bounded to a certain monitored activity. Other studies, such as the one presented in [142], made use of a smart meter, which periodically measured the global energy consumption in the house, combined with some smart plugs for punctually monitoring specific electrical devices. The goal was also to track elderly behaviour by detecting the usage of home appliances.

Another work on this topic is [131], which also proposes the use of only smart meter data to develop sustainable models for healthcare in smart homes, with very low intrusiveness, massive deployment and reduced cost. The usage patterns of appliances are used to evaluate the behaviour of elderly people and to determine when a subject has modified their routine. The system provides a daily score of normality regarding the regular behaviour (obtained from previous statistical analysis, using Dempster-Shafer theory on the disaggregated consumption of several homes). When this score was lower than a predetermined threshold, an alert could be derived. The same authors, in [141], present an interesting study of NILM classification, depending on disaggregation accuracy and sampling frequency, and of homecare monitoring system classification according to accuracy and scalability (where the systems based on NILM are less intrusive and more scalable, but at the cost of accuracy).

Despite the intrusiveness of these systems being low, privacy can still be an important issue. The authors of [143] analysed the electricity consumption of more than 5000 households over a 18-month period and deployed several machine learning methods to forecast home occupancy in the short and long term. The results revealed that the present and future occupancy status of households can only be established with high confidence based on smart meter data. In this context, it is also significant to secure the communication and storage techniques and equipment related to smart meter data, as well as to fulfil the corresponding legislation about how to treat such data.

5. Guidelines for Future Research

Although NILM methods are becoming recognized tools for home energy management systems and for ambient assisted living applications, several aspects still deserve further research.

NILM has been an active topic of research, mainly due to advances in computational intelligence and sensing technology. Although NILM has been around for 30 years now, only recently has the technology made its way into public domain, due to high equipment cost, which hinders the scalability, and a lack of disaggregation accuracy [6]. Future research and development in this area should focus on the solutions to these problems.

Current NILM methods work well for two-state appliances, but it is still difficult to identify some multi-state appliances, and even more challenging with continuous-state appliances. Typically, supervised methods are able to generalize better to unseen scenarios, e.g., different houses, than unsupervised techniques. However, they require a huge database and an off-line training phase. The use of semi-supervised algorithms, requiring some labelled training examples, might be a mechanism to achieve "low-cost" generalization accuracy. Another aspect would be using special features, such as time of day, temperature, frequency of appliance usage, and so on, together with more classical features obtained from steady and transient signatures. Notice that some of these features are already employed in HEMS applications.

Finally, the different techniques should be compared using common datasets. Nowadays, there are several public datasets available; however, these only cover developed countries. Regarding the established performance criteria, they should also consider the complexity of the solution, both from the software and hardware points of view, as well as the level of load usage and their usage patterns.

Focusing now on the application of NILM in HEMS, several aspects are worth mentioning. First, the performance of NILM techniques should be considered according to the final impact and cost; e.g., a classification accuracy improvement in appliance identification from 85% to 87% can be translated, in HEMS operations, into a much smaller reduction in electricity consumption (or in the electricity bill), requiring, however, much more complex hardware and/or software solutions.

Research in HEMS should also consider the absolute improvement that the different types of apparatus might achieve in the final electricity consumption. A 5% improvement in lighting, for instance, has much less impact than the same reduction in HVAC equipment consumption. For this reason, HVAC equipment should not only be efficiently scheduled by the HEMS, but its real-time control during their periods of operation should be as efficient as possible. The authors of [142], in a study of a large appliance consumption database in Sydney, Australia, studied the real influence of air-conditioners on summer demand peaks. By clustering the load profiles and proposing load control strategies, they estimated that 9% of the total peak demand could be reduced. Model-Based Predictive Control (MBPC) is the control technique that has the largest potential of energy reduction for HVAC systems [144]. By employing MBPC approaches such as the one detailed in [145], allowing user-defined schedules and thus being suitable for HEMS, and allowing different levels of occupants (thermal) comfort to be considered, the potential for savings in home electricity consumption is large.

As reported before, the concept of appliance flexibility and its calculation deserve further research. Usage patterns should take into account the type of day, such as weekday, weekend or bank holiday; season and/or outside weather information (HVAC systems usage is strongly correlated with average outside air temperature and, therefore, with the season); associations of appliances (for instance, cookers and range hoods are typically used together); and, obviously, occupancy and occupants' preferences. Taking all these factors into consideration is not, however, an easy task.

The existence of disaggregated energy achieved by NILM allows us to obtain better forecasts of energy consumption, which, together with the better forecasts of electricity produced by renewables, allows for better appliance scheduling and flexibility for DR schemes. In this way, improvements in HEMS also require research on the forecasting methods applied to the variables at stake. As examples, as equipment usage depends on household occupation, the authors of [146] proposed a method based on dynamic genetic programming to detect, and forecast, the occupancy of residential buildings, starting with their smart meter data. Several techniques for short-term load forecasting can be found in [147] and in the works included in [148]. Short-term forecasts of the electricity produced by PVs require forecasts of solar radiation and atmospheric air temperature, the former being the most difficult due to the existence of clouds. In this time range, machine learning methods are the most used techniques [149]. There is already commercial instrumentation available that is capable of producing not only measurements, but also forecasts of weather variables [150].

Regarding AAL, better activity recognition needs to be achieved, requiring smart meters or third-party devices with higher sampling frequency. The regulations on the way smart meter data are stored and shared with third parties in health contexts must be adapted from the current situation. The level of fault tolerance in critical health uses is much lower than in those applications about standard energy metering, and, consequently, possible responsibility should be clearly defined in case of failure.

In addition, in AAL more advances and contributions are necessary in the field of linking patterns of energy use to health conditions. That implies the work of multidisciplinary teams, involving computing and engineering people with specialists and practitioners working in health and care (with new ethical issues). Novel use cases should be proposed and tested with a representative population. Finally, it is important to consider issues concerning user acceptance of smart meters applied to health domains and compared with other tele-healthcare approaches.

6. Conclusions

Although it was proposed nearly 30 years ago, NILM technology has only made its way to public domain in more recent years, mainly due to advances in computational intelligence, sensing technology and the Internet of Things, smart grids and demand response energy programs. Since then, the NILM field and its applications to home energy management systems and ambient assisted living have evolved rapidly.

We hope that this review, focusing on proposals that appeared recently in the literature and pointing out new research issues related to the techniques and their applications in HEMS and AAL, is able to foster further interest in this technology.

Finally, we should remark that NILM techniques have the potential to be used for other applications that are outside of the scope of this paper. Examples are, for instance, recommender systems for energy efficiency, whether for individuals [151], companies or governments, or fault diagnosis applications [146,152].

Author Contributions: All authors contributed equally to the review.

Funding: This research was funded by Programa Operacional Portugal 2020 and Programa Operacional Regional do Algarve (grant 01/SAICT/2018/39578); Fundação para a Ciência e Tecnologia grants SFRH/BSAB/142998/2018, SFRH/BSAB/142997/2018 and UID/EMS/50022/2019, through IDMEC, under LAETA; Junta de Comunidades de Castilla-La-Mancha, Spain grant SBPLY/17/180501/000392; and the Spanish Ministry of Economy, Industry and Competitiveness (SOC-PLC project, ref. TEC2015-64835-C3-2-R MINECO/FEDER).

Conflicts of Interest: The authors declare no conflict of interest.

Acronyms

AAL	Ambient Assisted Living	ILM	Intrusive Load Monitoring
ADL	Activities of Daily Livings	IoT	Internet of Things
ANN	Artificial Neural Networks	k-NN	K-Nearest Neighbours
ARX	Auto-Regressive models with eXogeneous inputs	LDA	Latent Dirichlet Allocation
CNN	Convolutional Neural Network	LSTM	Long Short-Time Memory
D	Distortion power	MLP	Multilayer Perception
DBSCAN	Density-based spatial clustering of applications with noise	NILM	Non-Intrusive Load Monitoring
DR	Demand Response	PAR	Peak-to-Average Ratio
DT	Decision Trees	PCA	Principal Component Analysis
DWT	Discrete Wavelet Transform	PSQ	Active, Apparent and Reactive Powers
EMI	ElectroMagnetic Interference	PF	Power Factor
FSM	Finite-State-Machines	PV	Photovoltaics
GLR	Generalised Likelihood Ratio	QDA	Quadratic Discriminate Analysis
GOP	Goodness-Of Fit	RBF	Radial Basis Function Network
		RF	Random Forest
GSP	Graph Signal Processing	RMS	Root Mean Square
HMM	Hidden Markov Model	SG	Smart Grids
HVAC	Heating, Ventilating and Air Conditioning	SVM	Support Vector Machines
HEMS	Home Energy Management Systems	TCL	Thermostatically-Controlled Loads
HMM	Hidden Markov Methods	THD	Total Harmonic Distortion
IAW	Instantaneous Admittance Waveform	WS	Wave-Shape

Appendix A

Table A1. Main characteristics of selected NILM techniques with very low, low and medium sampling rates.

Ref #	Sampling Rate	Features	Load Identification	Contribution	Data Source	Application
55	Very Low	P	Source Separation via Tensor and Matrix Factorization (STMF)	Analysis of the seasonal trend patterns using	IHEPCD	HEMS
57	Very Low	P and I_{RMS}	Maximum a Posteriori (MAP) probability	Usage of MAP in NILM	AMPds	
75	Very Low	Power consumption and appliance consumption patterns	Fuzzy c-Means Clustering and Dynamic Time-Warping	Iterative disaggregation approach based on appliance consumption pattern	AMPds	HEMS
87	Very Low	V-I trajectory	Deep Neural Networks	Learning based on multiple layers	REDD and Pecan Street	
88	Very Low	Power features	Neural networks (autoencoders)	Unsupervised anomaly detection of building operational data	Experimental data	HEMS
49	Very Low and Low	S	Discriminative Disaggregation Sparse Coding (DDSC) and Source Separation Via Tensor and Matrix Factorizations (SMTF)	NILM interpreted as a source separation problem	REDD	HEMS
83	Very Low and Low	Power features	HMM with Viterbi decoding	Consider the identification problem as a segmented integer quadratic program, together with constraint programming	REDD	
10	Low	Voltage, current, power	DBSCAN followed by QDA	Correlates occupancy events and power changes	Private data	HEMS
36	Low	Power consumption and occupancy information	Modified Combinatorial Optimization	A location-aware energy disaggregation framework (LocED) proposed to derive accurate appliance level data.	DRED, REDD	HEMS

Table A1. *Cont.*

Ref #	Sampling Rate	Features	Load Identification	Contribution	Data Source	Application
41	Low	P	GSP compared with several other methods	Mitigates the effect of measurement noise and unknown loads in	REDD and REFIT	
50	Low	S	HMM	Hybrid approach, combining supervised and unsupervised methods	TRACEBASE and REDD	HEMS
52	Low	P and Q	MLP	MLP parameters tuned by PSO	Laboratory data	
53	Low	P and Q	LSTM, denoising autoencoders, specific Deep NN architecture	Comparison of 2 Deep NN architectures against combinatorial optimization and Factorial HNN	UK-DALE	
53	Low	P and V_{RMS}	Karhunen-Loève Spectral Decomposition	Real-Time NILM working under severe voltage fluctuations	Private Data	
61	Low	Powerlets	Optimization with several priors	Collects power signatures (powerlets) in a dictionary, using optimization to solve	REDD	
85	Low	Power features and WS	HMM, Deep Neural networks.	Disaggregation based on Long Short-Term Memory Recurrent Neural Network (LSTM-RNN) and advanced deep learning. Novel signature model based on multistate appliance case	UK-DALE and REDD	
99	Low	Power features	HMM	Adaptive approach for estimating devices based on HMM	REDD	
100	Low	Power features	Factorial Hidden Markov Models and Iterative Subsequence Dynamic Time Warping	Hybrid Signature-based Iterative Disaggregation (HSID)	AMPds	

Table A1. *Cont.*

Ref #	Sampling Rate	Features	Load Identification	Contribution	Data Source	Application
102	Low	Power features	Hierarchical HMM and particle filtering	Modelling of multi-mode appliances by HHMM	REDD	
104	Low	P	Graph signal processing Event detection	No training required for NILM	REDD and REFIT	
128	Low	P trace and usage pattern profiles	MAP criterion	Incorporates appliance usage patterns for load identification and forecasting	TRACEBASE and REDD	HEMS
15	Medium	Parameters of current transients	KNN applied to examples selected by Cross-Validation strategies	Identification considering the influence of voltage variations	Laboratory data	
24	Medium	Features obtained from the PSD of the power signal	Gaussian Process Classifier	Use of multiple models in a committee voting mechanism	Laboratory data	HEMS
59	Medium	Current Duty Cycle, Slope of On-State, Variance of On-State, Zero Crossing and combinations	K-NN and Naive Bayes, DT and Adaboost classifiers	Compares different features and different classifiers	Private Data	
63	Medium	7 features extracted from the I_{RMS} FSM representation	Several supervised classifiers	Efficient method to represent long-term raw current waveforms of electric loads by FSMs.	Public Database	

Table A2. Main characteristics of selected NILM techniques with high, very high and extremely high sampling rates.

Ref #	Sampling Rate	Features	Load Identification	Contribution	Data Source	Application
119	Low and High	I_{RMS}, Average Displacement power PF, Fundamental Phase angle Average THD, and the 3th and 5th current harmonics	Self-Organizing Mapping (SOM)	Integration of NILM into a DR system	Private data	HEMS
14	High	Real and Reactive Power and Current Harmonics	Variant of HMM, against Particle Swarm Optimization (PSO)	New HMM algorithm to detect appliances and their states, for DR applications	Laboratory data	HEMS
64	High	Steady-state current harmonics and the rate of change of the transient signal after an event	Rule-Based and Naïve Bayes Classifier	Method with small complexity	Laboratory Data	
65	High	Shannon and Renyi entropies and spectral band energy for specified frequency bands of the current spectrum	Linear search of a database	Simple method for the configuration of robust and distinct load signatures	Private data	
66	High	Wavelet Transform Coefficients (WTC)	MLP	Number of WTCs reduced using Parseval's theorem	Simulation and Laboratory Data	
67	High	Maximum magnitude of the first to eighth harmonics of current, obtained by Stockwell's Transform	Ant Colony Optimization	Delivers good results for Multiple Loads	Private Data	
123	High	P and D	ANN designed using PSO	Integration of NILM into a Demand-Side Management system	Private data	HEMS
5	High and Very High	V-I Trajectory Images	Siamese ANNs, followed by DBSCAN	Detects unidentified appliances	PLAID and WHITED	
139	High, Very High	P, Q, D trajectories	PQD-PCA	Excellent classifier performance, compared with other approaches.	PLAID and BLUED	AAL

Table A2. *Cont.*

Ref #	Sampling Rate	Features	Load Identification	Contribution	Data Source	Application
12	Very High	current *WS*, *P* and *Q*, harmonics, quantized waveforms, V-I binary image	K-NN, Gaussian Naive Bayes, logistic regression classifier, SVM, linear discriminant analysis/QDA, DT, RF,Adaptive Boosting	Compares the discriminative power of different features and the performance of different classifiers	PLAID	
13	Very High	V-I Trajectory	SVM	Introduces features based on the V-I trajectory	REDD and laboratory data	
42	Very High	Current Harmonics, V-I trajectory, and *PQD*	MLPs	Uses DBSCAN for Event Detection, followed by MLP classification	BLUED and laboratory data	
60	Very High	Current (*i*)	DWT and ensemble of DTs	Investigates the effect of DWT order and the DTs number in the ensemble	Simulated Data	
69	Very High	*WS* metrics	MLPs, SVM and AdaBoost	Applicable for challenging scenarios such as multiple near-identical appliances	REDD	
77	Very High	55 steady-state and 23 transient features	Random Forest	Proposes a feature selection algorithm	PLAID	
89	Very High, Extremely High	V-I trajectories	Convolutional neural networks	CNN applied to loads identification	PLAID and WHITED	
18	Extremely High	Current amplitudes of the fundamental frequency and the 3rd and 5th harmonics	Linear search of a database	Extension of [65] for simultaneous operation of various appliances	Private data	
19	Extremely High	Current vectors (phasors) of the fundamental frequency and the 3rd and 5th harmonics	Naïve Bayes Classifier	Able to identify simultaneous combinations of small nonlinear loads	Private data	
20	Extremely High	Current vectors (phasors) of the fundamental frequency and the 3rd and 5th harmonics	Linear search of a database	Extension of [18] considering Harmonics phasors, instead of amplitudes	Private data	
71	Extremely High	EMI signals	K-NN	Able to differentiate similar switching-mode power supplies	Private Data	

References

1. Zeifman, M.; Roth, K. Nonintrusive appliance load monitoring: Review and outlook. *IEEE Trans. Consum. Electron.* **2011**, *57*, 76–84. [CrossRef]
2. Esa, N.F.; Abdullah, M.P.; Hassan, M.Y. A review disaggregation method in Non-intrusive Appliance Load Monitoring. *Renew. Sustain. Energy Rev.* **2016**, *66*, 163–173. [CrossRef]
3. Hosseini, S.S.; Agbossou, K.; Kelouwani, S.; Cardenas, A. Non-intrusive load monitoring through home energy management systems: A comprehensive review. *Renew. Sustain. Energy Rev.* **2017**, *79*, 1266–1274. [CrossRef]
4. Tabatabaei, S.M.; Dick, S.; Xu, W. Toward Non-Intrusive Load Monitoring via Multi-Label Classification. *IEEE Trans. Smart Grid* **2017**, *8*, 26–40. [CrossRef]
5. De Baets, L.; Develder, C.; Dhaene, T.; Deschrijver, D. Detection of unidentified appliances in non-intrusive load monitoring using siamese neural networks. *Int. J. Electr. Power Energy Syst.* **2019**, *104*, 645–653. [CrossRef]
6. Haq, A.U.; Jacobsen, H.-A. Prospects of Appliance-Level Load Monitoring in Off-the-Shelf Energy Monitors: A Technical Review. *Energies* **2018**, *11*, 189. [CrossRef]
7. Neuro. *Neurio Sensor W1TM Overview. Product Specification*; Neurio Technology Inc.: Vancouver, BC, Canada, 2019.
8. Smappee. *Smappee Plus Manual. Product Specification and Reference Manual*; Smappee N.V.: Harelbeke, Belgium, 2018.
9. Jin, M.; Jia, R.; Spanos, C.J. Virtual Occupancy Sensing: Using Smart Meters to Indicate Your Presence. *IEEE Trans. Mob. Comput.* **2017**, *16*, 3264–3277. [CrossRef]
10. Rafsanjani, H.N.; Ahn, C.R.; Chen, J. Linking building energy consumption with occupants' energy-consuming behaviors in commercial buildings: Non-intrusive occupant load monitoring (NIOLM). *Energy Build.* **2018**, *172*, 317–327. [CrossRef]
11. Figueiredo, M. Contributions to Electrical Energy Disaggregation in a Smart Home. Ph.D. Thesis, University of Coimbra, Coimbra, Portugal, 2013.
12. Gao, J.; Kara, E.C.; Giri, S.; Bergés, M. A feasibility study of automated plug-load identification from high-frequency measurements. In Proceedings of the 2015 IEEE Global Conference on Signal and Information Processing (GlobalSIP), Orlando, FL, USA, 14–16 December 2015; pp. 220–224.
13. Wang, A.L.; Chen, B.X.; Wang, C.G.; Hua, D. Non-intrusive load monitoring algorithm based on features of V–I trajectory. *Electr. Power Syst. Res.* **2018**, *157*, 134–144. [CrossRef]
14. Agyeman, K.; Han, S.; Han, S. Real-Time Recognition Non-Intrusive Electrical Appliance Monitoring Algorithm for a Residential Building Energy Management System. *Energies* **2015**, *8*, 9029. [CrossRef]
15. Lin, Y.; Hung, S.; Tsai, M. Study on the Influence of voltage variations for Non-Intrusive Load Identifications. In Proceedings of the 2018 International Power Electronics Conference, Niigata, Japan, 20–24 May 2018; pp. 1575–1579.
16. Ribeiro, M.; Pereira, L.; Quintal, F.; Nunes, N. SustDataED: A Public Dataset for Electric Energy Disaggregation Research. In Proceedings of the ICT for Sustainability 2016, Amsterdam, The Netherlands, 30 August 2016.
17. Lee, D. Phase noise as power characteristic of individual appliance for non-intrusive load monitoring. *Electron. Lett.* **2018**, *54*, 993–995. [CrossRef]
18. Bouhouras, A.S.; Gkaidatzis, P.A.; Chatzisavvas, K.C.; Panagiotou, E.; Poulakis, N.; Christoforidis, G.C. Load Signature Formulation for Non-Intrusive Load Monitoring Based on Current Measurements. *Energies* **2017**, *10*, 538. [CrossRef]
19. Djordjevic, S.; Simic, M. Nonintrusive identification of residential appliances using harmonic analysis. *Turk. J. Electr. Eng. Comput. Sci.* **2018**, *26*, 780–791. [CrossRef]
20. Bouhouras, A.S.; Gkaidatzis, P.A.; Panagiotou, E.; Poulakis, N.; Christoforidis, G.C. A NILM algorithm with enhanced disaggregation scheme under harmonic current vectors. *Energy Build.* **2019**, *183*, 392–407. [CrossRef]
21. Anderson, K.; Ocneanu, A.; Benitez, D.; Carlson, D.; Rowe, A.; Berges, M. BLUED: A Fully Labeled Public Dataset for Event-Based Non-Intrusive Load Monitoring Research. In Proceedings of the 2nd Workshop on Data Mining Applications in Sustainability, Beijing, China, 12 August 2012.
22. Clark, M.S. Improving the Feasibility of Energy Disaggregation in Very High- and Low-Rate Sampling Scenarios. Master's Thesis, The University of British Columbia, Vancouver, BC, Canada, 2015.

23. Adabi, A.; Manovi, P.; Mantey, P. Cost-effective instrumentation via NILM to support a residential energy management system. In Proceedings of the 2016 IEEE International Conference on Consumer Electronics (ICCE), Las Vegas, NV, USA, 7–11 January 2016; pp. 107–110.
24. Sun, X.; Wang, X.; Liu, Y.; Wu, J. Non-intrusive sensing based multi-model collaborative load identification in cyber-physical energy systems. In Proceedings of the 2014 IEEE International Instrumentation and Measurement Technology Conference (I2MTC) Proceedings, Montevideo, Uruguay, 12–15 May 2014; pp. 1–6.
25. Garcia, F.C.C.; Creayla, C.M.C.; Macabebe, E.Q.B. Development of an Intelligent System for Smart Home Energy Disaggregation Using Stacked Denoising Autoencoders. *Procedia Comput. Sci.* **2017**, *105*, 248–255. [CrossRef]
26. Le, X.; Vrigneau, B.; Sentieys, O. l1-Norm minimization based algorithm for non-intrusive load monitoring. In Proceedings of the 2015 IEEE International Conference on Pervasive Computing and Communication Workshops (PerCom Workshops), St. Louis, MO, USA, 23–27 March 2015; pp. 299–304.
27. Noury, N.; Berenguer, M.; Teyssier, H.; Bouzid, M.; Giordani, M. Building an Index of Activity of Inhabitants from Their Activity on the Residential Electrical Power Line. *IEEE Trans. Inf. Technol. Biomed.* **2011**, *15*, 758–766. [CrossRef] [PubMed]
28. Ming-Chun, L.; Yung-Chi, C.; Shiao-Li, T.; Wenshiang, T. Design and implementation of a home and building gateway with integration of nonintrusive load monitoring meters. In Proceedings of the 2012 IEEE International Conference on Industrial Technology, Athens, Greece, 19–21 March 2012; pp. 148–153.
29. Pereira, L.; Nunes, N. Performance evaluation in non-intrusive load monitoring: Datasets, metrics, and tools—A review. *Wiley Interdiscip. Rev. Data Min. Knowl. Discov.* **2018**, *8*, 1265. [CrossRef]
30. Zico Kolter, J.; Johnson, M.J. REDD: A Public Data Set for Energy Disaggregation Research. In Proceedings of the SustKDD workshop on Data Mining Applications in Sustainability, San Diego, CA, USA, 21 August 2011.
31. Gao, J.; Giri, S.; Kara, E.C.; Berg, M. PLAID: A public dataset of high-resoultion electrical appliance measurements for load identification research: Demo abstract. In Proceedings of the 1st ACM Conference on Embedded Systems for Energy-Efficient Buildings, Memphis, Tennessee, 3–6 November 2014; pp. 198–199.
32. Firth, S.K.; Cockbill, S.; Dimitriou, V.; Hargreaves, T.; Hassan, T.M.; Hauxwell-Baldwin, R.; Kane, T.; Liao, J.; May, A.; Murray, D.; et al. *Smart Homes and Saving Energy: The REFIT Project Final Report for Industry and Government*; Loughborough University: Loughborough, UK, 2015.
33. Reinhardt, A.; Baumann, P.; Burgstahler, D.; Hollick, M.; Chonov, H.; Werner, M.; Steinmetz, R. On the accuracy of appliance identification based on distributed load metering data. In Proceedings of the 2012 Sustainable Internet and ICT for Sustainability (SustainIT), Pisa, Italy, 4–5 October 2012; pp. 1–9.
34. Kahl, M.; Haq, A.U.; Kriechbaumer, T.; Jacobsen, H.-A. WHITED-a worldwide household and industry transient energy data set. In Proceedings of the 3rd International Workshop on Non-Intrusive Load Monitoring (NILM), Vancouver, BC, Canada, 14–15 May 2016.
35. Kelly, J.; Knottenbelt, W. The UK-DALE dataset, domestic appliance-level electricity demand and whole-house demand from five UK homes. *Sci. Data* **2015**, *2*, 150007. [CrossRef]
36. Nambi, A.S.N.U.; Lua, A.R.; Prasad, V.R. LocED: Location-aware Energy Disaggregation Framework. In Proceedings of the 2nd ACM International Conference on Embedded Systems for Energy-Efficient Built Environments, Seoul, Korea, 4–5 November 2015; pp. 45–54.
37. Anderson, K.D.; Bergés, M.E.; Ocneanu, A.; Benitez, D.; Moura, J.M.F. Event detection for Non Intrusive load monitoring. In Proceedings of the 38th Annual Conference on IEEE Industrial Electronics Society, Montreal, QC, Canada, 25–28 October 2012; pp. 3312–3317.
38. Lucas, P.; Filipe, Q.; Rodolfo, G.; Nuno Jardim, N. SustData: A Public Dataset for ICT4S Electric Energy Research. In Proceedings of the ICT for Sustainability 2014 (ICT4S-14), Stockholm, Sweden, 24–27 Aug 2014.
39. Weiss, M.; Helfenstein, A.; Mattern, F.; Staake, T. Leveraging smart meter data to recognize home appliances. In Proceedings of the 2012 IEEE International Conference on Pervasive Computing and Communications, Lugano, Switzerland, 19–23 March 2012; pp. 190–197.
40. Alcalá, J.M.; Ureña, J.; Hernández, Á. Event-based detector for non-intrusive load monitoring based on the Hilbert Transform. In Proceedings of the 2014 IEEE Emerging Technology and Factory Automation (ETFA), Barcelona, Spain, 16–19 September 2014; pp. 1–4.
41. Zhao, B.; He, K.; Stankovic, L.; Stankovic, V. Improving Event-Based Non-Intrusive Load Monitoring Using Graph Signal Processing. *IEEE Access* **2018**, *6*, 53944–53959. [CrossRef]

42. Zheng, Z.; Chen, H.; Luo, X. A Supervised Event-Based Non-Intrusive Load Monitoring for Non-Linear Appliances. *Sustainability* **2018**, *10*, 1001. [CrossRef]
43. Yang, C.C.; Soh, C.S.; Yap, V.V. Comparative Study of Event Detection Methods for Non-intrusive Appliance Load Monitoring. *Energy Procedia* **2014**, *61*, 1840–1843. [CrossRef]
44. Alcalá, J.; Ureña, J.; Hernández, Á.; Gualda, D. Event-Based Energy Disaggregation Algorithm for Activity Monitoring from a Single-Point Sensor. *IEEE Trans. Instrum. Meas.* **2017**, *66*, 2615–2626. [CrossRef]
45. Meziane, M.N.; Ravier, P.; Lamarque, G.; Bunetel, J.L.; Raingeaud, Y. High accuracy event detection for Non-Intrusive Load Monitoring. In Proceedings of the 2017 IEEE International Conference on Acoustics, Speech and Signal Processing (ICASSP), New Orleans, LA, USA, 5–9 March 2017; pp. 2452–2456.
46. Ducange, P.; Marcelloni, F.; Marinari, D. An algorithm based on finite state machines with fuzzy transitions for non-intrusive load disaggregation. In Proceedings of the 2012 Sustainable Internet and ICT for Sustainability (SustainIT), Pisa, Italy, 4–5 October 2012; pp. 1–5.
47. Le, T.-T.-H.; Kim, H. Non-Intrusive Load Monitoring Based on Novel Transient Signal in Household Appliances with Low Sampling Rate. *Energies* **2018**, *11*, 3409. [CrossRef]
48. Liu, Q.; Kamoto, K.M.; Liu, X.; Sun, M.; Linge, N. Low-Complexity Non-Intrusive Load Monitoring Using Unsupervised Learning and Generalized Appliance Models. *IEEE Trans. Consum. Electron.* **2019**, *65*, 28–37. [CrossRef]
49. Figueiredo, M.; Ribeiro, B.; Almeida, A.d. Electrical Signal Source Separation Via Nonnegative Tensor Factorization Using On Site Measurements in a Smart Home. *IEEE Trans. Instrum. Meas.* **2014**, *63*, 364–373. [CrossRef]
50. Parson, O.; Ghosh, S.; Weal, M.; Rogers, A. An unsupervised training method for non-intrusive appliance load monitoring. *Artif. Intell.* **2014**, *217*, 1–19. [CrossRef]
51. Tomás, A.T.V. Inference Methods for Nonintrusive Load Monitoring Applications. Ph.D. Thesis, Instituto Superior Tecnico, Lisboa, Portugal, 2014.
52. Chang, H.-H.; Wiratha, P.W.; Chen, N. A Non-intrusive Load Monitoring System Using an Embedded System for Applications to Unbalanced Residential Distribution Systems. *Energy Procedia* **2014**, *61*, 146–150. [CrossRef]
53. Kelly, J.; Knottenbelt, W. Neural NILM: Deep Neural Networks Applied to Energy Disaggregation. In Proceedings of the 2nd ACM International Conference on Embedded Systems for Energy-Efficient Built Environments, Seoul, Korea, 4–5 November 2015; pp. 55–64.
54. Welikala, S.; Thelasingha, N.; Akram, M.; Ekanayake, P.B.; Godaliyadda, R.I.; Ekanayake, J.B. Implementation of a robust real-time non-intrusive load monitoring solution. *Appl. Energy* **2019**, *238*, 1519–1529. [CrossRef]
55. Figueiredo, M.; Ribeiro, B.; Almeida, A.d. Analysis of trends in seasonal electrical energy consumption via non-negative tensor factorization. *Neurocomputing* **2015**, *170*, 318–327. [CrossRef]
56. Bache, K.; Lichman, M. UCI Machine Learning Repository. Available online: http://archive.ics.uci.edu/ml (accessed on 20 April 2019).
57. Makonin, S.; Popowich, F.; Bartram, L.; Gill, B.; Bajić, I.V. AMPds: A public dataset for load disaggregation and eco-feedback research. In Proceedings of the 2013 IEEE Electrical Power & Energy Conference, Halifax, NS, Canada, 21–23 August 2013; pp. 1–6.
58. Makonin, S. Real-Time Embedded Low-Frequency Load Disaggregation. Ph.D. Thesis, Simon Fraser University, Burnaby, BC, Canada, 2014.
59. Lu-Lulu, L.-L.; Park, S.-W.; Wang, B.-H. Electric Load Signature Analysis for Home Energy Monitoring System. *Int. J. Fuzzy Log. Intell. Syst.* **2012**, *12*, 193–197. [CrossRef]
60. Alshareef, S.; Morsi, W.G. Application of wavelet-based ensemble tree classifier for non-intrusive load monitoring. In Proceedings of the 2015 IEEE Electrical Power and Energy Conference (EPEC), London, ON, Canada, 26–28 October 2015; pp. 397–401.
61. Elhamifar, E.; Sastry, S. Energy disaggregation via learning 'Powerlets' and sparse coding. In Proceedings of the Twenty-Ninth AAAI Conference on Artificial Intelligence, Austin, TX, USA, 25–30 January 2015; pp. 629–635.
62. Patri, O.P.; Panangadan, A.V.; Chelmis, C.; Prasanna, V.K. Extracting discriminative features for event-based electricity disaggregation. In Proceedings of the 2014 IEEE Conference on Technologies for Sustainability (SusTech), Portland, OR, USA, 24–26 July 2014; pp. 232–238.

63. Du, L.; Yang, Y.; He, D.; Harley, R.G.; Habetler, T.G. Feature Extraction for Load Identification Using Long-Term Operating Waveforms. *IEEE Trans. Smart Grid* **2015**, *6*, 819–826. [CrossRef]
64. Meehan, P.; McArdle, C.; Daniels, S. An Efficient, Scalable Time-Frequency Method for Tracking Energy Usage of Domestic Appliances Using a Two-Step Classification Algorithm. *Energies* **2014**, *7*, 7041. [CrossRef]
65. Bouhouras, A.S.; Milioudis, A.N.; Labridis, D.P. Development of distinct load signatures for higher efficiency of NILM algorithms. *Electr. Power Syst. Res.* **2014**, *117*, 163–171. [CrossRef]
66. Chang, H.; Lian, K.; Su, Y.; Lee, W. Power-Spectrum-Based Wavelet Transform for Nonintrusive Demand Monitoring and Load Identification. *IEEE Trans. Ind. Appl.* **2014**, *50*, 2081–2089. [CrossRef]
67. Lin, Y.; Tsai, M. Development of an Improved Time–Frequency Analysis-Based Nonintrusive Load Monitor for Load Demand Identification. *IEEE Trans. Instrum. Meas.* **2014**, *63*, 1470–1483. [CrossRef]
68. Nakajima, H.; Nagasawa, K.; Shishido, Y.; Kagiya, Y.; Takagi, Y. The state estimation of existing home appliances using signal analysis technique. In Proceedings of the SICE Annual Conference, Sapporo, Japan, 9–12 September 2014; pp. 1247–1252.
69. Hassan, T.; Javed, F.; Arshad, N. An Empirical Investigation of V-I Trajectory Based Load Signatures for Non-Intrusive Load Monitoring. *IEEE Trans. Smart Grid* **2014**, *5*, 870–878. [CrossRef]
70. Liang, J.; Ng, S.K.K.; Kendall, G.; Cheng, J.W.M. Load Signature Study—Part I: Basic Concept, Structure, and Methodology. *IEEE Trans. Power Deliv.* **2010**, *25*, 551–560. [CrossRef]
71. Gupta, S.; Reynolds, M.S.; Patel, S.N. ElectriSense: Single-point sensing using EMI for electrical event detection and classification in the home. In Proceedings of the 12th ACM International Conference on Ubiquitous Computing, Copenhagen, Denmark, 26–29 September 2010; pp. 139–148.
72. Elbe, C.; Schmautzer, E. Appliance-specific usage patterns for load disaggregation methods. In Proceedings of the Internationale Energiewirtschaftstagung an der TU Wien, Vienna, Austria, 13–15 February 2013; pp. 1–4.
73. Zeifman, M. Disaggregation of home energy display data using probabilistic approach. *IEEE Trans. Consum. Electron.* **2012**, *58*, 23–31. [CrossRef]
74. Piga, D.; Cominola, A.; Giuliani, M.; Castelletti, A.; Rizzoli, A.E. Sparse Optimization for Automated Energy End Use Disaggregation. *IEEE Trans. Control Syst. Technol.* **2016**, *24*, 1044–1051. [CrossRef]
75. Wang, H.; Yang, W. An Iterative Load Disaggregation Approach Based on Appliance Consumption Pattern. *Appl. Sci.* **2018**, *8*, 542. [CrossRef]
76. Zhai, S.; Wang, Z.; Yan, X.; He, G. Appliance Flexibility Analysis Considering User Behavior in Home Energy Management System Using Smart Plugs. *IEEE Trans. Ind. Electron.* **2019**, *66*, 1391–1401. [CrossRef]
77. Wytock, M.; Zico Kolter, J. Contextually Supervised Source Separation with Application to Energy Disaggregation. In Proceedings of the AAAI'14 Twenty-Eighth AAAI Conference on Artificial Intelligence, Québec, QC, Canada, 27–31 July 2013; pp. 486–492.
78. Liang, J.; Ng, S.K.K.; Kendall, G.; Cheng, J.W.M. Load Signature Study—Part II: Disaggregation Framework, Simulation, and Applications. *IEEE Trans. Power Deliv.* **2010**, *25*, 561–569. [CrossRef]
79. Sadeghianpourhamami, N.; Ruyssinck, J.; Deschrijver, D.; Dhaene, T.; Develder, C. Comprehensive feature selection for appliance classification in NILM. *Energy Build.* **2017**, *151*, 98–106. [CrossRef]
80. Mayhorn, E.T.; Sullivan, G.P.; Petersen, J.M.; Butner, R.S.; Johnson, E.M. *Load Disaggregation Technologies: Real World and Laboratory Performance*; American Council for an Energy-Efficient Economy: Washington, DC, USA; Pacific Northwest National Lab. (PNNL): Richland, WA, USA, 2016.
81. Kong, W.; Dong, Z.Y.; Hill, D.J.; Luo, F.; Xu, Y. Improving Nonintrusive Load Monitoring Efficiency via a Hybrid Programing Method. *IEEE Trans. Ind. Inform.* **2016**, *12*, 2148–2157. [CrossRef]
82. Egarter, D.; Sobe, A.; Elmenreich, W. Evolving Non-Intrusive Load Monitoring. In *Lecture Notes in Computer Science*; Esparcia-Alcázar, A.I., Ed.; Springer: Berlin/Heidelberg, Germany, 2013; Volume 7835, pp. 182–191.
83. Kong, W.; Dong, Z.Y.; Ma, J.; Hill, D.J.; Zhao, J.; Luo, F. An Extensible Approach for Non-Intrusive Load Disaggregation with Smart Meter Data. *IEEE Trans. Smart Grid* **2018**, *9*, 3362–3372. [CrossRef]
84. Chang, H.-H. Non-Intrusive Demand Monitoring and Load Identification for Energy Management Systems Based on Transient Feature Analyses. *Energies* **2012**, *5*, 4569. [CrossRef]
85. Wu, Q.; Wang, F. Concatenate Convolutional Neural Networks for Non-Intrusive Load Monitoring across Complex Background. *Energies* **2019**, *12*, 1572. [CrossRef]
86. Kim, J.; Le, T.-T.-H.; Kim, H. Nonintrusive Load Monitoring Based on Advanced Deep Learning and Novel Signature. *Comput. Intell. Neurosci.* **2017**, *2017*, 4216281. [CrossRef]

87. Devlin, M.; Hayes, B. Non-Intrusive Load Monitoring Using Electricity Smart Meter Data: A Deep Learning Approach. 2018. Available online: https://www.researchgate.net/publication/328784204_Non-Intrusive_Load_Monitoring_using_Electricity_Smart_Meter_Data_A_Deep_Learning_Approach (accessed on 1 March 2019).
88. Singh, S.; Majumdar, A. Deep Sparse Coding for Non–Intrusive Load Monitoring. *IEEE Trans. Smart Grid* **2018**, *9*, 4669–4678. [CrossRef]
89. Fan, C.; Xiao, F.; Zhao, Y.; Wang, J. Analytical investigation of autoencoder-based methods for unsupervised anomaly detection in building energy data. *Appl. Energy* **2018**, *211*, 1123–1135. [CrossRef]
90. De Baets, L.; Ruyssinck, J.; Develder, C.; Dhaene, T.; Deschrijver, D. Appliance classification using VI trajectories and convolutional neural networks. *Energy Build.* **2018**, *158*, 32–36. [CrossRef]
91. Xia, M.; Liu, W.A.; Wang, K.; Zhang, X.; Xu, Y. Non-intrusive load disaggregation based on deep dilated residual network. *Electr. Power Syst. Res.* **2019**, *170*, 277–285. [CrossRef]
92. Figueiredo, M.B.; de Almeida, A.; Ribeiro, B. An Experimental Study on Electrical Signature Identification of Non-Intrusive Load Monitoring (NILM) Systems. In Proceedings of the 10th ICANNGA, Ljubljana, Slovenia, 14–16 April 2011; pp. 31–40.
93. Kramer, O.; Wilken, O.; Beenken, P.; Hein, A.; Hüwel, A.; Klingenberg, T.; Meinecke, C.; Raabe, T.; Sonnenschein, M. *On Ensemble Classifiers for Nonintrusive Appliance Load Monitoring*; Springer: Berlin/Heidelberg, Germany, 2012; pp. 322–331.
94. Giri, S.; Bergés, M.; Rowe, A. Towards automated appliance recognition using an EMF sensor in NILM platforms. *Adv. Eng. Inform.* **2013**, *27*, 477–485. [CrossRef]
95. Barker, S.; Musthag, M.; Irwin, D.; Shenoy, P. Non-intrusive load identification for smart outlets. In Proceedings of the 2014 IEEE International Conf. on Smart Grid Communications, Venice, Italy, 3–6 November 2014; pp. 548–553.
96. He, H.; Liu, Z.; Jiao, R.; Yan, G. A Novel Nonintrusive Load Monitoring Approach based on Linear-Chain Conditional Random Fields. *Energies* **2019**, *12*, 1797. [CrossRef]
97. Srinivasan, D.; Ng, W.S.; Liew, A.C. Neural-network-based signature recognition for harmonic source identification. *IEEE Trans. Power Deliv.* **2006**, *21*, 398–405. [CrossRef]
98. Wang, Z.; Zheng, G. Residential Appliances Identification and Monitoring by a Nonintrusive Method. *IEEE Trans. Smart Grid* **2012**, *3*, 80–92. [CrossRef]
99. Yang, C.C.; Soh, C.S.; Yap, V.V. A systematic approach to ON-OFF event detection and clustering analysis of non-intrusive appliance load monitoring. *Front. Energy* **2015**, *9*, 231–237. [CrossRef]
100. Wong, Y.F.; Şekercioğlu, Y.A.; Drummond, T.; Wong, V.S. Recent approaches to non-intrusive load monitoring techniques in residential settings. In Proceedings of the 2013 IEEE Computational Intelligence Applications in Smart Grid (CIASG), Singapore, 16–19 April 2013; pp. 73–79.
101. Aiad, M.; Lee, P.H. Non-intrusive load disaggregation with adaptive estimations of devices main power effects and two-way interactions. *Energy Build.* **2016**, *130*, 131–139. [CrossRef]
102. Cominola, A.; Giuliani, M.; Piga, D.; Castelletti, A.; Rizzoli, A.E. A Hybrid Signature-based Iterative Disaggregation algorithm for Non-Intrusive Load Monitoring. *Appl. Energy* **2017**, *185*, 331–344. [CrossRef]
103. Cutsem, O.V.; Lilis, G.; Kayal, M. Automatic multi-state load profile identification with application to energy disaggregation. In Proceedings of the 2017 22nd IEEE International Conference on Emerging Technologies and Factory Automation (ETFA), Limassol, Cyprus, 12–15 September 2017; pp. 1–8.
104. Kong, W.; Dong, Z.Y.; Hill, D.J.; Ma, J.; Zhao, J.H.; Luo, F.J. A Hierarchical Hidden Markov Model Framework for Home Appliance Modeling. *IEEE Trans. Smart Grid* **2018**, *9*, 3079–3090. [CrossRef]
105. Mueller, J.A.; Kimball, J.W. Accurate Energy Use Estimation for Nonintrusive Load Monitoring in Systems of Known Devices. *IEEE Trans. Smart Grid* **2018**, *9*, 2797–2808. [CrossRef]
106. Zhao, B.; Stankovic, L.; Stankovic, V. On a Training-Less Solution for Non-Intrusive Appliance Load Monitoring Using Graph Signal Processing. *IEEE Access* **2016**, *4*, 1784–1799. [CrossRef]
107. He, K.; Stankovic, L.; Liao, J.; Stankovic, V. Non-Intrusive Load Disaggregation Using Graph Signal Processing. *IEEE Trans. Smart Grid* **2018**, *9*, 1739–1747. [CrossRef]
108. DOE. *Quadrennial Technology Review: An assessment of Energy Technologies and Research Opportunities—Chapter 1: Energy Challenges*; Department of Energy: Washington, DC, USA, 2015.
109. Beaudin, M.; Zareipour, H. Home energy management systems: A review of modelling and complexity. *Renew. Sustain. Energy Rev.* **2015**, *45*, 318–335. [CrossRef]

110. Liu, Y.; Qiu, B.; Fan, X.; Zhu, H.; Han, B. Review of Smart Home Energy Management Systems. *Energy Procedia* **2016**, *104*, 504–508. [CrossRef]
111. Manic, M.; Wijayasekara, D.; Amarasinghe, K.; Rodriguez-Andina, J.J. Building Energy Management Systems: The Age of Intelligent and Adaptive Buildings. *IEEE Ind. Electron. Mag.* **2016**, *10*, 25–39. [CrossRef]
112. Shareef, H.; Ahmed, M.S.; Mohamed, A.; Hassan, E.A. Review on Home Energy Management System Considering Demand Responses, Smart Technologies, and Intelligent Controllers. *IEEE Access* **2018**, *6*, 24498–24509. [CrossRef]
113. Zairi, A.; Chaabene, M. A review on home energy management systems. In Proceedings of the 9th International Renewable Energy Congress, Hammamet, Tunisia, 20–22 March 2018; pp. 1–6.
114. Zhou, B.; Li, W.; Chan, K.W.; Cao, Y.; Kuang, Y.; Liu, X.; Wang, X. Smart home energy management systems: Concept, configurations, and scheduling strategies. *Renew. Sustain. Energy Rev.* **2016**, *61*, 30–40. [CrossRef]
115. Pau, G.; Collotta, M.; Ruano, A.; Qin, J. Smart Home Energy Management. *Energies* **2017**, *10*, 382. [CrossRef]
116. Yan, X.; Ozturk, Y.; Hu, Z.; Song, Y. A review on price-driven residential demand response. *Renew. Sustain. Energy Rev.* **2018**, *96*, 411–419. [CrossRef]
117. Kelly, J.; Knottenbelt, W. Does disaggregated electricity feedback reduce domestic electricity consumption? A systematic review of the literature. In Proceedings of the 3rd International NILM Workshop, Vancouver, BC, Canada, 14–15 May 2016.
118. Zhuang, M.; Shahidehpour, M.; Li, Z. An Overview of Non-Intrusive Load Monitoring: Approaches, Business Applications, and Challenges. In Proceedings of the 2018 International Conference on Power System Technology, Guangzhou, China, 6–8 November 2018.
119. D'hulst, R.; Labeeuw, W.; Beusen, B.; Claessens, S.; Deconinck, G.; Vanthournout, K. Demand response flexibility and flexibility potential of residential smart appliances: Experiences from large pilot test in Belgium. *Appl. Energy* **2015**, *155*, 79–90. [CrossRef]
120. Kohlhepp, P.; Harb, H.; Wolisz, H.; Waczowicz, S.; Müller, D.; Hagenmeyer, V. Large-scale grid integration of residential thermal energy storages as demand-side flexibility resource: A review of international field studies. *Renew. Sustain. Energy Rev.* **2019**, *101*, 527–547. [CrossRef]
121. He, D.; Lin, W.; Liu, N.; Harley, R.G.; Habetler, T.G. Incorporating Non-Intrusive Load Monitoring Into Building Level Demand Response. *IEEE Trans. Smart Grid* **2013**, *4*, 1870–1877. [CrossRef]
122. Lin, Y.; Tsai, M. An Advanced Home Energy Management System Facilitated by Nonintrusive Load Monitoring with Automated Multiobjective Power Scheduling. *IEEE Trans. Smart Grid* **2015**, *6*, 1839–1851. [CrossRef]
123. Lin, Y.-H.; Hu, Y.-C. Residential Consumer-Centric Demand-Side Management Based on Energy Disaggregation-Piloting Constrained Swarm Intelligence: Towards Edge Computing. *Sensors* **2018**, *18*, 1365. [CrossRef] [PubMed]
124. Khan, W.Z.; Ahmed, E.; Hakak, S.; Yaqoob, I.; Ahmed, A. Edge computing: A survey. *Future Gener. Comput. Syst.* **2019**, *97*, 219–235. [CrossRef]
125. Lin, Y.-H.; Hu, Y.-C. Electrical Energy Management Based on a Hybrid Artificial Neural Network-Particle Swarm Optimization-Integrated Two-Stage Non-Intrusive Load Monitoring Process in Smart Homes. *Processes* **2018**, *6*, 236. [CrossRef]
126. Xia, C.; Li, W.; Chang, X.; Delicato, F.; Yang, T.; Zomaya, A. Edge-based Energy Management for Smart Homes. In Proceedings of the 2018 IEEE 16th International Conference on Dependable, Autonomic and Secure Computing, 16th International Conference on Pervasive Intelligence and Computing, 4th International Conference on Big Data Intelligence and Computing and Cyber Science and Technology Congress(DASC/PiCom/DataCom/CyberSciTech), Athens, Greece, 12–15 August 2018; pp. 849–856.
127. Sculley, D. Web-scale k-means clustering. In Proceedings of the 19th international conference on World wide web, Raleigh, NC, USA, 26–30 April 2010; pp. 1177–1178.
128. Pelleg, D.; Moore, A.W. X-means: Extending K-means with Efficient Estimation of the Number of Clusters. In Proceedings of the 17th International Conference on Machine Learning, Pittsburgh, PA, USA, 29 June–2 July 2000; pp. 727–734.
129. Dinesh, C.; Welikala, S.; Liyanage, Y.; Ekanayake, M.P.B.; Godaliyadda, R.I.; Ekanayake, J. Non-intrusive load monitoring under residential solar power influx. *Appl. Energy* **2017**, *205*, 1068–1080. [CrossRef]

130. Welikala, S.; Dinesh, C.; Ekanayake, M.P.B.; Godaliyadda, R.I.; Ekanayake, J. Incorporating Appliance Usage Patterns for Non-Intrusive Load Monitoring and Load Forecasting. *IEEE Trans. Smart Grid* **2019**, *10*, 448–461. [CrossRef]
131. Alcalá, J.M.; Ureña, J.; Hernández, Á.; Gualda, D. Assessing Human Activity in Elderly People Using Non-Intrusive Load Monitoring. *Sensors* **2017**, *17*, 351. [CrossRef]
132. Corbishley, P.; Rodriguez-Villegas, E. Breathing Detection: Towards a Miniaturized, Wearable, Battery-Operated Monitoring System. *IEEE Trans. Biomed. Eng.* **2008**, *55*, 196–204. [CrossRef] [PubMed]
133. Wang, A.; Chen, G.; Yang, J.; Zhao, S.; Chang, C. A Comparative Study on Human Activity Recognition Using Inertial Sensors in a Smartphone. *IEEE Sens. J.* **2016**, *16*, 4566–4578. [CrossRef]
134. Massot, B.; Noury, N.; Gehin, C.; McAdams, E. On designing an ubiquitous sensor network for health monitoring. In Proceedings of the 2013 IEEE 15th International Conference on e-Health Networking, Applications and Services (Healthcom 2013), Lisbon, Portugal, 9–12 October 2013; pp. 310–314.
135. Zhang, X.; Kato, T.; Matsuyama, T. Learning a context-aware personal model of appliance usage patterns in smart home. In Proceedings of the IEEE Innovative Smart Grid Technologies, Kuala Lumpur, Malaysia, 20–23 May 2014; pp. 73–78.
136. Patrono, L.; Rametta, P.; Meis, J. Unobtrusive Detection of Home Appliance's Usage for Elderly Monitoring. In Proceedings of the 2018 3rd International Conference on Smart and Sustainable Technologies (SpliTech), Split, Croatia, 26–29 June 2018; pp. 1–6.
137. Alcalá, J.; Parson, O.; Rogers, A. Detecting Anomalies in Activities of Daily Living of Elderly Residents via Energy Disaggregation and Cox Processes. In Proceedings of the 2nd ACM International Conference on Embedded Systems for Energy-Efficient Built Environments, Seoul, Korea, 4–5 November 2015; pp. 225–234.
138. Chalmers, C.; Hurst, W.; Mackay, M.; Fergus, P. Smart Monitoring: An Intelligent System to Facilitate Health Care across an Ageing Population. In Proceedings of the The Eighth International Conference on Emerging Networks and Systems Intelligence, Venice, Italy, 9–13 October 2016.
139. Belley, C., Gaboury, S.; Bouchard, B.; Bouzouane, A. An efficient and inexpensive method for activity recognition within a smart home based on load signatures of appliances. *Pervasive Mob. Comput.* **2014**, *12*, 58–78. [CrossRef]
140. Patrono, L.; Primiceri, P.; Rametta, P.; Sergi, I.; Visconti, P. An innovative approach for monitoring elderly behavior by detecting home appliance's usage. In Proceedings of the 2017 25th Int. Conference on Software, Telecommunications and Computer Networks (SoftCOM), Split, Croatia, 21–23 September 2017; pp. 1–7.
141. Alcalá, J.M.; Ureña, J.; Hernández, Á.; Gualda, D. Sustainable Homecare Monitoring System by Sensing Electricity Data. *IEEE Sens. J.* **2017**, *17*, 7741–7749. [CrossRef]
142. Malik, A.; Haghdadi, N.; MacGill, I.; Ravishankar, J. Appliance level data analysis of summer demand reduction potential from residential air conditioner control. *Appl. Energy* **2019**, *235*, 776–785. [CrossRef]
143. Razavi, R.; Gharipour, A.; Fleury, M.; Akpan, I.J. Occupancy detection of residential buildings using smart meter data: A large-scale study. *Energy Build.* **2019**, *183*, 195–208. [CrossRef]
144. Ferreira, P.M.; Ruano, A.E.; Silva, S.; Conceicao, E.Z.E. Neural Networks based predictive control for thermal comfort and energy savings in public buildings. *Energy Build.* **2012**, *55*, 238–251. [CrossRef]
145. Ruano, A.E.; Pesteh, S.; Silva, S.; Duarte, H.; Mestre, G.; Ferreira, P.M.; Khosravani, H.R.; Horta, R. The IMBPC HVAC system: A complete MBPC solution for existing HVAC systems. *Energy Build.* **2016**, *120*, 145–158. [CrossRef]
146. Rashid, H.; Singh, P.; Stankovic, V.; Stankovic, L. Can non-intrusive load monitoring be used for identifying an appliance's anomalous behaviour? *Appl. Energy* **2019**, *238*, 796–805. [CrossRef]
147. Khosravani, H.; Castilla, M.; Berenguel, M.; Ruano, A.; Ferreira, P. A Comparison of Energy Consumption Prediction Models Based on Neural Networks of a Bioclimatic Building. *Energies* **2016**, *9*, 57. [CrossRef]
148. Hong, W.-C.; Li, M.-W.; Fan, G.-F. (Eds.) *Short-Term Load Forecasting by Artificial Intelligent Technologies*; MDPI: Basel, Switzerland, 2019.
149. Voyant, C.; Notton, G.; Kalogirou, S.; Nivet, M.-L.; Paoli, C.; Motte, F.; Fouilloy, A. Machine learning methods for solar radiation forecasting: A review. *Renew. Energy* **2017**, *105*, 569–582. [CrossRef]
150. Mestre, G.; Ruano, A.; Duarte, H.; Silva, S.; Khosravani, H.; Pesteh, S.; Ferreira, P.; Horta, R. An Intelligent Weather Station. *Sensors* **2015**, *15*, 31005–31022. [CrossRef] [PubMed]

151. Luo, F.; Ranzi, G.; Kong, W.; Dong, Z.Y.; Wang, S.; Zhao, J. Non-intrusive energy saving appliance recommender system for smart grid residential users. *IET Gener. Transm. Distrib.* **2017**, *11*, 1786–1793. [CrossRef]
152. Lindahl, P.A.; Green, D.H.; Bredariol, G.; Aboulian, A.; Donnal, J.S.; Leeb, S.B. Shipboard Fault Detection Through Nonintrusive Load Monitoring: A Case Study. *IEEE Sens. J.* **2018**, *18*, 8986–8995. [CrossRef]

Article

Energy Sustainability in Smart Cities: Artificial Intelligence, Smart Monitoring, and Optimization of Energy Consumption

Kwok Tai Chui [1,*], Miltiadis D. Lytras [2,3] and Anna Visvizi [2,4]

1 Department of Electronic Engineering, City University of Hong Kong, Hong Kong SAR, China
2 School of Business & Economics, Deree College—The American College of Greece, Gravias 6, 153 42 Aghia Paraskevi, Greece; mlytras@acg.edu (M.D.L.); avisvizi@acg.edu (A.V.)
3 Effat College of Engineering, Effat University, P.O. Box 34689, Jeddah 21478, Saudi Arabia
4 Effat College of Business, Effat University, P.O. Box 34689, Jeddah 21478, Saudi Arabia
* Correspondence: ktchui3-c@my.cityu.edu.hk

Received: 10 October 2018; Accepted: 19 October 2018; Published: 23 October 2018

Abstract: Energy sustainability is one of the key questions that drive the debate on cities' and urban areas development. In parallel, artificial intelligence and cognitive computing have emerged as catalysts in the process aimed at designing and optimizing smart services' supply and utilization in urban space. The latter are paramount in the domain of energy provision and consumption. This paper offers an insight into pilot systems and prototypes that showcase in which ways artificial intelligence can offer critical support in the process of attaining energy sustainability in smart cities. To this end, this paper examines smart metering and non-intrusive load monitoring (NILM) to make a case for the latter's value added in context of profiling electric appliances' electricity consumption. By employing the findings in context of smart cities research, the paper then adds to the debate on energy sustainability in urban space. Existing research tends to be limited by data granularity (not in high frequency) and consideration of about six kinds of appliances. In this paper, a hybrid genetic algorithm support vector machine multiple kernel learning approach (GA-SVM-MKL) is proposed for NILM, with consideration of 20 kinds of appliance. Genetic algorithm helps to solve the multi-objective optimization problem and design the optimal kernel function based on various kernel properties. The performance indicators are sensitivity (S_e), specificity (S_p) and overall accuracy (OA) of the classifier. First, the performance evaluation of proposed GA-SVM-MKL achieves S_e of 92.1%, S_p of 91.5% and OA of 91.8%. Second, the percentage improvement of performance indicators using proposed method is more than 21% compared with traditional kernel. Third, results reveal that by keeping different modes of electric appliance as identical class label, the performance indicators can increase to about 15%. Forth, tunable modes of GA-SVM-MKL classifier are proposed to further enhance the performance indicators up to 7%. Overall, this paper is a bold and novel contribution to the debate on energy utilization and sustainability in urban spaces as it integrates insights from artificial intelligence, IoT, and big data analytics and queries them in a context defined by energy sustainability in smart cities.

Keywords: artificial intelligence; demand response; energy; policy making; genetic algorithm; multiple kernel learning; non-intrusive load monitoring; smart grid; smart metering; support vector machine; smart cities; smart villages

---

1. Introduction

Cities are the major consumers of electricity today. Considering the correlation that exists between energy consumption, the environmental footprint it leaves, and the implications for and of global warming [1,2], energy sustainability emerges as one of the key questions that beholds the stakeholders, including the industry, decision-makers and the society. Consensus has emerged that replacing old electrical infrastructure by smart grid might be the most effective way of addressing the challenge worldwide. Microgrid applications like transactive energy framework [3,4], energy management [5–7] and advanced retail electricity market [8], play an important role in context of smart grid development. Microgrids are typically supported by generators or renewable wind and solar energy resources and are often used to provide backup power or to supplement the main power grid during periods of heavy demand. A microgrid strategy that integrates local wind or solar resources can provide redundancy for essential services and make the main grid less susceptible to localized disaster. Smart metering is one of the key features that conditions the functioning of a smart grid [9]. By 2020, worldwide, the estimated number of smart meters will exceed 800 million, while the penetration rate will be 50% [10,11]. The question is to what extent and how smart metering may contribute to attaining greater efficiency of a smart grid, e.g., by optimizing it. To address this question, this paper employs advances in artificial intelligence and big data analytics to query in which ways their integrated use in context of smart metering and smart grid optimization may yield positive results in the form of decreased energy consumption and greater energy sustainability. Inserting the discussion in context of smart cities, adds an additional twist to this discussion. The argument is structured as follows. In the first section, a review of load monitoring methods is discussed briefly to highlight the value added of non-intrusive load monitoring. Next, the research methodology is outlined, which is followed by overview of empirical testing and analysis. Section 5 evaluates the performance of proposed method and its comparison with related work. Finally a conclusion is drawn.

2. Related Works—Non-Intrusive Load Monitoring (NILM) and Its Value Added

The evolution of modern advanced computational forecasting methods provides new tools for electricity forecasting and pattern recognition. According to individual smart data and smart metering techniques will have a great impact in the efficiency of smart energy solutions. In addition, artificial intelligence techniques and smart grid approaches can set up sophisticated services for the optimization of energy consumption. Toward this direction advanced demand modelling using machine learning algorithms will offer new predicting capabilities. Furthermore, Big Data context increases the complexity of the problem and also requires novel mining techniques based on energy time series for behavioral analytics. Therefore, user behavior and analysis is directly linked, as is integrated behavioral analytics and smart energy modelling, metering and solutions.

Recent research focused on intrusive load monitoring (ILM) and non-intrusive load monitoring (NILM). A study concluded that load monitoring can reduce 20% electricity consumption [12]. In contrast, ILM is distributed sensing, whereas NILM is single-point sensing. ILM uses more than one smart meter per apartment (could be one smart meter per power outlet), but NILM uses only one smart meter in the apartment. Theoretically, more smart meters can yield higher accuracy for the detection of appliance consumption, because the number of appliances that need to be disaggregated is lower [13]. However, disadvantages exist. These include: High cost, complex smart metering network configuration, and management. This paper focuses specifically on NILM and its value added.

Figure 1 shows the general architecture of NILM for electricity suppliers, companies and users. The NILM benefits to electricity suppliers, manufacturers and users. Electricity suppliers can achieve a more accurate demand response by understanding the electricity consumption profile of each electric appliance. Therefore, a better energy demand prediction model can be achieved using the usage pattern. Furthermore, it tries to lower the gap between total electricity supply and demand, in other words, the electricity wastage attributable to unused electricity decreases (it is worth mentioning that the energy supply is always larger than energy demand to ensure it can still fulfill the demand requirement if abrupt increase in demand occurs). For manufacturers, they would be able to develop a better understanding of the relationship between appliances and their usage patterns. One may focus on increasing the energy efficiency of frequently used and power hungry appliances. Last, the electricity consumption pattern of each appliance may correct the misunderstanding of end users whom normally have no idea on electricity consumption. They can formulate a direction to reduce the electricity bill, especially in power hungry appliance.

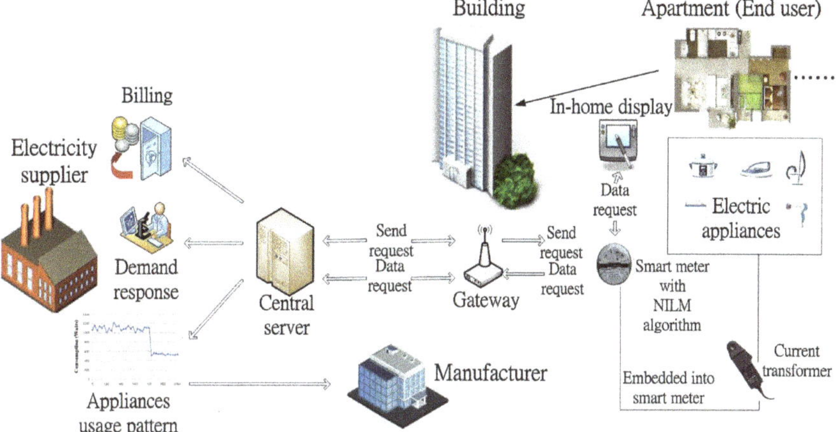

Figure 1. Non-intrusive load monitoring (NILM)—general architecture: Electricity suppliers, companies and end users.

Various approaches for NILM have been proposed. For instance, decision tree [14,15], graph signal processing [16], hidden Markov model [17,18], k-nearest neighbor [19], clustering [20] and cepstrum-smoothing [21]. It can be seen that the detection interval for some works is not real-time, 8 s in [15] and 1 min in [16–18,20]. This is often impractical because the actual operation time for an electric appliance is usually not a divider of 1-min or 8 s. When it comes to NILM, unsupervised or supervised classification is required. It is invalid to define the class label when the operation time is not a divider of the detection interval. Thus, a real-time detection interval 50 Hz or 60 Hz is required, which depends on the line voltage standard of the district. The works in [16,20,21] adopt detection interval of 60 Hz, 50 Hz and 0.5 s respectively. However, these works focused on NILM of 4 or 6 electric appliances, which are far from adequate in the practical situation. The details of [16–21], as well as comparison between proposed work and these works will be discussed in Section 5.5.

In this paper, a hybrid generic algorithm support vector machine multiple kernel learning (GA-SVM-MKL) approach has been proposed for NILM of 20 electric appliances. Genetic algorithm helps to solve the multi-objective optimization problem and design the optimal kernel function based on various kernel properties. SVM is adopted owning to the fact that it takes key advantages in (i) avoid over-fitting; (ii) kernel trick; (iii) convex optimization problem; and (iv) good out-of-sample generalization. The contribution is as follows (i) GA-SVM-MKL is capable of analyzing and disaggregating the energy profile of single point into list of 20 common types of operating electric appliances, which is far more than that in existing works; (ii); GA-SVM-MKL achieves Sensitivity (S_e) of 92.1–98.4%, Specificity (S_p) of 91.5–98.8% and overall accuracy (OA) of 91.8–98.6% and (iii) Tunable modes of GA-SVM-MKL is introduced to enhance the classification performance by 7% because we can reduce the number of types of appliances in certain period in order to reduce the complexity of model and thus increase the performance of classification model.

The rest of the paper is organized as follows. The methodology and formulation of the proposed algorithm is presented in Section 2. Section 3 carries out performance evaluation of the proposed algorithm and comparison is made with existing methods. Finally, a conclusion is drawn in Section 4.

3. Research Methodology and Research Problem Formulation

This paper examines to what extent and how smart metering may contribute to attaining greater efficiency of smart grid, for example by optimizing it by deploying advances from the fields of artificial intelligence and big data analytics. To address this question, several hypotheses have been made, as well as corresponding research, including literature review and primary research. Figure 2 depicts the methodology and the workflow. In brief, the research presented here draws from insights from three converging fields of scientific inquiry to rethink the question of smart grid optimization. These insights include:

- Insights from artificial intelligence (AI) and cognitive computing and the value added they bring into the process of designing, managing and utilizing smart energy systems
- Insights from smart cities and smart villages research, as well as considerations specific to the debate on sustainability, including the SDGs, and their value added consistent with an emphasis on wellbeing and inclusive socio-economic growth and development
- Insights from the broad field pertinent to energy supply and demand and related questions the value added if ICT-driven coherent and effective policymaking

It is at the intersection of these three broad domains that our research question is located. Accordingly, the more specific research questions that this paper will address include: In which ways novel ICT-enhanced solutions, including algorithms and data integration, can contribute to efficient and sustainable consumption of resources, like energy.

"What is the optimal design of classification model for NILM application". The multiple objectives optimization problem will be solved by multi-objective genetic algorithm.

"Can we reduce the number of types of appliances in certain period in order to reduce the complexity of model and thus increase the performance of classification model". This will be addressed in Section 5.4.

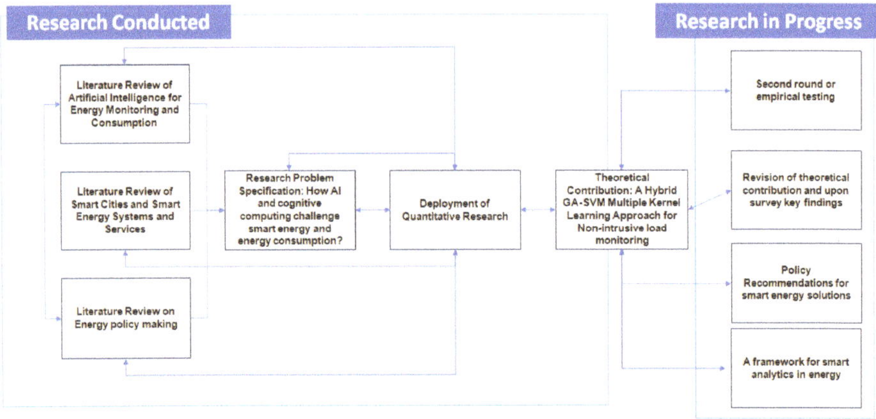

Figure 2. Flow chart of research methodology. AI, artificial intelligence; GA-SVM, generic algorithm support vector machine.

4. Overview of Empirical Testing and Analysis

The general flow of GA-SVM-MKL classifier for NILM is given in Figure 3. The smart meter will measure the current and voltage waveform of the apartment continuously. Both waveforms are carried out signal preprocessing includes dc offset elimination, interval segmentation. In this paper, 0.2 s interval is selected as of the line voltage standards in Hong Kong, 220 V/50 Hz. Features of each interval segment are then computed. The features act as input for the embedded and trained GA-SVM-MKL classifier. The training of classifier includes signal preprocessing and features extraction. Then, formulation of different multi-objective SVM classifiers is carried out by various combinations of typical kernels. The multi-objective optimization problems are solved by genetic algorithm. The optimal designs of classifier under different combinations of typical kernels can be concluded. It is worth mentioning that a well-known 10-fold cross-validation is adopted for the training of classifier [22–24]. The outputs of the classifier are types of operating electric appliances and electricity consumption of operating electric appliances. Based on the outputs of the classifier, three major applications, billing, demand response and appliance usage pattern can be obtained. Electricity suppliers may utilize all applications whereas companies and end users may only utilize the appliance usage pattern.

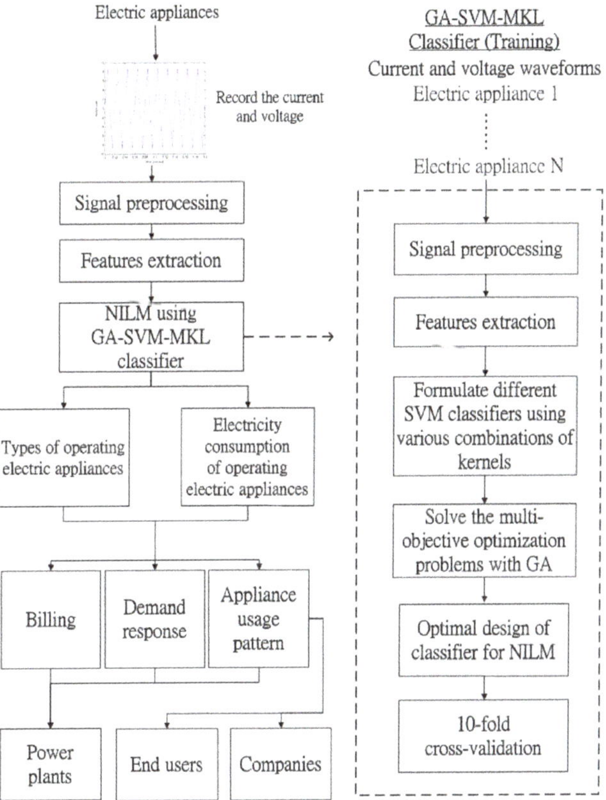

Figure 3. Flow chart for the operation of generic algorithm support vector machine multiple kernel learning (GA-SVM-MKL) classifier for NILM.

This section comprises of three subsections. First, the measurement and preparation of datasets for training and validation of GA-SVM-MKL classifier are discussed. Second, possible features for constructing the GA-SVM-MKL classifier are presented. At last, the formulation of optimal design of GA-SVM-MKL classifier is explained.

4.1. Datasets of Electric Appliances

Figure 4 shows the measurement set up for obtaining the voltage and current waveforms of all electric appliances. The voltage of 220 Vrms is measured with differential probe using cathode ray oscilloscope (CRO) with sampling frequency F_s = 10 kHz. A current transformer with ratio 50/5 is utilized for measuring the current waveform. The resistor R1 is chosen to be large value (10 MΩ), which because it has negligible effect to the circuit. The resistor R2 of 10 MΩ is connected in series with the secondary winding of current transformer to avoid open circuit.

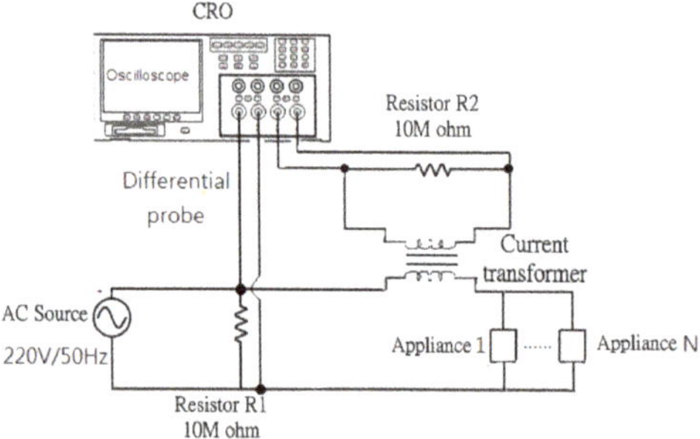

Figure 4. Measurement set up for capturing voltage and current waveforms of electric appliances. CRO, cathode ray oscilloscope.

The datasets consist of 20 electric appliances that are commonly used in typical households. The measurement allows multiple operation of electric appliances, in other words, the current waveforms may be superimposed by multiple electric appliances. Table 1 summarizes the electric appliances along with their type of activity, modes and number of brands being considered. The electric appliances can be divided into six activities, lighting, cooking, home living, computing, renovating and audio and video. There is limitation that the measurement cannot cover all brands of each electric appliance, each electric appliance has at least two brands for consideration. Likewise, there is a maximum number for each electric appliance operating at any instant in a typical household. For every combination of electric appliances, the corresponding voltage and current waveforms are recorded for 30 s (equivalent to 30 × 50 = 1500 samples). Each combination is assigned with a unique class label. It is noted that in Section 5.1, different modes of electric appliances will be assumed as identical class label and that in Section 5.2 will be assumed as different class labels.

Table 1. List of electric appliances that have been analyzed.

Type of Activity	Electric Appliance	Modes	No. of Brands	Maximum No. of Appliances at One Time
Cooking	Electric stove	2	3	2
	microwave oven	3	2	1
	cooker	3	2	1
Home living	Ironbrush	4	2	1
	Vacuum cleaner	1	2	1
	Fan	3	3	2
	Hair dryer	2	3	2
	Electric heater	2	2	1
Renovation	Electric drill	1	2	1
	Electric sander	1	2	1

Type of Activity	Electric Appliance	Modes	No. of Brands	Maximum No. of Appliances at One Time
Lighting	Fluorescent light	1	2	3
	Light bulb	1	2	3
	LED tube	1	3	3
	LED light bulb	1	2	3
Computing	Notebook	1	6	3
	desktop	1	3	3
	All in one printer and scanner	1	2	1
	Mobile charger	1	5	3
Audio and video	Radio	1	2	1
	Television	1	2	1

After measuring the voltage and current waveforms of electric appliances, the waveforms perform dc offset elimination, whichIndividual samples can be obtained by segmentation of signals with interval of 0.2 s.

4.2. Features Extraction

F The individual samples $I(n)$ and $V(n)$ are transformed to feature vector. The proposed GA-SVM-MKL adopts seven features: Maximum current ($I_{\max}$), root-mean-square current (I_{rms}), average current (I_{avg}), active power (P_{act}), apparent power (P_{app}), reactive power (P_{rea}) and power factor (PF). The features can be computed by:

$$I_{\max} = \max\{I(n)\}, \tag{1}$$

$$I_{rms} = \sqrt{[I^2(1) + I^2(2) + \ldots + I^2(F_s/50)]/(F_s/50)}, \tag{2}$$

$$I_{avg} = \frac{1}{F_s/50} \sum_{a=1}^{a=F_s/50} I(a), \tag{3}$$

$$P_{act} = \frac{1}{F_s/50} \sum_{a=1}^{a=F_s/50} I(a)V(a), \tag{4}$$

$$\begin{array}{c} P_{app} = \sqrt{[V^2(1) + \ldots + V^2(F_s/50)]/(F_s/50)} \\ \times\sqrt{[I^2(1) + \ldots + I^2(F_s/50)]/(F_s/50)} \end{array}, \tag{5}$$

$$P_{rea} = \sqrt{{P_{app}}^2 - {P_{act}}^2}, \tag{6}$$

$$PF = P_{act}/P_{app}. \tag{7}$$

It is worth mentioning that dimensionality reduction (e.g., in [11]) is not adopted because all of these features are essential for distinguishing between electric appliances in nature. The focus will be devoted on the optimal design of kernel function for building SVM classifier.

4.3. Optimal Design of GA-SVM-MKL Classifier

Denote electric appliances samples by $X_{ij}(n)$ with current $I_{ij}(n)$ and $V_{ij}(n)$ for class $i = 1, \ldots, N_c$ and $j = 1, \ldots, N_i$ where $N_i = 1500$ is the total number of samples in class i. Let feature vector be $f_{ij} = \{I_{max,ij}, I_{rms,ij}, I_{avg,ij}, P_{act,ij}, P_{app,ij}, P_{rea,ij}, PF_{ij}\}$ corresponds to $X_{ij}(n)$.

When it comes to the selection of kernels, there are five typical kernels $k(x_1, x_2)$ with inner product $\langle x_1, x_2\rangle$. They are linear kernel, qth degree polynomial kernel, complete polynomial kernel, radial basis function (RBF) kernel and sigmoid kernel. The expressions of these kernels can be summarized as follows:

$$\begin{array}{ll} \text{Linear kernel}: & k_1(x_1, x_2) = \langle x_1, x_2\rangle. \quad (8) \\ \text{qth degree polynomial kernel}: & k_2(x_1, x_2) = (\langle x_1, x_2\rangle)^q. \quad (9) \\ \text{Complete polynomial kernel}: & k_3(x_1, x_2) = (\langle x_1, x_2\rangle + c)^q. \quad (10) \\ \text{RBF kernel}: & k_4(x_1, x_2) = \exp(||x_1 - x_2||_2/2\sigma). \quad (11) \\ \text{Sigmoid kernel}: & k_5(x_1, x_2) = \tanh(\langle x_1, x_2\rangle + c), \quad (12) \end{array}$$

where $c, \sigma \in \Re, q \in N^+$.

Different kernels possess different characteristics where there is no single kernel that works well in all applications. In this paper, the proposed GA-SVM-MKL classifier adopts the idea that by combining multiple kernels (namely multiple kernel learning), the classifier can achieve better performance for

NILM after taking the advantages from each kernel. In order to combine kernels to form a new one, the kernel should obey Mercer's Theorem. According to [25], there are four properties (P):

$$P_1: \quad k(x_1, x_2) = k_i(x_1, x_2) + k_j(x_1, x_2). \tag{13}$$

$$P_2: \quad k(x_1, x_2) = c \cdot k_i(x_1, x_2), \forall c \in \Re^+. \tag{14}$$

$$P_3: \quad k(x_1, x_2) = k_i(x_1, x_2) + c, \forall c \in \Re^+. \tag{15}$$

$$P_4: \quad k(x_1, x_2) = k_i(x_1, x_2) \cdot k_j(x_1, x_2), \tag{16}$$

where $k_i : \chi \times \chi \to \Re$ and $k_j : \chi \times \chi \to \Re$ are any two Mercer kernels. It is noted that properties 1 and 4 can be further extended to infinite number of Mercer kernels.

The optimal design of classifier for NILM is formulated as a multi-objective optimization problem and solved by genetic algorithm. Multi-objective optimization is an integral part of optimization activities and has tremendous practical importance, since almost all real-world optimization problems are ideally suited to be modeled using multiple conflicting objectives [26]. Compared with single objective optimizations, which usually scalarizing multiple-objectives into one single objective, multi-objective optimization can give trade-off optimal solutions more accurately. Besides, the multi-objective optimization has multiple cardinalities of the optimal set, multiple objectives and different search spaces [27]. The objective functions constitute a multidimensional space, which is known as objective spaces [28]. The optimal solutions presented in objective spaces are referred to as Pareto optimal solutions and the set of such solutions are called Pareto Front.

As the objectives conflict with each other, it is usually impossible to obtain one single optimal objective. Therefore, for obtaining the optimal solutions in multi-objective optimizations, the most used concept is domination. Assuming for an M-objective minimization problem, candidate solution u is dominated by another candidate solution v if and only if function values of u is partially less than v, which is formulated as [26]:

$$\begin{cases} f_m(u) \geq f_m(v) & \forall m = 1, 2, \ldots, M \\ f_m(u) > f_m(v) & \exists n = 1, 2, \ldots, M \end{cases}. \tag{17}$$

Based on the concept of domination, what we prefer are the non-dominated solutions, which compose the Pareto Front. In this paper, in order to give the optimal design of classifier for NILM, multi-objective optimization genetic algorithm (MOGA) [27] for solving the multiple kernels is designed. The flow of the MOGA for the optimal design of kernel functions is shown in Figure 5. The procedures are as follows: (i) The population size and values of objective function are initialized; (ii) the values of objective function of individuals in the population are computed using the values of objective function defined in (i); (iii) ranking the individuals according to the values of objective function; (iv) the population convergence is dependent on small group of pareto optimal solutions, but not all optimal solutions attributable to the nature of the stochastic selection errors, given a limited population size; (v) niche count is introduced to enhance the population diversity by lengthening the distance between two optimal solutions along the axis of objective functions. The convergence to small group solutions will be avoided; (vi) a new offspring is generated and the values of objective functions are evaluated; (vii) ranks assignment and niche count calculation are carried out repeatedly in the new offspring; and (viii) the algorithm is terminated if it attains the maximum number of generations or if the output reaches the pareto front. It is noted that there exist other stopping criteria in literature for stochastic optimization algorithm and can be referred to [29–31].

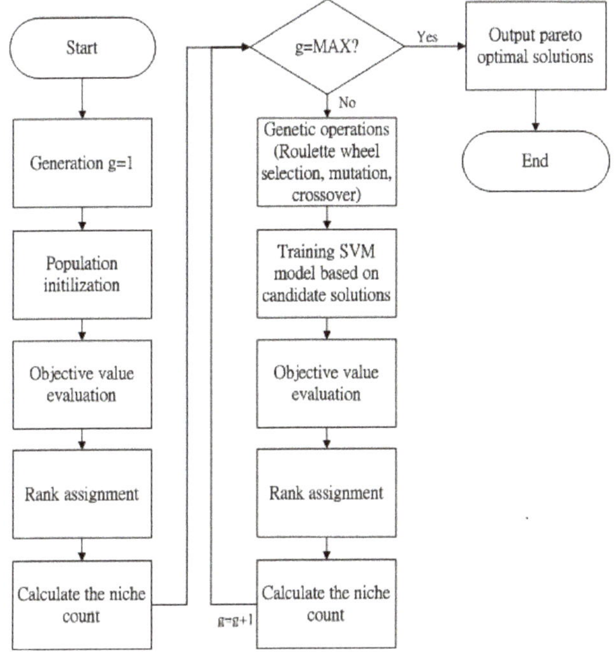

Figure 5. Flow chart of the optimal design of the classifier using GA-SVM-MKL.

The multi-objective optimization problem for NILM can be formulated as:

$$\begin{array}{ll} Max & Se \\ Max & Sp \\ Max & \widetilde{D} \end{array}, \tag{18}$$

$$s.t. \quad \alpha_j \geq 0, \sum_{j=1}^{N} \alpha_j y_j = 0, i = 1, \ldots, N, \tag{19}$$

where S_e is the sensitivity of the classifier, S_p is the specificity of the classifier, $\widetilde{D}$ is a margin equals to distance of closest samples from the hyperplane, α_j is the Lagrange multiplier and $y_j \in \{-1, +1\}$ is the output of the classifier. The three objective functions S_e, S_p and $\widetilde{D}$ are defined as:

$$S_e = TP/N_p, \tag{20}$$

$$S_p = TN/N_n, \tag{21}$$

$$\widetilde{D} = \sum_{i=1}^{N} \alpha_i - \frac{1}{2} \sum_{i=1}^{N} \sum_{j=1}^{N} \alpha_i \alpha_j y_i y_j k_{NILM}(x_i, x_j), \tag{22}$$

where TP is the number of true positive samples, TN is the number of true negative samples, N_p is the total number of positive samples, N_n is the total number of negative samples. The customized and optimized kernel for NILM, k_{NILM} varies by different combination of typical kernels in (8)–(12) using Properties 1–4 in (13)–(16). These scenarios are summarized in Appendix A Table A1, which have been studied and analyzed. It is noted that due to there are infinite scenarios settings, only property combinations of property (P), P_1, P_2, P_3, P_4, P_1P_2, P_1P_3, P_1P_4, P_1P_5, P_2P_3, P_2P_4, P_2P_5, P_3P_4,

P_3P_5, P_4P_5 are illustrated and analyzed. These 285 scenario settings cover adequate analysis for taking the advantages from individual kernel to form a multiple kernel for k_{NILM}.

The proof of combinations of property P_1P_2, P_1P_3, P_1P_4, P_1P_5, P_2P_3, P_2P_4, P_2P_5, P_3P_4, P_3P_5, P_4P_5 is shown below:

For all $r \in$ N and all sequences $(x_1, \ldots, x_r) \in X^r$ let K_1, K_2, K_3, K_4, K_{P1P2}, K_{P1P3}, K_{P1P4}, K_{P2P3}, K_{P2P4} and K_{P2P3} be the $r \times r$ matrices whose i, j-th element is given by $k_1(x_i, x_j)$, $k_2(x_i, x_j)$, $k_3(x_i, x_j)$, $k_4(x_i, x_j)$, $c_1k_1(x_i, x_j) + c_2k_2(x_i, x_j)$, $k_1(x_i, x_j) + k_2(x_i, x_j) + c$, $(k_1(x_i, x_j) + k_2(x_i, x_j))(k_3(x_i, x_j)k_4(x_i, x_j))$, $c_1k_1(x_i, x_j) + c_2$, $ck_1(x_i, x_j)k_2(x_i, x_j)$ and $(k_1(x_i, x_j) + c)k_2(x_i, x_j)$ respectively. It is required to show that K_{P1P2}, K_{P1P3}, K_{P1P4}, K_{P2P3}, K_{P2P4} and K_{P2P3} are positive semidefinite using only that K_1, K_2, K_3 and K_4 are positive semidefinite, i.e., for all $r \in \Re^r$, $\alpha' K_1 \alpha \geq 0$, $\alpha' K_2 \alpha \geq 0$, $\alpha' K_3 \alpha \geq 0$ and $\alpha' K_4 \alpha \geq 0$.

(i)
$$\alpha' K_{P1P2} \alpha = \alpha'(c_1 K_1 + c_2 K_2)\alpha \geq 0, \tag{23}$$

(ii)
$$\alpha' K_{P1P3} \alpha = \alpha'(K_1 + K_2 + c11')\alpha, \tag{24}$$
$$= \alpha' K_1 \alpha + \alpha' K_2 \alpha + c||1'\alpha||_2 \geq 0. \tag{25}$$

(iii) The $r^2 \times r^2$ matrix $H = K_1 \otimes (K_3K_4)$ and $G = K_2 \otimes (K_3K_4)$ are positive semidefinite, that is, for all $a \in \Re^{r^2}$, $a'Ha \geq 0$ and $a'a \geq 0$. Given any $\alpha \in \Re^r$, consider $a = (\alpha_1 e_1', \ldots, \alpha_r e_r') \in \Re^{r^2}$. Then

$$a'Ha = \sum_{i=1}^{r^2}\sum_{j=1}^{r^2} a_i a_j H_{ij} = \sum_{i=1}^{r}\sum_{j=1}^{r} \alpha_i \alpha_j H_{i+(i-1)r, j+(j-1)r}, \tag{26}$$

$$= \sum_{i=1}^{r}\sum_{j=1}^{r} \alpha_i \alpha_j k_1(x_i, x_j) k_3(x_i, x_j) k_4(x_i, x_j). \tag{27}$$

Similarly, it can be derived that

$$a'a = \sum_{i=1}^{r^2}\sum_{j=1}^{r^2} a_i a_j G_{ij} = \sum_{i=1}^{r}\sum_{j=1}^{r} \alpha_i \alpha_j G_{i+(i-1)r, j+(j-1)r}. \tag{28}$$

$$= \sum_{i=1}^{r}\sum_{j=1}^{r} \alpha_i \alpha_j k_2(x_i, x_j) k_3(x_i, x_j) k_4(x_i, x_j). \tag{29}$$

Thus, $a'Ha + a'a$

$$= \sum_{i=1}^{r}\sum_{j=1}^{r} \alpha_i \alpha_j (k_1(x_i, x_j) + k_2(x_i, x_j))(k_3(x_i, x_j) k_4(x_i, x_j)), \tag{30}$$

$$= \alpha' K_{P1P4} \alpha \geq 0. \tag{31}$$

(iv)
$$\alpha' K_{P2P3} \alpha = \alpha'(c_1 K_1 + c_2 11')\alpha, \tag{32}$$
$$= c_1 \alpha' K_1 \alpha + c_2 ||1'\alpha||_2 \geq 0. \tag{33}$$

(v)
$$\alpha' K_{P2P4} \alpha = \alpha'(c K_1 K_2)\alpha, \tag{34}$$
$$= c\alpha' K_1 K_2 \alpha \geq 0. \tag{35}$$

(vi)
$$\alpha' K_{P3P4} \alpha = \alpha'(K_1 + K_2 + c11')\alpha, \tag{36}$$
$$= \alpha' K_1 \alpha + \alpha' K_2 \alpha + c||1'\alpha||_2. \tag{37}$$

A 10-fold cross validation is used for the for performance evaluation of the kernels [22–24]. The classifiers are deduced using 1-against-1 multi-class SVM. This is because 1-against-1 multi-class SVM approach was generally performed better than 1-against-all multi-class SVM [25–28].

5. Performance Evaluation and Comparisons

This section is divided into five subsections. Section 5.1 discusses the performance of the proposed GA-SVM-MKL classifiers. In Section 5.2, in order to show the effectiveness of k_{NILM} using multiple kernels, the performance of classifier using k_{NILM} is compared with either single kernel is used. The feasibility study of breaking down electric appliances into different modes is discussed in Section 5.3. Intuitively, some activities like cooking and renovating are carried out in certain period. Thus, the number of classes for classifier can be reduced when these electric appliances are not in-use and the classifier is then retrained. Results in Section 5.4 support this hypothesis. Finally, comparison between proposed GA-SVM-MKL classifier and related works is carried out in Section 5.5.

5.1. Performance Evaluation of GA-SVM-MKL Classifier

285 scenarios for k_{NILM} using P_1, P_2, P_3, P_4, P_1P_2, P_1P_3, P_1P_4, P_2P_3, P_2P_4 and P_3P_4, with typical kernels k_1, k_2, k_3, k_4 and k_5 are optimally designed. The S_e, S_p and overall accuracy (OA) of the GA-SVM-MKL in each scenario are recorded as shown in Appendix A Table A2. OA is defined as the average of S_e and S_p given that the identical sample size in each class of the classifier. Probability distribution of the OAs for 285 scenarios is shown in Appendix A, as in Figure A1. The skewness and kurtosis of the OA for all scenarios are −0.0902 (left skewed) and 1.547 (heavy-tailed) respectively.

$$OA = (S_e + S_p)/2. \tag{38}$$

All results are obtained using 10-fold cross-validation. Scenario 178 using P_1P_2 achieves the best performance with S_e of 92.1%, S_p of 91.5% and OA of 91.8%. The average OA using different properties can be ranked by $OA_{P2P3} > OA_{P3P4} > OA_{P2P4} > OA_{P1P2} > OA_{P1P3} > OA_{P1P4} > OA_{P2} > OA_{P1} > OA_{P3} > OA_{P4}$ with accuracies 87.3%, 86.7%, 85.8%, 83.4%, 76.7%, 76.6%, 75.8%, 75.6%, 75.3%, and 74.7% respectively.

Results reveal that merging kernel properties and adopting multiple kernel learning can achieve better performance than using single property.

5.2. Comparisons to Single Kernel Based SVM Classifier

The performance of proposed GA-SVM-MKL classifier is compared to traditional SVM classifier using single kernel k_1, k_2, k_3, k_4 and k_5. It is noted that this SVM classifier deals with single objective maximization problem, which maximizes the margin $\tilde{D}$ which has been defined in (22). The comparison is shown in Table 2. The proposed GA-SVM-MKL classifier increases the Se, Sp and OA by 21.3–28.6%, 21.5–26.7% and 21.4–27.7% respectively. Among five scenarios using traditional SVM with k_1–k_5, the best performance is using k4, which follows by k_5, k_3, k_2 and k_1. The better performance of proposed GA-SVM-MKL can be explained by two reasons. First, GA-SVM-MKL adopts optimal kernel using multiple kernel learning with kernel properties in which it takes the advantages from each individual kernel for customization to NILM. Second, traditional SVM aims at single objective optimization, which maximizes the margin, but not S_e and S_p.

Table 2. Comparisons between Proposed and Traditional SVM Classifier.

Method	S_e (%)	S_p (%)	OA (%)
GA-SVM-MKL	92.1	91.5	91.8
SVM using k1	71.6	72.2	71.9
SVM using k2	72.3	72.8	72.6
SVM using k3	73.5	74.2	73.9
SVM using k4	75.9	75.3	75.6
SVM using k5	74.9	75.1	75

5.3. Feasibility Study of Assignment a Class Label for Different Modes of Electric Appliance

Among 20 electric appliances in this study, seven electric appliances, electric stove, microwave oven, cooker, ironbrush, fan, hair dryer and electric heater have more than one mode. These are activities of cooking and home living. In Section 5.1, it is assumed that different modes of the same electric appliances are of the same class. In this section, analysis has been made to assign different modes of the same electric appliances to be different classes. Thus, the 20 electric appliances can be extended to 32 electric appliances. Table 3 shows four scenarios S1, S2, S3 and S4 for the performance comparisons of GA-SVM-MKL classifier between before and after the assignment of new classes.

Compared between S1 and S2, the assignment of new class label for different modes of electric appliances decreases the Se, Sp and OA by 15.7%, 14.4% and 15.0% respectively. Scenarios S3 and S4 reveal that the decrease in Se, Sp and OA are mainly due to the introduction of new class labels for activities of cooking and home living. Therefore, the original assumption that different modes of same electric appliances should be considered as identical electric appliance is verified.

Table 3. Performance evaluation of assignment a class label for different modes in electric appliances.

Scenario (S1 to S4)	S_e (%)	S_p (%)	OA (%)
S1: 20 appliances (Different modes, same class)	92.1	91.5	91.8
S2: 32 appliances (Different modes, different classes)	79.6	80.0	79.8
S3: 32 appliances (only cooking related combinations)	77.4	77.1	77.3
S4: 32 appliances (only home living related combinations)	75.3	76.6	76.0

5.4. Tunable Mode for GA-SVM-MKL Classifier

Aforementioned, the 20 electric appliances for study can be divided into six activities, lighting, cooking, home living, computing, renovating and audio and video. All activities except renovating are daily used. For cooking, it is periodic activities in which users turn on the electric appliances in breakfast, lunch or dinner. Thus, it is proposed that GA-SVM-MKL classifier can be tuned for different electric appliances detection with five tunable modes (TMs).

(i) TM 1 assumes a full range classifier, in which all 20 electric appliances in six activities can be detected.

(ii) TM 2 can be selected when it is breakfast, lunch or dinner so that electric appliances of cooking should be detected by the classifier. Provided that there is no renovating, five activities, lighting, cooking, home living, computing and audio and video can be detected.

(iii) TM 3 is a non-eating period where electric appliances of cooking are not necessary. However, there is small-scale renovating activity, which allows normal activities inside the house. Five activities, including lighting, home living, computing, renovating, and audio and video, can be detected.

(iv) TM 4 assumes electric appliances related to cooking and renovating activities will not be operated. Only four activities, lighting, home living, computing and audio and video will be operated and detected.

(v) TM 5 assumes a large-scale renovating, in which only electric appliances of renovating (1 activity) are detected.

Table 4 summarizes the modes and the activities of GA-SVM-MKL classifier. For each mode, a GA-SVM-MKL classifier is trained using 10-fold cross-validation. Practically, end users can enter the period for breakfast, lunch and dinner during weekday and weekend so that GA-SVM-MKL classifier can detect electric appliances of cooking in specific time interval. Also, the ability to detect electric appliances of renovating is turned off until end users specify there is a renovation activity in their apartment.

Table 4. Various Modes in GA-SVM-MKL classifier.

Activity	Mode of Classifier				
	Mode 1	Mode 2	Mode 3	Mode 4	Mode 5
Lighting	✓	✓	✓	✓	X
Cooking	✓	✓	X	X	X
Home living	✓	✓	✓	✓	X
Computing	✓	✓	✓	✓	X
Renovating	✓	X	✓	X	✓
Audio and video	✓	✓	✓	✓	X

The S_e, S_p and OA for the classifier in TM 1 to 5 have been recorded in Table 5. A finding is observed, the S_e, S_p and OA of the classifier increase when the number of activities (or classes) decreases. This may be explained by fewer classes, the classification problem is less complex. Thus, it is shown that the proposed mode tunable GA-SVM-MKL classifier can help improving the Se, Sp and OA for NILM. Compared between TM1 and TM2-TM5, the percentage improvement using tunable mode is ranged (1.85%, 6.84%), (2.84%, 7.98%), (2.51%, 7.41%) for Se, Sp and OA respectively.

Table 5. Performance Evaluation of GA-SVM-MKL classifier in each tunable modes (TM).

S_e ǀ S_p and OA (%) of GA-SVM-MKL Classifier									
TM 1		TM 2		TM 3		TM 4		TM 5	
Se	Sp	Se	Sp	Se	Sp	Se	Sp	Se	Sp
92.1	91.5	93.8	94.3	94.6	94.1	96.6	96.1	98.4	98.8
OA	91.8	OA	94.1	OA	94.4	OA	96.4	OA	98.6

5.5. Comparisons to Related Works

Related works for NILM include different methods like decision tree [14,15], graph signal processing [16], hidden Markov model [17,18], k-nearest neighbor [19], clustering [20] and cepstrum-smoothing [21]. The features, datasets, cross-validation, detection interval, and OA of each method have been summarized in Table 6. It should be noted that related work in [18] focused on building a probabilistic appliance model which has been generalized to match previously unseen households; thus, it did not involve any classifier for NILM.

Table 6. Performance comparisons between GA-SVM-MKL and related works.

Work	Features	Dataset	Cross-Validation	Detection Interval	OA
Decision tree and wavelet transform [14]	approximation level and detail level	Four electric appliances—battery charger, compact fluorescent lamp, personal computer and incandescent light bulb (total 864 samples)	No	0.0167 s (60 Hz)	96.65%
Decision tree method [15]	the first increasing edge at the start of the event and the last decreasing edge at the end of the event	Ten Activities—cooking, washing, laundering, cleaning, watching TV, listening to radio, games, computing, hobbies and socializing (unknown sample size)	N/A	8 s	59%
Graph signal processing [16]	Active power edges	Nine electric appliances—416 microwave oven, 311 washer dryer, 61 oven, 330 lighting, 2228 refrigerator, 264 dishwasher, 138 stove, 62 heater and 54 air conditioner	No	1 min	77.2%
Factorial Hidden Markov Model [17]	Factorial main factors	Five electric appliances—90 microwave oven, 121 electric stove, 883 refrigerator, 58 dishwasher and 189 lighting	N/A	1 min	70.84%

Table 6. *Cont.*

Work	Features	Dataset	Cross-Validation	Detection Interval	OA
Hidden Markov model [18]	log-odds score	Four electric appliances—2228 refrigerator, 416 microwave oven, 311 washing machine and 264 dishwasher	50-fold	1 min	N/A
k-nearest neighbor and artificial neural network [19]	maximum, average and root mean square of the current wave in transient stage	Four electric appliances—27 Fan, 30 Fluorescent light, 19 radio and 18 microwave oven	No	0.02 s (50 Hz)	97.87%
Clustering [20]	real power, reactive power, apparent power and voltage features	Seven electric appliances—800 oven, 56 refrigerator, 452 dishwasher, 65 lighting, 154 washer, 443 microwave oven, 236 dryer	No	1 min	77.6%
Cepstrum-smoothing [21]	Frequency and amplitude of the dominant peaks in the smoothed cepstrum	Six electric appliances—television, computer, monitor, refrigerator, washer and vacuum cleaner (unknown sample size)	N/A	0.5 s	96.37%
Proposed GA-SVM-MKL	$I_{\max}$, I_{rms}, I_{avg}, P_{act}, P_{app}, P_{rea} and PF	20 electric appliances—Fluorescent light, light bulb, LED tube, LED light bulb, electric stove, microwave oven, cooker, ironbrush, vacuum cleaner, fan, hair dryer, electric heater, notebook, desktop, all in one printer and scanner, mobile charger, electric drill, electric sander, radio and television (each of 1500 samples)	10-fold	0.02 s (50 Hz)	91.8%, 94.1%, 94.4%, 96.4%, 98.6% for TM1 to TM5

It can be seen that the existing works [15–17,20] using detection interval of 8 s or 1-min interval, which is far from using real-time data. There are two concerns for using these detection intervals. First, the operation time of electric appliances is generally not a divider of 8 s or 1-min. It is difficult to define the class label. On the other hand, it increases the difficulty for the classification, because (i) detection interval of 8 s, researchers are expected to find out whether the actual operation time of electric appliance is 1 s, 2 s, ... or 8 s. (ii) detection interval of 1-min, likewise, the determination of operation time of electric appliance equals 1 s, 2 s, ... or 60 s is required. Thus, related works in [15–17,20] achieve S_e, S_p and OA less than 80%.

The detection intervals in [14,19,21] are 60 Hz, 0.5 s and 50 Hz respectively. For OA, these works achieve 96.65% [14], 94.87% [19] and 96.37% [21]. However, these works only consider the NILM of 4 or 6 electric appliances, which is much less than that in this paper (20 electric appliances). Also, previous works are lack of or without mentioned one of the most important part in the performance evaluation, cross-validation. One can pick up a bias training dataset to train the classifier so that the results are not convincing and reliable. In aforementioned related works, S_e and S_p are not given, which is believed to be important criteria to evaluate both the accuracies in determining the true positive and true negative samples. It is noted that when S_e and S_p are far from each other, the chance of having bias in some classifiers (toward specific classes) is high.

By comparing the GA-SVM-MKL TM 1-TM5 with [14,19,21] their OAs are similar. Thus, it can be concluded that the proposed method achieves good performance in NILM when the number of electric appliances is extended to 20.

We have to comment on the adoption of this method in the real world. Smart metering on real time basis is quite complicated research problem. The evolution of machine learning techniques along with real time sensors and big data capabilities will increase our capacity to model, meter and analyze behavioral patterns over energy consumption. This will help us a lot to understand the linkages between behavior and energy consumption. From a decision support point of view, irrelevant of the programing and development environments, e.g., smart grid, the key challenge is to be capable of aggregating smart energy data for advanced computational processing. Within this context some of the most challenging future research directions can be:

- Standardization of Smart Energy data sets;

- Interoperability in the Energy Smart Grid;
- Adoption of machine learning techniques for the provision and measurement of Behavioral analytics;
- Integration of Smart Grid approaches in Energy Sector with a new era of Key Performance Indicators (KPIs) and Energy Analytics;
- Large scale experimentation with millions of electrical devices for pattern analysis;
- Optimization of electricity consumption on real time basis based on smart energy data;
- Ontological Engineering and Semantic Annotation of smart energy data.

6. Conclusions

Considering the energy sustainability challenge cities/urban areas are exposed to today, the objective of this paper was to examine ways of optimizing the use of electricity consumption and suggest ways of employing these solutions in cities'/urban areas' context. Specifically, the research presented in this paper focused on the question of to what extent and how smart metering may contribute to attaining greater efficiency of smart grid. The hypothesis underlying the research was that an integrated approach consistent with engaging insights from (i) artificial intelligence, cognitive computing and big data analytics, (ii) smart cities and smart villages research, and (iii) energy sustainability debate, may yield novel findings. In fact, having employed a complex methodology, as a result of research discussed in this paper a genetic algorithm support vector machine multiple kernel learning (GA-SVM-MKL) approach has been proposed for NILM. A customized kernel has been designed using typical kernel functions with kernel properties. This approach is customized to specific problem, which is NILM for energy disaggregation. Applying kernel properties in various types of kernels can increase the performance of the classifier. Three objective functions have been solved for the optimal design of the classifier to detect 20 common household electric appliances with five tunable modes. The effectiveness of GA-SVM-MKL has been demonstrated. To this end, (i) 20 common types of of electric appliances have been considered, which is far more than that in existing works (at most 10 as in Table 6); (ii) it achieves S_e of 92.1–98.4%, S_p of 91.5–98.8% and OA of 91.8–98.6%; and (iii) tunable modes of GA-SVM-MKL is introduced to enhance the classification performance by 7%.

The authors are aware of the limitations of this research. The consideration of the number of types of appliance, the number of modes and brands, as well as the maximum number of appliance is limited. The coverage of the dataset could be extended when it comes to large-scale study. In addition, investigation of the feature extraction could be one of the solutions to further improve the accuracy of the classifier.

The contribution of this paper to the research agenda outlined in the Special Issue titled Artificial Intelligence for Smart Grid is multifold:

First, from a technical point of view it demonstrates the capacity of AI techniques to model complex problems and to simulate optimized solutions. Furthermore, it proves the new era of computational problems where the creation and consumption of big data requires efficient and coherent approaches integrating IoT, big data analytics and AI algorithms:

- Insights from artificial intelligence (AI) and cognitive computing and the value added they bring into the process of smart systems [32]
- Insights from smart cities as well as considerations specific to the debate on sustainability, including the SDGs, and their value added consistent with an emphasis on wellbeing and inclusive socio-economic growth and development [33,34]
- Insights from the broad field pertinent to energy supply and demand and related questions the value added if ICT-driven coherent and effective policymaking [35–37].

Second, from a strategic management and sustainability point of view, this paper heralds the onset of a new era of energy-focused data-driven decision-making. This new era defined by the imperative of energy sustainability requires dynamic real time distributed infrastructure and techniques to manage and utilize data flows from millions of devices (IoT), It also requires high speed networks that can bring together all stakeholders, including energy produces, providers, businesses, end-users, decisionmakers. This suggests that new research is needed that would focus on the question of how blockchain technology may effectively serve this role [37]. Indeed, this is subject of our research in-progress.

Additionally, the decision-making point of view, the arguments outlined in this paper suggest that more attention needs to be devoted to the work in progress undertaken by key stakeholders involved in efforts geared toward optimizing electricity consumption. This includes the key electric appliances producers, as well as key actors involved in devising regulatory frameworks, incl. the Organization for Economic Cooperation and Development (OECD) and the European Union (EU). Arguably, several of actions undertaken by these actors would benefit from the findings discussed in this paper.

In the direction of future research, several interesting new research areas promote the interdisciplinary nature of sustainable smart energies research: Based on [38,39] the evolution of individual smart data and smart metering techniques together with advanced Artificial Intelligence and Machine Learning approaches will set up new challenges for intelligent energy agents. Sophisticated and complicated modelling of energy consumption will also allow new analytical processing and predicting capabilities [38]. The evolution of Data Mining, multidimensional data based and distributed DataWarehouses, together with Cloud Services will promote the vision of Enengies' Software, Platform and Infrastructure as a Service [39,40]. In this direction, user behavior and a behavioral analysis is directly linked, as is integrated behavioral analytics and smart energy modelling, metering and solutions [41]. We plan very shortly to present a global survey on the social impact of Big Data for Sustainable Energy.

Author Contributions: Conceptualization, K.T.C., M.D.L. and A.V.; Methodology, K.T.C.; Validation, K.T.C., M.D.L. and A.V.; Formal Analysis, K.T.C.; Writing-Original Draft Preparation, K.T.C., M.D.L. and A.V.; Writing-Review & Editing, K.T.C., M.D.L. and A.V.

Funding: This research received no external funding. M.D.L. and A.V. would like to thank Effat University in Jeddah, Saudi Arabia, for offering resources for conducting research through the Research and Consultancy Institute.

Conflicts of Interest: The authors declare no conflict of interest.

Appendix A

Table A1. Scenario setting for k_{NILM} using properties 1–4 with typical kernels.

No.	P	k_{NILM}	No.	P	k_{NILM}	No.	P	k_{NILM}	No.	P	k_{NILM}	No.	P	k_{NILM}
1	1	$k_1 + k_1$	2	1	$k_1 + k_2$	3	1	$k_1 + k_3$	4	1	$k_1 + k_4$	5	1	$k_1 + k_5$
6	1	$k_2 + k_2$	7	1	$k_2 + k_3$	8	1	$k_2 + k_4$	9	1	$k_2 + k_5$	10	1	$k_3 + k_3$
11	1	$k_3 + k_4$	12	1	$k_3 + k_5$	13	1	$k_4 + k_4$	14	1	$k_4 + k_5$	15	1	$k_5 + k_5$
16	2	ck_1	17	2	ck_2	18	2	ck_3	19	2	ck_4	20	2	ck_5
21	3	$k_1 + c$	22	3	$k_2 + c$	23	3	$k_3 + c$	24	3	$k_4 + c$	25	3	$k_5 + c$
26	4	k_1k_1	27	4	k_1k_2	28	4	k_1k_3	29	4	k_1k_4	30	4	k_1k_5
31	4	k_2k_2	32	4	k_2k_3	33	4	k_2k_4	34	4	k_2k_5	35	4	k_3k_3
36	4	k_3k_4	37	4	k_3k_5	38	4	k_4k_4	39	4	k_4k_5	40	4	k_5k_5
41	1.2	$c_1k_1 + c_2k_2$	42	1.2	$ck_1 + k_2$	43	1.2	$k_1 + ck_2$	44	1.2	$c_1k_1 + c_2k_3$	45	1.2	$ck_1 + k_3$
46	1.2	$k_1 + ck_3$	47	1.2	$c_1k_1 + c_2k_4$	48	1.2	$ck_1 + k_4$	49	1.2	$k_1 + ck_4$	50	1.2	$c_1k_1 + c_2k_5$
51	1.2	$ck_1 + k_5$	52	1.2	$k_1 + ck_5$	53	1.2	$c_1k_2 + c_2k_3$	54	1.2	$ck_2 + k_3$	55	1.2	$k_2 + ck_3$
56	1.2	$c_1k_2 + c_2k_4$	57	1.2	$ck_2 + k_4$	58	1.2	$k_2 + ck_4$	59	1.2	$c_1k_2 + c_2k_5$	60	1.2	$ck_2 + k_5$
61	1.2	$k_2 + ck_5$	62	1.2	$c_1k_3 + c_2k_4$	63	1.2	$ck_3 + k_4$	64	1.2	$k_3 + ck_4$	65	1.2	$c_1k_3 + c_2k_5$
66	1.2	$ck_3 + k_5$	67	1.2	$k_3 + ck_5$	68	1.2	$c_1k_4 + c_2k_5$	69	1.2	$ck_4 + k_5$	70	1.2	$k_4 + ck_5$
71	1.3	$k_1 + k_1 + c$	72	1.3	$k_1 + k_2 + c$	73	1.3	$k_1 + k_3 + c$	74	1.3	$k_1 + k_4 + c$	75	1.3	$k_1 + k_5 + c$
76	1.3	$k_2 + k_2 + c$	77	1.3	$k_2 + k_3 + c$	78	1.3	$k_2 + k_4 + c$	79	1.3	$k_2 + k_5 + c$	80	1.3	$k_3 + k_3 + c$
81	1.3	$k_3 + k_4 + c$	82	1.3	$k_3 + k_5 + c$	83	1.3	$k_4 + k_4 + c$	84	1.3	$k_4 + k_5 + c$	85	1.3	$k_5 + k_5 + c$
86	1.4	$k_1k_1 + k_1$	87	1.4	$k_1k_1 + k_2$	88	1.4	$k_1k_1 + k_3$	89	1.4	$k_1k_1 + k_4$	90	1.4	$k_1k_1 + k_5$
91	1.4	$k_1k_2 + k_1$	92	1.4	$k_1k_2 + k_2$	93	1.4	$k_1k_2 + k_3$	94	1.4	$k_1k_2 + k_4$	95	1.4	$k_1k_2 + k_5$
96	1.4	$k_1k_3 + k_1$	97	1.4	$k_1k_3 + k_2$	98	1.4	$k_1k_3 + k_3$	99	1.4	$k_1k_3 + k_4$	100	1.4	$k_1k_3 + k_5$
101	1.4	$k_1k_4 + k_1$	102	1.4	$k_1k_4 + k_2$	103	1.4	$k_1k_4 + k_3$	104	1.4	$k_1k_4 + k_4$	105	1.4	$k_1k_4 + k_5$
106	1.4	$k_1k_5 + k_1$	107	1.4	$k_1k_5 + k_2$	108	1.4	$k_1k_5 + k_3$	109	1.4	$k_1k_5 + k_4$	110	1.4	$k_1k_5 + k_5$
111	1.4	$k_2k_2 + k_1$	112	1.4	$k_2k_2 + k_2$	113	1.4	$k_2k_2 + k_3$	114	1.4	$k_2k_2 + k_4$	115	1.4	$k_2k_2 + k_5$
116	1.4	$k_2k_3 + k_1$	117	1.4	$k_2k_3 + k_2$	118	1.4	$k_2k_3 + k_3$	119	1.4	$k_2k_3 + k_4$	120	1.4	$k_2k_3 + k_5$
121	1.4	$k_2k_4 + k_1$	122	1.4	$k_2k_4 + k_2$	123	1.4	$k_2k_4 + k_3$	124	1.4	$k_2k_4 + k_4$	125	1.4	$k_2k_4 + k_5$
126	1.4	$k_2k_5 + k_1$	127	1.4	$k_2k_5 + k_2$	128	1.4	$k_2k_5 + k_3$	129	1.4	$k_2k_5 + k_4$	130	1.4	$k_2k_5 + k_5$
131	1.4	$k_3k_3 + k_1$	132	1.4	$k_3k_3 + k_2$	133	1.4	$k_3k_3 + k_3$	134	1.4	$k_3k_3 + k_4$	135	1.4	$k_3k_3 + k_5$
136	1.4	$k_3k_4 + k_1$	137	1.4	$k_3k_4 + k_2$	138	1.4	$k_3k_4 + k_3$	139	1.4	$k_3k_4 + k_4$	140	1.4	$k_3k_4 + k_5$
141	1.4	$k_3k_5 + k_1$	142	1.4	$k_3k_5 + k_2$	143	1.4	$k_3k_5 + k_3$	144	1.4	$k_3k_5 + k_4$	145	1.4	$k_3k_5 + k_5$
146	1.4	$k_4k_4 + k_1$	147	1.4	$k_4k_4 + k_2$	148	1.4	$k_4k_4 + k_3$	149	1.4	$k_4k_4 + k_4$	150	1.4	$k_4k_4 + k_5$

Table A1. *Cont.*

No.	P	k_{NILM}	No.	P	k_{NILM}	No.	P	k_{NILM}	No.	P	k_{NILM}	No.	P	k_{NILM}
151	1.4	$k_4k_5 + k_1$	15yh72	1.4	$k_4k_5 + k_2$	153	1.4	$k_4k_5 + k_3$	154	1.4	$k_4k_5 + k_4$	155	1.4	$k_4k_5 + k_5$
156	1.4	$k_5k_5 + k_1$	157	1.4	$k_5k_5 + k_2$	158	1.4	$k_5k_5 + k_3$	159	1.4	$k_5k_5 + k_4$	160	1.4	$k_5k_5 + k_5$
161	2.3	$c_1k_1(k_1 + c_2)$	162	2.3	$c_1k_1(k_2 + c_2)$	163	2.3	$c_1k_1(k_3 + c_2)$	164	2.3	$c_1k_1(k_4 + c_2)$	165	2.3	$c_1k_1(k_5 + c_2)$
166	2.3	$c_1k_2(k_1 + c_2)$	167	2.3	$c_1k_2(k_2 + c_2)$	168	2.3	$c_1k_2(k_3 + c_2)$	169	2.3	$c_1k_2(k_4 + c_2)$	170	2.3	$c_1k_2(k_5 + c_2)$
171	2.3	$c_1k_3(k_1 + c_2)$	172	2.3	$c_1k_3(k_2 + c_2)$	173	2.3	$c_1k_3(k_3 + c_2)$	174	2.3	$c_1k_3(k_4 + c_2)$	175	2.3	$c_1k_3(k_5 + c_2)$
176	2.3	$c_1k_4(k_1 + c_2)$	177	2.3	$c_1k_4(k_2 + c_2)$	178	2.3	$c_1k_4(k_3 + c_2)$	179	2.3	$c_1k_4(k_4 + c_2)$	180	2.3	$c_1k_4(k_5 + c_2)$
181	2.3	$c_1k_5(k_1 + c_2)$	182	2.3	$c_1k_5(k_2 + c_2)$	183	2.3	$c_1k_5(k_3 + c_2)$	184	2.3	$c_1k_5(k_4 + c_2)$	185	2.3	$c_1k_5(k_5 + c_2)$
186	2.4	$ck_1(k_1k_2)$	187	2.4	$ck_1(k_1k_3)$	188	2.4	$ck_1(k_1k_4)$	189	2.4	$ck_1(k_1k_5)$	190	2.4	$ck_1(k_2k_3)$
191	2.4	$ck_1(k_2k_4)$	192	2.4	$ck_1(k_2k_5)$	193	2.4	$ck_1(k_3k_4)$	194	2.4	$ck_1(k_3k_5)$	195	2.4	$ck_1(k_4k_5)$
196	2.4	$ck_2(k_1k_2)$	197	2.4	$ck_2(k_1k_3)$	198	2.4	$ck_2(k_1k_4)$	199	2.4	$ck_2(k_1k_5)$	200	2.4	$ck_2(k_2k_3)$
201	2.4	$ck_2(k_2k_4)$	202	2.4	$ck_2(k_2k_5)$	203	2.4	$ck_2(k_3k_4)$	204	2.4	$ck_2(k_3k_5)$	205	2.4	$ck_2(k_4k_5)$
206	2.4	$ck_3(k_1k_2)$	207	2.4	$ck_3(k_1k_3)$	208	2.4	$ck_3(k_1k_4)$	209	2.4	$ck_3(k_1k_5)$	210	2.4	$ck_3(k_2k_3)$
211	2.4	$ck_3(k_2k_4)$	212	2.4	$ck_3(k_2k_5)$	213	2.4	$ck_3(k_3k_4)$	214	2.4	$ck_3(k_3k_5)$	215	2.4	$ck_3(k_4k_5)$
216	2.4	$ck_4(k_1k_2)$	217	2.4	$ck_4(k_1k_3)$	218	2.4	$ck_4(k_1k_4)$	219	2.4	$ck_4(k_1k_5)$	220	2.4	$ck_4(k_2k_3)$
221	2.4	$ck_4(k_2k_4)$	222	2.4	$ck_4(k_2k_5)$	223	2.4	$ck_4(k_3k_4)$	224	2.4	$ck_4(k_3k_5)$	225	2.4	$ck_4(k_4k_5)$
226	2.4	$ck_5(k_1k_2)$	227	2.4	$ck_5(k_1k_3)$	228	2.4	$ck_5(k_1k_4)$	229	2.4	$ck_5(k_1k_5)$	230	2.4	$ck_5(k_2k_3)$
231	2.4	$ck_5(k_2k_4)$	232	2.4	$ck_5(k_2k_5)$	233	2.4	$ck_5(k_3k_4)$	234	2.4	$ck_5(k_3k_5)$	235	2.4	$ck_5(k_4k_5)$
236	3.4	$(k_1 + c)k_1k_2$	237	3.4	$(k_1 + c)k_1k_3$	238	3.4	$(k_1 + c)k_1k_4$	239	3.4	$(k_1 + c)k_1k_5$	240	3.4	$(k_1 + c)k_2k_3$
241	3.4	$(k_1 + c)k_2k_4$	242	3.4	$(k_1 + c)k_2k_5$	243	3.4	$(k_1 + c)k_3k_4$	244	3.4	$(k_1 + c)k_3k_5$	245	3.4	$(k_1 + c)k_4k_5$
246	3.4	$(k_2 + c)k_1k_2$	247	3.4	$(k_2 + c)k_1k_3$	248	3.4	$(k_2 + c)k_1k_4$	249	3.4	$(k_2 + c)k_1k_5$	250	3.4	$(k_2 + c)k_2k_3$
251	3.4	$(k_2 + c)k_2k_4$	252	3.4	$(k_2 + c)k_2k_5$	253	3.4	$(k_2 + c)k_3k_4$	254	3.4	$(k_2 + c)k_3k_5$	255	3.4	$(k_2 + c)k_4k_5$
256	3.4	$(k_3 + c)k_1k_2$	257	3.4	$(k_3 + c)k_1k_3$	258	3.4	$(k_3 + c)k_1k_4$	259	3.4	$(k_3 + c)k_1k_5$	260	3.4	$(k_3 + c)k_2k_3$
261	3.4	$(k_3 + c)k_2k_4$	262	3.4	$(k_3 + c)k_2k_5$	263	3.4	$(k_3 + c)k_3k_4$	264	3.4	$(k_3 + c)k_3k_5$	265	3.4	$(k_3 + c)k_4k_5$
266	3.4	$(k_4 + c)k_1k_2$	267	3.4	$(k_4 + c)k_1k_3$	268	3.4	$(k_4 + c)k_1k_4$	269	3.4	$(k_4 + c)k_1k_5$	270	3.4	$(k_4 + c)k_2k_3$
271	3.4	$(k_4 + c)k_2k_4$	272	3.4	$(k_4 + c)k_2k_5$	273	3.4	$(k_4 + c)k_3k_4$	274	3.4	$(k_4 + c)k_3k_5$	275	3.4	$(k_4 + c)k_4k_5$
276	3.4	$(k_5 + c)k_1k_2$	277	3.4	$(k_5 + c)k_1k_3$	278	3.4	$(k_5 + c)k_1k_4$	279	3.4	$(k_5 + c)k_1k_5$	280	3.4	$(k_5 + c)k_2k_3$
281	3.4	$(k_5 + c)k_2k_4$	282	3.4	$(k_5 + c)k_2k_5$	283	3.4	$(k_5 + c)k_3k_4$	284	3.4	$(k_5 + c)k_3k_5$	285	3.4	$(k_5 + c)k_4k_5$

Table A2. Optimal Design of GA-SVM-MKL Classifier in 285 Scenario using Various Kernel and Kernel Properties.

No.	Performance (%)			No.	Performance (%)			No.	Performance (%)			No.	Performance (%)			No.	Performance (%)		
	S_e	S_p	OA		S_e	S_p	OA		S_e	S_p	OA		S_e	S_p	OA		S_e	S_p	OA
1	71.8	72.3	72.1	2	73.1	72.7	72.9	3	75.3	76.3	75.8	4	76.4	75.9	76.2	5	72.9	73.6	73.3
6	72.4	72.7	72.6	7	75.7	76.4	76.1	8	78.5	78.8	78.7	9	75.7	76.1	75.9	10	74.8	75.4	75.1
11	79.3	80.1	79.7	12	76.9	77.8	77.4	13	76.5	77.1	76.8	14	77.1	76.2	76.7	15	75.4	74.2	74.8
16	73.4	72.9	73.2	17	74.9	75.3	75.1	18	75.9	76.1	76	19	78.2	78.8	78.5	20	76.2	75.8	76
21	71.9	72.6	72.3	22	74.6	75.1	74.9	23	75.3	76.3	75.8	24	76.7	77.3	77	25	76.8	76.0	76.4
26	70.8	71.5	71.2	27	72.6	72.8	72.7	28	73.6	73.4	73.5	29	75.7	76.3	76	30	73.3	72.9	73.1
31	70.3	71.9	71.1	32	75.1	74.5	74.8	33	78.2	77.6	77.9	34	75.3	76.8	76.1	35	72.9	73.7	73.3
36	77.4	78.4	77.9	37	76.4	77.5	77.0	38	76.3	75.6	76.0	39	76.8	75.3	76.1	40	72.8	73.1	73.0
41	80.3	81.4	80.9	42	79.4	78.8	79.1	43	79.5	78.5	79	44	82.7	83.7	83.2	45	79.4	78.6	79
46	80.4	81.1	80.8	47	84.3	85.1	84.7	48	81.8	82.3	82.1	49	82.4	82.9	82.7	50	81.5	82.4	82.0
51	78.6	79.5	79.1	52	80.1	81.4	80.8	53	83.9	82.9	83.4	54	82.7	81.6	82.2	55	83.5	84.2	83.9
56	85.6	84.9	85.3	57	84.5	84.9	84.7	58	85.3	86.2	85.8	59	84.3	83.6	84.0	60	82.5	83.1	82.8
61	83.5	84.0	83.8	62	86.8	87.2	87	63	85.7	86.4	86.1	64	85.3	85.7	85.5	65	85.2	86.3	85.8
66	84.8	84.2	84.5	67	84.3	85.4	84.9	68	87.3	86.9	87.1	69	85.4	86.7	86.1	70	86.1	86.6	86.4
71	73.4	72.5	73.0	72	74.5	73.8	74.2	73	77.1	76.2	76.7	74	77.5	76.9	77.2	75	75.6	74.9	75.3
76	73.8	74.5	74.2	77	76.3	77.1	76.7	78	79.9	78.5	79.2	79	77.3	78.4	77.9	80	76.3	77.6	77.0
81	81.2	80.6	80.9	82	78.5	79.1	78.8	83	76.8	77.7	77.3	84	76.3	77.4	76.9	85	76.4	75.3	75.9
86	72.4	73.4	72.9	87	73.5	74.2	73.9	88	74.8	75.6	75.2	89	75.6	76.2	75.9	90	74.6	75.1	74.9
91	74.2	73.6	73.9	92	74.1	75.9	75	93	74.3	74.9	74.6	94	76.3	77.4	76.9	95	75.2	74.6	74.9
96	74.5	75.6	75.1	97	75.3	76.2	75.8	98	75.6	76.5	76.1	99	77.8	78.2	78	100	77.4	76.3	76.9
101	75.8	76.4	76.1	102	77.4	78.4	77.9	103	77.5	76.3	76.9	104	80.1	79.7	79.9	105	78.4	79.6	79
106	76.1	75.9	76	107	76.7	77.9	77.3	108	75.4	76.6	76	109	75.7	76.1	75.9	110	74.2	75.8	75
111	73.4	74.8	74.1	112	75.1	74.7	74.9	113	75.1	76.3	75.7	114	75.9	76.7	76.3	115	76.8	75.3	76.1
116	74.8	75.6	75.2	117	75.5	76.1	75.8	118	77.5	76.2	76.9	119	78.1	78.9	78.5	120	77.3	78.2	77.8
121	75.1	75.2	75.2	122	77.4	76.7	77.1	123	77.3	78.9	78.1	124	79.5	78.9	79.2	125	76.3	75.8	76.1
126	75.3	74.5	74.9	127	76.4	77.1	76.8	128	76.1	76.4	76.3	129	78.3	77.3	77.8	130	76.8	75.9	76.4
131	75.4	74.6	75	132	75.8	76.8	76.3	133	76.2	77.3	76.8	134	77.3	76.4	76.9	135	76.8	76.4	76.6
136	76.4	76.2	76.3	137	78.4	77.9	78.2	138	78.1	79.3	78.7	139	81.2	80.4	80.8	140	80.9	80.4	80.7
141	76.4	75.3	75.9	142	76.8	77.4	77.1	143	78.4	79.5	79.0	144	79.6	80.1	79.9	145	79.9	78.9	79.4
146	75.6	76.3	76.0	147	76.2	76.1	76.2	148	77.5	78.4	78.0	149	79.4	78.8	79.1	150	77.9	78.4	78.2
151	75.6	74.8	75.2	152	76.4	76.7	76.6	153	75.6	75.3	75.5	154	78.6	79.1	78.9	155	77.5	78.1	77.8
156	74.6	73.5	74.1	157	74.9	75.8	75.4	158	74.3	75.9	75.1	159	77.6	75.9	76.8	160	75.3	76.3	75.8
161	81.9	82.4	82.2	162	81.8	82.3	82.1	163	83.6	84.6	84.1	164	84.1	84.6	84.4	165	83.3	84.1	83.7
166	83.4	82.7	83.1	167	82.3	83.8	83.1	168	83.5	84.1	83.8	169	84.8	85.7	85.3	170	85.1	84.9	85
171	86.1	85.7	85.9	172	86.8	87.2	87	173	88.9	87.5	88.2	174	91.5	90.3	90.9	175	91.2	90.8	91
176	88.9	89.5	89.2	177	88.6	89.8	89.2	178	92.1	91.5	91.8	179	91.2	91.5	91.4	180	90.3	90.1	90.2
181	88.5	89.2	88.9	182	88.8	89.6	89.2	183	88.3	89.2	88.8	184	88.4	89.7	89.1	185	88.5	89.1	88.8
186	83.1	82.5	82.8	187	83.4	84.1	83.8	188	85.6	84.7	85.2	189	83.9	84.2	84.1	190	83.9	84.5	84.2
191	85.3	86.1	85.7	192	83.7	84.1	83.9	193	85.2	84.6	84.9	194	83.8	82.9	83.4	195	84.1	83.7	83.9
196	83.4	82.7	83.1	197	84.3	85.1	84.7	198	86.3	85.8	86.1	199	84.8	85.7	85.3	200	85.5	86.2	85.9
201	85.7	86.3	86	202	84.9	85.1	85	203	86.1	87.3	86.7	204	84.9	85.3	85.1	205	85.2	86.2	85.7
206	84.5	85.1	84.8	207	85.7	86.8	86.3	208	86.2	86.4	86.3	209	85.9	84.8	85.4	210	84.6	84.9	84.8
211	85.2	86.4	85.8	212	85.7	86.5	86.1	213	86.7	87.1	86.9	214	86.3	87.4	86.9	215	86.7	87.2	87.0

Table A2. *Cont.*

No.	Performance (%)			No.	Performance (%)			No.	Performance (%)			No.	Performance (%)			No.	Performance (%)		
	S_e	S_p	OA		S_e	S_p	OA		S_e	S_p	OA		S_e	S_p	OA		S_e	S_p	OA
216	85.8	86.7	86.3	217	88.2	87.1	87.7	218	87.8	88.5	88.2	219	86.3	87.2	86.8	220	85.1	86.3	85.7
221	86.6	87.3	87.0	222	87.3	86.4	86.9	223	88.9	88.2	88.6	224	86.7	87.5	87.1	225	87.5	88.2	87.9
226	85.6	86.8	86.2	227	86.3	87.3	86.8	228	86.1	85.9	86	229	85.4	85.8	85.6	230	85.9	84.3	85.1
231	84.6	85.1	84.9	232	86.1	87.4	86.8	233	87.1	86.4	86.8	234	86.3	87.8	87.1	235	85.2	86.3	85.8
236	83.4	84.9	84.2	237	83.3	84.1	83.7	238	84.9	85.6	85.3	239	84.7	85.1	84.9	240	85.8	86.6	86.2
241	86.1	86.7	86.4	242	85.6	84.8	85.2	243	86.7	85.9	86.3	244	85.3	84.1	84.7	245	85.1	85.9	85.5
246	86.1	85.3	85.7	247	86.2	85.6	85.9	248	87.1	86.3	86.7	249	86.2	85.9	86.1	250	86.1	86.8	86.5
251	87.2	86.4	86.8	252	86.7	85.7	86.2	253	87.8	88.3	88.1	254	86.7	86.5	86.6	255	86.3	87.3	86.8
256	86.4	87.3	86.9	257	86.9	85.8	86.4	258	86.3	87.3	86.8	259	87.1	86.3	86.7	260	87.5	88.1	87.8
261	86.9	87.6	87.3	262	87.4	88.1	87.8	263	89.4	88.4	88.9	264	87.4	86.7	87.1	265	87.4	87.9	87.7
266	86.3	87.4	86.9	267	86.9	87.5	87.2	268	88.2	87.1	87.7	269	85.7	86.4	86.1	270	86.7	87.5	87.1
271	87.5	86.7	87.1	272	87.4	88.3	87.9	273	88.4	87.5	88.0	274	88.6	89.2	88.9	275	86.9	87.2	87.1
276	86.7	86.8	86.8	277	87.1	86.5	86.8	278	88.5	87.5	88	279	86.3	87.2	86.8	280	86.5	86.8	86.7
281	86.4	87.1	86.8	282	86.7	87.8	87.3	283	87.5	88.6	88.1	284	87.1	87.5	87.3	285	86.7	87.9	87.3

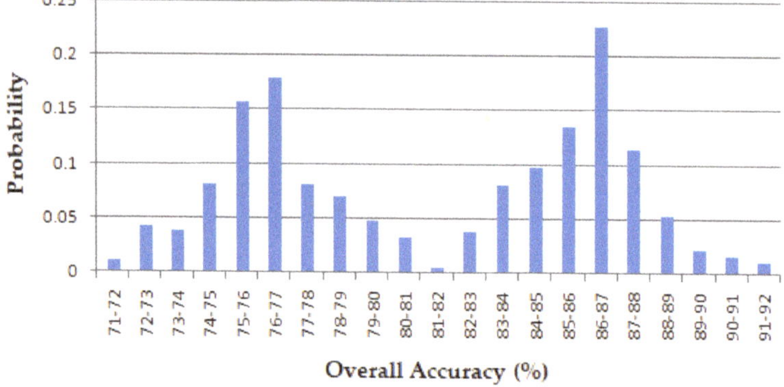

Figure A1. Probability distribution of overall accuracy for optimal design of GA-SVM-MKL classifier in 285 scenario using various kernel and kernel properties.

References

1. Yao, S.L.; Luo, J.J.; Huang, G.; Wang, P. Distinct global warming rates tied to multiple ocean surface temperature changes. *Nat. Clim. Chang.* **2017**, *7*, 486. [CrossRef]
2. Voigt, C.; Lamprecht, R.E.; Marushchak, M.E.; Lind, S.E.; Novakovskiy, A.; Aurela, M.; Martikainen, P.J.; Biasi, C. Warming of subarctic tundra increases emissions of all three important greenhouse gases–carbon dioxide, methane, and nitrous oxide. *Glob. Chang. Biol.* **2017**, *23*, 3121–3138. [CrossRef] [PubMed]
3. Sharma, K.; Saini, L.M. Performance analysis of smart metering for smart grid: An overview. *Renew. Sustain. Energy Rev.* **2015**, *49*, 720–735. [CrossRef]
4. *The Smart Meter Revolution Towards a Smarter Future*; Telefonica: London, UK, 2014.
5. Malvoni, M.; De Giorgi, M.G.; Congedo, P.M. Photovoltaic forecast based on hybrid PCA–LSSVM using dimensionality reducted data. *Neurocomputing* **2016**, *211*, 72–83. [CrossRef]
6. Alahakoon, D.; Yu, X. Smart electricity meter data intelligence for future energy systems: A survey. *IEEE Trans. Ind. Inform.* **2016**, *12*, 425–436. [CrossRef]
7. Armel, K.C.; Gupta, A.; Shrimali, G.; Albert, A. Is disaggregation the holy grail of energy efficiency? The case of electricity. *Energy Policy* **2013**, *52*, 213–234. [CrossRef]
8. Zoha, A.; Gluhak, A.; Imran, M.A.; Rajasegarar, S. Non-intrusive load monitoring approaches for disaggregated energy sensing: A survey. *Sensors* **2012**, *12*, 16838–16866. [CrossRef] [PubMed]
9. Gillis, J.M.; Alshareef, S.M.; Morsi, W.G. Nonintrusive load monitoring using wavelet design and machine learning. *IEEE Trans. Smart Grid* **2016**, *7*, 320–328. [CrossRef]
10. Stankovic, L.; Stankovic, V.; Liao, J.; Wilson, C. Measuring the energy intensity of domestic activities from smart meter data. *Appl. Energy* **2016**, *183*, 1565–1580. [CrossRef]

11. Zhao, B.; Stankovic, L.; Stankovic, V. On a training-less solution for non-intrusive appliance load monitoring using graph signal processing. *IEEE Access* **2016**, *4*, 1784–1799. [CrossRef]
12. Aiad, M.; Lee, P.H. Non-intrusive load disaggregation with adaptive estimations of devices main power effects and two-way interactions. *Energy Build.* **2016**, *130*, 131–139. [CrossRef]
13. Parson, O.; Ghosh, S.; Weal, M.; Rogers, A. An unsupervised training method for non-intrusive appliance load monitoring. *Artif. Intell.* **2014**, *217*, 1–19. [CrossRef]
14. Tsai, M.S.; Lin, Y.H. Modern development of an adaptive non-intrusive appliance load monitoring system in electricity energy conservation. *Appl. Energy* **2012**, *96*, 55–73. [CrossRef]
15. Giri, S.; Berges, M. An energy estimation framework for event-based methods in non-intrusive load monitoring. *Energy Convers. Manag.* **2015**, *90*, 488–498. [CrossRef]
16. Kong, S.; Kim, Y.; Ko, R.; Joo, S.K. Home appliance load disaggregation using cepstrum-smoothing-based method. *IEEE Trans. Consum. Electron.* **2015**, *61*, 24–30. [CrossRef]
17. McLachlan, G.J.; Do, K.A.; Ambroise, C. Supervised classification of tissue samples. In *Analyzing Microarray Gene Expression Data*; John Wiley & Sons Inc.: Hoboken, NJ, USA, 2004; pp. 221–251.
18. Refaeilzadeh, P.; Tang, L.; Liu, H. Cross-validation. In *Encyclopedia of Database Systems*; Springer: Boston, MA, USA, 2009; pp. 532–538.
19. Ritchie, M.D.; Hahn, L.W.; Roodi, N.; Bailey, L.R.; Dupont, W.D.; Parl, F.F.; Moore, J.H. Multifactor-dimensionality reduction reveals high-order interactions among estrogen-metabolism genes in sporadic breast cancer. *Am. J. Hum. Genet.* **2001**, *69*, 138–147. [CrossRef] [PubMed]
20. Deb, K. *Multi-Objective, Search Methodologies*; Springer: Berlin, Germany, 2014.
21. Wang, Y.; Cai, Z. Combining Multiobjective Optimization with Differential Evolution to Solve Constrained Optimization Problems. *IEEE Trans. Evol. Comput.* **2012**, *16*, 117–134. [CrossRef]
22. Shim, V.A.; Tan, K.C.; Tang, H. Adaptive Memetic Computing for Evolutionary Multiobjective Optimization. *IEEE Trans. Cybern.* **2015**, *45*, 610–621. [CrossRef] [PubMed]
23. Hsu, C.W.; Lin, C.J. A comparison of methods for multiclass support vector machine. *IEEE Trans. Neural Netw.* **2002**, *13*, 415–425. [PubMed]
24. Herbrich, R. *Learning Kernel Classifiers Theory and Algorithms*; The MIT Press: London, UK, 2002.
25. Marzband, M.; Sumper, A.; Domínguez-García, J.L.; Gumara-Ferret, R. Experimental validation of a real time energy management system for microgrids in islanded mode using a local day-ahead electricity market and MINLP. *Energy Convers. Manag.* **2013**, *76*, 314–322. [CrossRef]
26. Marzband, M.; Azarinejadian, F.; Savaghebi, M.; Pouresmaeil, E.; Guerrero, J.M.; Lightbody, G. Smart transactive energy framework in grid-connected multiple home microgrids under independent and coalition operations. *Renew. Energy* **2018**, *126*, 95–106. [CrossRef]
27. Marzband, M.; Fouladfar, M.H.; Akorede, M.F.; Lightbody, G.; Pouresmaeil, E. Framework for smart transactive energy in home-microgrids considering coalition formation and demand side management. *Sustain. Cities Soc.* **2018**, *40*, 136–154. [CrossRef]
28. Tavakoli, M.; Shokridehaki, F.; Akorede, M.F.; Marzband, M.; Vechiu, I.; Pouresmaeil, E. CVaR-based energy management scheme for optimal resilience and operational cost in commercial building microgrids. *Int. J. Electr. Power Energy Syst.* **2018**, *100*, 1–9. [CrossRef]
29. Marzband, M.; Javadi, M.; Pourmousavi, S.A.; Lightbody, G. An advanced retail electricity market for active distribution systems and home microgrid interoperability based on game theory. *Electr. Power Syst. Res.* **2018**, *157*, 187–199. [CrossRef]
30. Tavakoli, M.; Shokridehaki, F.; Marzband, M.; Godina, R.; Pouresmaeil, E. A Two Stage Hierarchical Control Approach for the Optimal Energy Management in Commercial Building Microgrids Based on Local Wind Power and PEVs. *Sustain. Cities Soc.* **2018**, *41*, 332–340. [CrossRef]
31. Selvakumar, A.I.; Thanushkodi, K. A new particle swarm optimization solution to nonconvex economic dispatch problems. *IEEE Trans. Power Syst.* **2007**, *22*, 42–51. [CrossRef]
32. Lytras, M.D.; Raghavan, V.; Damiani, E. Big data and data analytics research: From metaphors to value space for collective wisdom in human decision making and smart machines. *Int. J. Semant. Web Inf. Syst. (IJSWIS)* **2017**, *13*, 1–10. [CrossRef]
33. Lytras, M.; Visvizi, A. Who uses smart city services and what to make of it: Toward interdisciplinary smart cities research. *Sustainability* **2018**, *10*, 1998. [CrossRef]

34. Visvizi, A.; Lytras, M.D. Rescaling and refocusing smart cities research: From mega cities to smart villages. *J. Sci. Technol. Policy Manag.* **2018**, *9*, 134–145. [CrossRef]
35. Visvizi, A.; Lytras, M.D. *Smart Cities: Issues and Challenges*; Elsevier-US: New York, NY, USA, 2019.
36. Visvizi, A.; Lytras, M.D. Editorial: Policy Making for Smart Cities: Innovation and Social Inclusive Economic Growth for Sustainability. *J. Sci. Technol. Policy Manag.* **2018**, *9*, 126–133. [CrossRef]
37. Sicilia, M.A.; Visvizi, A. Blockchain and OECD data repositories: Opportunities and policymaking implications. *Libr. Hi Tech* **2018**. [CrossRef]
38. Malvoni, M.; De Giorgi, M.G.; Congedo, P.M. Forecasting of PV Power Generation using weather input data-preprocessing techniques. *Energy Procedia* **2017**, *126*, 651–658. [CrossRef]
39. Gajowniczek, K.; Ząbkowski, T. Two-stage electricity demand modelling using machine learning algorithms. *Energies* **2017**, *10*, 1547. [CrossRef]
40. Singh, S.; Yassine, A. Big Data Mining of Energy Time Series for Behavioral Analytics and Energy Consumption Forecasting. *Energies* **2018**, *11*, 452. [CrossRef]
41. Gajowniczek, K.; Ząbkowski, T. Short term electricity forecasting based on user behavior using individual smart meter data. *Intell. Fuzzy Syst.* **2015**, *30*, 223–234. [CrossRef]

Article

Non-Intrusive Load Monitoring Based on Novel Transient Signal in Household Appliances with Low Sampling Rate

Thi-Thu-Huong Le † **and Howon Kim** *,†

School of Computer Science and Engineering, Pusan National University, Busan 609-735, Korea; lehuong7885@gmail.com

* Correspondence: howonkim@pusan.ac.kr

† Current address: Information Security & IoT Lab, Building A06, School of Computer Science & Engineering, Pusan National University, San-30, JangJeon–dong, Geumjeong–gu, Busan 609-735, Korea.

Received: 31 October 2018; Accepted: 3 December 2018; Published: 5 December 2018

Abstract: Nowadays climate change problems have been more and more concerns and urgent in the real world. Especially, the energy power consumption monitoring is a considerate trend having positive effects in decreasing affecting climate change. Non-Intrusive Load Monitoring (NILM) is the best economic solution to solve the electrical consumption monitoring issue. NILM captures the electrical signals from the aggregate energy consumption, feature extraction from these signals and then learning and predicting the switch ON/OFF of appliances used these feature extracted. This paper proposed a NILM framework including data acquisition, data feature extraction, and classification model. The main contribution is to develop a new transient signal in a different aspect. The proposed transient signal is extracted from the active power signal in the low-frequency sampling rate. This transient signal is used to detect the event of household appliances. In household appliances event detection, we applied to Decision Tree and Long Short-Time Memory (LSTM) models. The average accuracies of these models achieved 92.64% and 96.85%, respectively. The computational and result experiments present the solution effectiveness for the accurate transient signal extraction in the electrical input signals.

Keywords: NILM; energy disaggregation; MCP39F511; Jetson TX2; transient signature; decision tree; LSTM

---

1. Introduction

Developing countries with rapid urbanization in high buildings construction and the high power demand are a reason for the need for conversation and efficient energy program. The program requires monitoring of customer appliances energy consumption in real-time. Using smart meter had led to NILM enables estimation of individual power consumption used for aggregate power consumption in energy management recently.

In the factory field, researchers are working on Factory Energy Management System (FEMS) for efficient electric energy use. In recently, FEMS has been linked with to the Cyber-Physical Systems (CPS) of Industry 4.0. And the related researches will be more significant in this area [1]. At resident homes, NILM provides the households understand their consumption usage via a cost-effective real-time monitoring appliances system. The customers need giving up unwanted activities to avoid producing unnecessary energy consumption such as the appropriate appliances usage time and appliances usage optimization. These activities can be obtained showing to customers the consumption of each appliance in the sum of the total billing to detect excesses or malfunction [2]. Furthermore, it could be possible notify users of possible savings in their billing electricity. In contrary to this,

Kelly et al. [3] have argued that it is not proven yet that above activity and the additional feedback which become saving energy. In addition, there is an increase in micro-grids and renewable energy facility installation also continuous growth in recently. In purpose of increasing efforts these saving, automating energy measurement, energy monitoring, and the power management system are needed. In the load measurement of a power system, load monitoring has the main role that the process of acquiring and identifying the load [4]. This load monitoring will determine the status of appliances and their consumption. Besides, it supports to understand the behavior of each load in the whole power system. There are two types of the load monitoring including Intrusive Load Monitoring (ILM) and Non-Intrusive Monitoring (NILM).

In ILM, the term "intrusive" means that there is the meter device in the resident house and close to the appliances to monitor. ILM deploys a measurement of the energy consumption of one or more household appliances using meter devices. In the ILM ecosystem, more low-end meter devices are needed. This makes hard to install, maintain, expand as well as expensive. In contrary to ILM, the term "non-intrusive" in NILM means that no extra equipment is installed in the house. NILM is a process which gives data from whole house energy consumption. This process includes installing a sensor device at the panel level and then the appliances will be inferred with being used. NILM preferred using than ILM because it is cheaper and easier installation. Instead of at least one-meter device per room, this technique requires only one-meter device for each energy entrance to the house. Energy disaggregation is another synonym for NILM. This technique estimates the power demand for each appliance from a single meter which contains the overall demand for several appliances.

A NILM system has three roles including capturing the signals from the aggregate consumption, extracting the feature uniquely from the load signal and classifying to identify which appliances are turned ON based on these features extracted. To identify the individual signature of each device, the NILM system requires several steps such as signal sampling (data acquisition), feature extraction based on signal analysis from the electrical signals. This paper presents a NILM framework including data acquisition, data feature extraction, and classification model. The contributions in this NILM study are: (1) propose a NILM framework, (2) collect the household appliances energy consumption data set in low-frequency sampling rate, (3) propose the algorithm to extract a new transient signal in low-frequency sampling rate, (4) improve the performance of NILM model in event detection as well as load identification.

The main contribution to be presented in this paper is a new approach to extract the transient uniquely. The proposed transient signal is extracted from the active power signal in the low-frequency sampling rate. This transient signal is used to detect the event of household appliances. Besides, an embedded board, Jetson TX2, is used to build the proposed NILM application system. This board has two roles to use in this research. The first role is to connect with the sensor to request and storage energy data (data acquisition). The second one is to build machine learning and deep learning models with high-performance training and testing. It is integrated Graphic Processing Unit (GPU) related to deep learning and its performance such as LSTM model. Furthermore, NILM Web application can be built for visualizing NILM result at this board. On the other hand, it can be replaced Jetson TX2 by other hardware that integrated GPU or normal personal computer (PC) without accelerated computing. Hence, the roles of Jetson TX2 is the same role with PC that integrated GPU to perform deep learning model. In summary, Jetson TX2 can connect to the sensor to storage data. Besides, it can build the machine learning, deep learning models, training, predicting and visualizing the result Jetson board or the PC has GPU with support high computation in a deep neural networks.

This paper is organized as follows. The next Section introduces the NILM system concept and presents a related literature works. In Section 3, a NILM framework is proposed. The core of this NILM framework is to propose the algorithm to develop a new transient signal from power signal in low-frequency sampling rate. The experimental setup is described and experimental results are presented in Section 4. Finally, conclusions and future works are drawn in Section 5.

2. Related Works

The initial NILM approach was proposed by Hart in 1990s [5]. The author introduced the Non-Intrusive Appliance Load Monitor(NALM) is a software which was able to analyze single point electrical data and then obtained energy used correspond to individual electrical appliances. Figure 1 shows the first concept of general NILM.

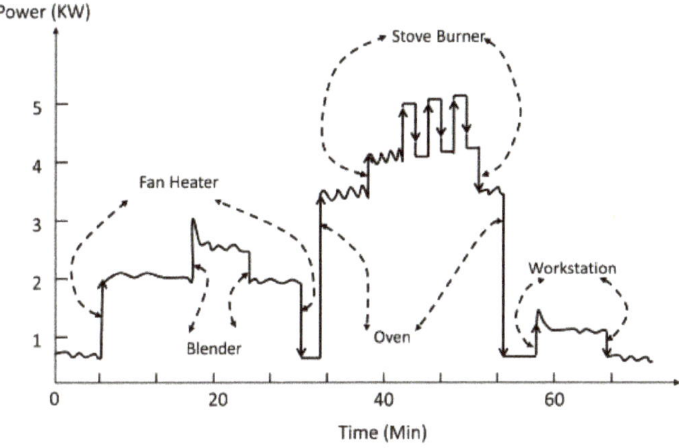

Figure 1. The first concept of NILM based on a single point measurement by Hart [5].

In recently, there are a number of studies have applied and extended this approach by using directly sampled which is increased the resolving power of the $\Delta P - \Delta Q$ space such in [6,7]. This approach has a limitation which cannot distinct appliances has similar in power signal and their operational principles, for example, hairdryer and iron. Hence, Laughman et al. [8] proposed another advanced solution. The capable of this idea based on the transient shapes of appliances to recognize individual load. Besides, they proposed the analysis of the spectral envelopes for continuously variable loads. However, they have not solved the electric noise of appliance usage yet. Therefore, Patel et al. [9] proposed a method to avoid the electric noise via combination software and hardware in household-level current sampling task at 1 MHz. They applied Support Vector Machine (SVM) model to trained the data to achieve accuracy at 90%. However, this technique has a limitation which requires high sampling rate with more kHz. Because of the meter device limitation, it is hard to apply in the real world environment. Furthermore, the need for an adjusting the prediction models on different data on training data is another drawback. The NILM algorithm needs to detect the appliance operation status (ON and OFF) from the power measurements. The NILM approaches can be classified as *event-based* or *state-based* based on different event detection strategies.

Event-based methods generated the state transient edges of appliances. This approach uses a change detection algorithm to determine the start and end of an event such in [10,11]. The significant information needs to extract to identification the event has occurred, for example, appliance signatures such as active power, rising or falling edge etc. These extracted features are analyzed to detect the event based on the appliances and their power consumption estimated. The researcher [12] has used different classification methods such as K-Means, K-Nearest Neighbour (KNN), SVM, Hidden Markov Model (HMM) and HMM's variations to detect event appliance. However, the performance of the event-based approach is not high because of the fixed threshold in the change detection algorithm, the large noise, and the similar among steady state signatures. Furthermore, false positive rate or false alarm rate may arise in the detection of edges methods.

State-based methods do not base on event detection. The idea is how to determine each appliance operation based on a state machine which is a different state transient using usage pattern of appliances [13]. When the appliance is turned ON/OFF or is changed running states, the method creates new state power signal through a probability distribution to match to the original power signal of the appliance. HMM and its variants [14–17] are used in state-based NILM. However, state-based approaches have several limitations. Firstly, the need for expert knowledge to set a prior value for each appliance by long training periods. Secondly, they have the complexity to compute [13,18]. Finally, there is not a good way to handle states may stable unchanged for long time intervals [19]. The requirement for an effective NILM algorithm is unique features or signatures have to characterize appliance behavior. Appliance signatures are a unique energy consumption pattern of all appliances. Ahmed Z., et al. [17] used appliance signatures to uniquely identify and recognized appliance operations from the aggregated load measurements. In feature selection, two main appliance features are used by NILM research to identify loads including *steady-state* and *transient state* [20].

Steady state is extracted when an appliance changes its running states related to sustained changes in power characteristics. The factors are used in this method including active power, reactive power, current, and voltage. The steady state signatures extraction of current and voltage do not demand high-end meter devices. Features at low frequency are used in the most commonly in steady state features in advance researches. However, the performance of this approach is limited by similarities among steady state signatures. In recently, deep neural network (DNN) in deep learning field becomes more attractive and widely system recognition applied in several areas such as handwritten recognition, speed recognition etc. Specifically, LSTM model is a kind of DNN model which is applied to classification applications have time series data. In NILM, this model is applied to load identification on the UK Domestic Appliance-Level Electricity (UK-DALE) dataset using active power by Kelly et al. [21]. However, the author pointed out the performance limitation on the appliances have informative events in power signal can be many time steps such as washing machine. Besides, Kim et al. [22] applied LSTM model on the variant power signals are generated from active power in low frequency on several public datasets. This method overcomes the long gaps between event may present a challenge in LSTM. In future NILM works, LSTM model may become more and more promising and effecting learning method for researchers.

Contrary to the *steady state* features, before settling into a steady state value, *transient state* features are short-term fluctuations in current or power. To create transient feature uniquely, the authors [17] defined appliance state transients which are shapes, size, duration, and harmonics by sampling current and voltage waveforms at high frequency. Hence, these transient features can achieve signal uniquely to a high degree. Besides, they capture all operation cycles in high sampling rates in longer monitoring time [23]. For example, Patel et al. [9] proposed a custom hardware built to detect the transient noise in range 0.001 kHz to 100 kHz frequency. Then, the authors used the fact that each appliance in-state operation transmits noise back to the power line. However, the high sampling rate required is the major drawback to obtain transient features such as current spikes, transient response time, repeatable transient power profiles, spectral envelopes, etc. [24]. When using a high sampling rate, the system demands a costly hardware and complicated to be installed in the home to detect these transient features [24]. The cause is smart meters report only low-frequency power. In the transient state analysis, the researchers [25,26] analyzed and captured the load signatures based on wavelet transform and transient energy algorithm. Artificial Neural Network (ANN) and HMM are used to improve the performance of NILM in these researches, respectively. Although the results were very much significant, there are still the little drawbacks. For example, the authors [25] sampled at high frequency for current and voltage waveform data to capture the transient effect. Nevertheless, doing this will increase the cost of energy meters because modern energy meters are not equipped with such functionality. Besides, the authors [26] used the data which have the repeatable transient energy signature for load identification because of the varying transient with these waveforms. Therefore, the sampling of the instantaneous load profile for each turn ON transient is required much diligence. In addition, the authors [27] applied the Wavelet Transform Coefficient (WTC) based on ON/OFF transient signal identification in data acquisition. Although WTC works better than Fourier Transform, WTC requires much longer computation time. Besides, it needs larger machine resource like memory usage.

In summary, the low-frequency or high-frequency data collected is used in classifying appliance recognition systems based on signature feature extraction. A low-frequency data sampling rate is implemented in without additional installation by using existing meter infrastructure. Contrary to this, high-frequency data sampling rate needs adding more hardware installing in data acquisition. Especially, the limitations of high sampling rate data acquisition are more expensive cost and more complex in signature database management [28,29]. However, a more accurate and precise analysis can be provided with more information and assistance [30]. Hence, using a low-frequency sampling rate is more promising in event detection appliances based on analysis the active power and/or reactive power. A new necessary technique of NILM has valid three most important factors to introduce into services for end-users. First thing is an ability to widely appropriately applied in household. The second thing is the usage of the low-cost device to retrieve the energy consumption. The third thing is able to recognition appliances with the same power signal and appliances with variable power signals in the low-frequency sampling data. Hence, in this work, new transient signal is extracted from the active power signal in the low-frequency sampling rate to overcome the problem in advance studies.

3. The Proposed NILM Framework

The proposed NILM framework is introduced in this Section. Figure 2 shows the proposed NILM system. The proposed NILM system includes several components such as energy stream data, sensing device MCP39F511 (Microchip Technology Inc., Chandler, AZ, USA), an embedded board Jetson TX2 (NVIDIA Corporation, Santa Clara, CA, USA), transient signal extraction and recognition, and energy monitoring and consumption prediction system.

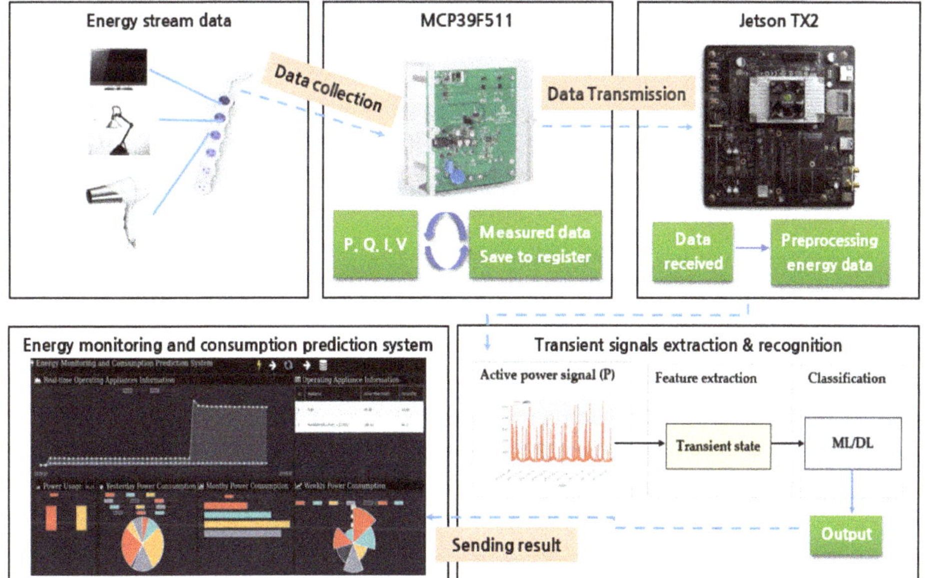

Figure 2. The proposed NILM framework.

3.1. Energy Stream Data

In data acquisition, several household appliances and the multi-tab power (HJ04009-10010C, KC Electric Company, Seoul, Korea) to collect energy data are prepared. Five appliances are collected including Air-purifier (LG Electronics Inc., Seoul, Korea), Fan (Cixi Xiongsheng Electrical Appliances Co., Ltd., Ningbo, China), Hairdryer (Korea Hanil Electronics, Seoul, Korea), LG monitor (LG Electronics Inc., Seoul, Korea), and Samsung monitor (Samsung Electronics Inc., Suwon, Korea). These appliances are connected to sensing device MCP39F511 via a multi-tab and the sensor's port. Table 1 describes models and power consumption of five appliances which are used to collect energy data.

Table 1. Specifications of five appliances.

Appliance	Model	Power Consumption (W)
Air-purifier	LG AS181DRWT	40
Fan	TESS-S1060	40
Hairdryer	Patech PH-3050	1300
LG monitor	24MP57VQ	23
Samsung monitor	S27D850T	90

Besides, twenty test cases data are also collected. The purpose of testing data collection is to obtain variant transient signals of five appliances. Each test case corresponds to set up that require the number of appliances is different. To each appliance, four combinations between this appliance and 1, 2, 3, or 4 other appliances are create in this work. Therefore, each appliance has four different transient signals in this collecting process. The process to collect testing data includes two steps as follows. The first step is to turn ON simultaneously all appliances required for each test. The second step is to make one appliance transient by turn OFF and then turn ON while other appliances are still ON. Table 2 presents the process to collect twenty test cases in detail.

Table 2. Twenty tests data information.

No	Test	Description
1	Test 1	[Air-purifier, LG monitor, Samsung monitor, Fan, Hairdryer] are ON simultaneously. Turn OFF and then turn ON [Air-purifier] while the other appliances are still ON.
2	Test 2	[Air-purifier, LG monitor, Samsung monitor, and Fan] are ON simultaneously. Turn OFF and then turn ON [Air-purifier] while the other appliances are still ON.
3	Test 3	[Air-purifier, LG monitor, Samsung monitor] are ON simultaneously. Turn OFF and then turn ON [Air-purifier] while the other appliances are still ON.
4	Test 4	[Air-purifier and LG monitor] are ON simultaneously. Turn OFF and then turn ON [Air-purifier] while the other appliance are still ON.
5	Test 5	[Fan and Air-purifier] are ON simultaneously. Turn OFF and then ON [Fan] while the other appliance are still ON.
6	Test 6	[Fan, Samsung monitor, and Air-purifier] are ON simultaneously. Turn OFF and then ON [Fan] while the other appliance are still ON.
7	Test 7	[Fan, Samsung monitor, LG monitor, Air-purifier, and Hairdryer] are ON simultaneously. Turn OFF and then ON [Fan] while the other appliance are still ON.
8	Test 8	[Fan, Samsung monitor, LG monitor, and Air-purifier] are ON simultaneously. Turn OFF and then ON [Fan] while the other appliance are still ON.
9	Test 9	[Hairdryer, Air-purifier, LG monitor, Fan, and Samsung monitor] are ON simultaneously. Turn OFF and turn ON [Hairdryer] while the other appliance are still ON.
10	Test 10	[Hairdryer, Air-purifier, LG monitor, and Fan] are ON simultaneously. Turn OFF and then turn ON [Hairdryer] while the other appliance are still ON.
11	Test 11	[Hairdryer, Air-purifier, and LG monitor] are ON simultaneously. Turn OFF and then tun ON [Hairdryer] while the other appliance are still ON.
12	Test 12	[Hairdryer and Air-purifier] are ON simultaneously. Turn OFF and then turn ON [Hairdryer] while the other appliance are still ON.
13	Test 13	[LG monitor, Samsung monitor, Fan, Air-purifier, and Hairdryer] are ON simultaneously. Turn OFF and then turn ON [LG monitor] while the other appliance are still ON.
14	Test 14	[LG monitor, Samsung monitor, Fan, and Air-purifier] are ON simultaneously. Turn OFF and then turn ON [LG monitor] while the other appliance are still ON.
15	Test 15	[LG monitor, Samsung monitor, and Fan] are ON simultaneously. Turn OFF and then turn ON [LG monitor] while the other appliance are still ON.
16	Test 16	[LG monitor and Samsung monitor] are ON simultaneously. Turn OFF and then turn ON [LG monitor] while the other appliance are still ON.

Table 2. *Cont.*

No	Test	Description
17	Test 17	[Hairdryer, Fan, Samsung monitor, LG monitor, and Air-purifier] are ON simultaneously. Turn OFF and then turn ON [Samsung monitor] while the other appliance are still ON.
18	Test 18	[Samsung monitor, Fan, LG monitor, and Air-purifier] are ON simultaneously. Turn OFF and then turn ON [Samsung monitor] while the other appliance are still ON.
19	Test 19	[Samsung monitor, LG monitor, and Air-purifier] are ON simultaneously. Turn OFF and then turn ON [Samsung monitor] while the other appliance are still ON.
20	Test 20	[Samsung monitor and Air-purifier] are ON simultaneously. Turn OFF and then turn ON [Samsung monitor] while the other appliance are still ON.

3.2. MCP39F511

The MCP39F511 is a power monitoring device that can measure input power in real time for the consumer, power distribution units, AC/DC power supplies. This sensor supports 2-wire serial protocols and Universal Asynchronous Receiver/Transmitter (UART)with enabling select full speed at up to 115.2 kbps. This sensor has a Power Monitor Demonstration Board which is a fully functional single-phase power. The system calculates and displays active power, reactive power, RMS current, RMS voltage, active energy (both import and export), and four quadrants reactive energy. MCP39F511 changes data acquisition mode compare to a conventional method. In the conventional method, data acquisition mode is getting energy data stored in registers by sending a command from PC. In this sensor device, the mode is getting energy data via connecting to Jetson TX2 (see Figure 3) via some steps as follows.

- Step 1: Jetson TX2 sends a command to switch to single wire mode. This single wire mode includes twenty bytes such as Header Byte ($0 \times$ AB), Header Byte 2 ($0 \times$ CD), Header Byte 3 ($0 \times$ EF), Current RMSs with Byte 0 to Byte 3, Voltage RMSs with Byte 0 to Byte 1, Active Power with Byte 0 to Byte 3, Reactive power with Byte 0 to Byte 3, Line Frequency with Byte 0 to Byte 1, and final is check sum.
- Step 2: Single wire mode is automatically sent whenever the sensing device updates energy data.
- Step 3: The sampling rate is in 15 Hz (see Figure 4).

There is a coherent sampling algorithm to phase lock the sampling rate to the line frequency based on an integer number of sample per line cycle in the computation cycle of MCP39F511. After that, it reports all power output quantities at a 2^N number of line cycles. The power outputs include RMS current, RMS voltage, apparent power, active power. The accumulation interval is defined as an 2^N number of line cycles, where N is the value in the Accumulation Interval Parameter register. Equations (1)–(5) system calculate and display RMS current (I_{RMS}), RMS voltage (V_{RMS}), Apparent power (S), Active power (P), Reactive power (Q), and Power factor as follows.

RMS current equation (I_{RMS}) with unit is Amps (A):

$$I_{RMS} = \sqrt{\frac{\sum_{n=0}^{2N-1}(i_n)^2}{2^N}} \tag{1}$$

RMS voltage equation (V_{RMS}) with unit is Volt (V):

$$V_{RMS} = \sqrt{\frac{\sum_{n=0}^{2N-1}(v_n)^2}{2^N}} \tag{2}$$

Apparent power equation (S) with unit is Volt-Amps (VA):

$$S = I_{RMS} \times V_{RMS} \tag{3}$$

Active power equation (P) with unit is Watts (W):

$$P = \frac{1}{2^N} \sum_{k=0}^{2^N - 1} V_k \times I_k \tag{4}$$

In the MCP39F511, Reactive power (Q) with Volt-Amps-Reactive unit (VAR) is measured based on a θ-degree phase shift in the voltage channel. The common degree is 90-degree phase shift. Accumulator Unit (ACCU) acts as the accumulator where has the similar accumulation principles applied to Active power (P). In the Gain Reactive power register, Gain is corrected. In the Reactive power register, an unsigned 32-bit value is located which is the final output. P is measured by the formula below.

$$Q = V_{RMS} \times I_{RMS} \times sin(\theta) \tag{5}$$

The ratio of P to S or Active power divided by Apparent power is Power factor (PF) measurement.

$$PF = \frac{P}{S} \tag{6}$$

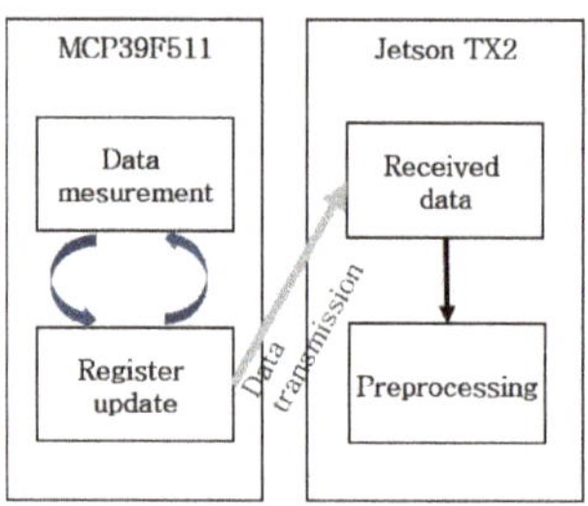

Figure 3. Data transfer process.

A	B	C	D	E	F	G	H	I	J	K	L
Year	Month	Day	Hour	Minute	Second	Millisecond	Current	Voltage	ActivePwr	ReactvPwr	Frequency
2018	3	5	9	21	20	50928	0.0835	222.9	3.82	16.33	60.048
2018	3	5	9	21	20	117481	0.0819	222.9	3.57	16.38	60.074
2018	3	5	9	21	20	184210	0.0827	222.7	3.56	16.43	60.124
2018	3	5	9	21	20	250923	0.0827	222.7	3.78	16.32	60.098
2018	3	5	9	21	20	317605	0.0835	222.8	3.87	16.28	60.098
2018	3	5	9	21	20	384271	0.0827	222.7	3.69	16.32	60.098
2018	3	5	9	21	20	450959	0.0827	222.7	3.69	16.32	60.098
2018	3	5	9	21	20	517666	0.0843	222.8	3.94	16.31	60.098
2018	3	5	9	21	20	584307	0.0835	222.9	3.85	16.31	60.098
2018	3	5	9	21	20	650940	0.0835	222.9	3.81	16.36	60.098
2018	3	5	9	21	20	717508	0.0835	222.8	3.65	16.46	60.098
2018	3	5	9	21	20	784372	0.0827	222.9	3.69	16.4	60.048
2018	3	5	9	21	20	850993	0.0835	222.9	3.73	16.36	60.048
2018	3	5	9	21	20	917845	0.0835	222.8	3.68	16.41	60.048
2018	3	5	9	21	20	984467	0.0835	222.9	3.73	16.4	60.048
2018	3	5	9	21	21	50976	0.0835	222.9	3.77	16.38	60.048
2018	3	5	9	21	21	117763	0.0843	223	3.82	16.35	60.074
2018	3	5	9	21	21	184411	0.0933	223	5.62	16.06	60.098
2018	3	5	9	21	21	251133	0.1014	223	7.36	15.75	60.098
2018	3	5	9	21	21	317812	0.1014	223	7.42	15.8	60.098
2018	3	5	9	21	21	384489	0.099	223	7.1	15.84	60.074
2018	3	5	9	21	21	451071	0.099	222.9	7.1	15.82	60.048
2018	3	5	9	21	21	517752	0.1249	222.6	11.33	15.27	60.074
2018	3	5	9	21	21	584576	0.1371	222.6	13.33	14.98	60.098
2018	3	5	9	21	21	651324	0.1379	222.8	13.53	14.98	60.048
2018	3	5	9	21	21	718006	0.1371	222.9	13.36	14.98	60.048
2018	3	5	9	21	21	784595	0.1371	222.8	13.38	15.01	60.074
2018	3	5	9	21	21	851286	0.1387	222.8	13.66	14.96	60.098
2018	3	5	9	21	21	917898	0.1379	222.8	13.49	14.97	60.098
2018	3	5	9	21	21	984733	0.1395	222.8	13.78	14.95	60.048

Figure 4. The low sampling rate format data collected in 15 Hz of Air-purifier device.

3.3. Jetson TX2

Jetson is the low-power embedded platform in the world's leading. Besides, it enables server-class AI to compute performance everywhere. Jetson's features include an integrated 256-core NVIDIA Pascal GPU, a hex-core ARMv8 64-bit CPU complex, and 8GB of LPDDR4 memory with a 128-bit interface. Figure 5 shows the CPU complex which combines a dual-core NVIDIA Denver 2 alongside a quad-core ARM Cortex-A57.

Table 3 shows Jetson TX2 technical specifications in detail. The installation files of Jetson TX2 are set up including,

- GPU includes Cuda, cudnn
- Machine learning/ deep learning with Tensorflow 1.3
- Python 3rd party lib with Pandas, numpy, jupyter, pyserial, matplotlib etc.
- Power Sensor Interlock is CDC ACM module
- Server includes Node.js 6.11.3, Npm 3.10.10, MongoDB-enterprise

Figure 5. Jetson TX2.

Table 3. Jetson TX2 Technical Specifications.

Feature	Byte
CPU	NVIDIA Pascal^{TM}, 256 CUDA cores
GPU	HMP Dual Denver 2/2 MB L2 + Quad ARM A57/2 MB L2
Monitor	8 GB 128 bit LPDDR4 59.7 GB/s
Data capacity	32 GB eMMC, SDIO, SATA
Connectivity	1 Gigabit Ethernet, 802.11ac WLAN, Bluetooth
Etc.	CAN, UART, SPI, I2C, I2S, GPIOs

It is a necessary to configure the system for collecting energy data from the appliances. The stream energy data is calculated in the power sensing device (MCP39F511). When the embedded board (Jetson TX2) sends a power data request, the sensing device will transfer power data via a single wire mode transmission frame 15 times per second. After that, Jetson TX2's processor operates to sort data received from MCP39F511 and save the data result to database management (MongoDB). For instance, Figure 6 shows the system configuration diagram to collect data from five household appliances (Air-purifier, Fan, Hairdryer, LG monitor, and Samsung monitor).

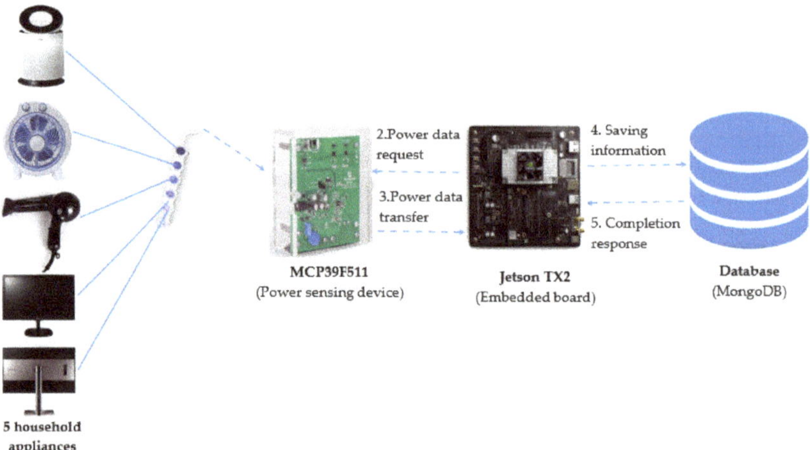

Figure 6. System configuration diagram collection data.

To collecting and storing the energy data, this paper proposes Algorithm 1. The algorithm consists of four steps as follows.

- Step 1. Setting up serial connect from NVIDIA Jetson TX2 to MCP39F511
- Step 2. Checking frame with 20-byte frame
- Step 3. Calculating power data including current, voltage, active power, reactive power, frequency
- Step 4. Storing data into NVIDIA Jetson TX2

Algorithm 1 The requesting and storing energy data of Jetson TX2

```
Require: single_wire_frame
Ensure: current (I), voltage (V), activepwr (P), reactivepwr (Q), frequency (F)
if Serial.isConnected () then
    Serial.write (single_wire_frame)
else
    print ("serial connection error!")
    exit()
end if
while Serial.isConnected() do
    single_wire_mode =Serial.read()
    check = check_frame (single_wire_frame)
    if check = True then
        I, V, P, Q, F = calculate_power_data (single_wire_mode)
        save_to_database (I,V,P,Q,F)
    end if
end while
```

In this algorithm , it defines two functions. The first function is **check_frame**() to check frame with 20-byte frame in single wire mode. The second one is **calculate_power_data**() to measure energy data information in Root Mean Square (RMS).

3.4. *Transient Signal Extraction and Recognition*

Based on literature observation above, this paper states three problems needing solve as follows.

Problem 1. Analyzing and extracting new transient signal from original power signal in low sampling rate data in feature extraction.

Problem 2. Labeling ON/OFF data on new transient signals extracted.

Problem 3. Improving the performance for event detection as well as load identification using a new transient signal results and machine learning/deep learning models.

This study selects active power factor which is a unique input in this approach. Figure 7 shows in detail of the proposed solution. The Figure 7a displays the process of this solution. The Figure 7b is to illustrate the process in Figure 7a, respectively. The proposed method includes three processes, such as feature extraction, labeling, and classification. The first, the feature extraction task is to generate the state of the appliance and to extract the transient signal of appliances. The second, labeling task is to label ON/OFF data with state and transient signals of appliance after extracting. The final, classification task is to learn and classify the ON/OFF appliance based on the transient signal and ON/OFF label signal.

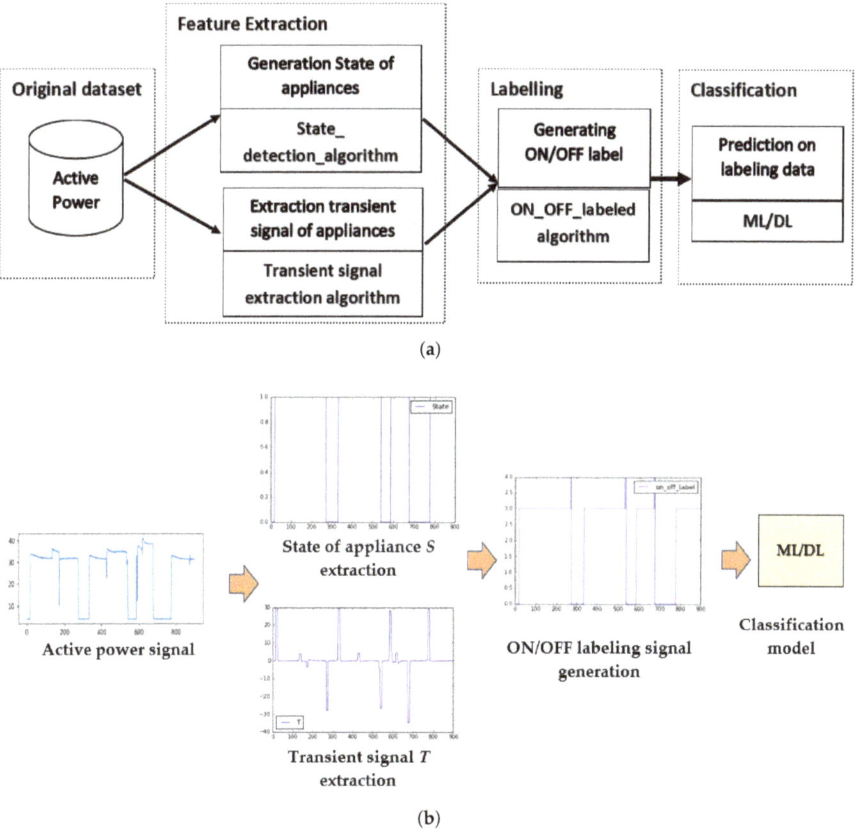

Figure 7. The proposed technique for transient signal extraction and recognition. (**a**) The proposed concept; (**b**) The illustration of the proposed process.

In feature extraction task, to label ON/OFF for appliance data, this work generates two signals including State of appliance signal and Transient signal. Hence, this paper proposes two algorithms, Algorithms 2 and 3.

The main idea of this task is how to detect whether or not rising and falling signals from the active power signal. The rising signal means appliance is operating (ON state). The falling signal means appliance is changing state from ON to OFF. Sliding window size of time series data is used to trace power signal to detect changing signals. Hence, the performance of the changing detection algorithm does not depend on the fixed or adaptive threshold, the large measure noise, and similarities among steady state signatures. Figure 8 shows the changing signals from active power signal of the appliance.

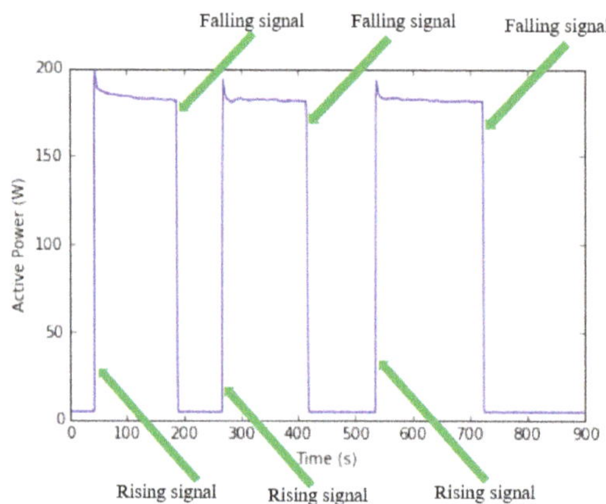

Figure 8. The analysis of active power signal with rising and falling signal.

Algorithm 2 is to generate State of the appliances, denoted by S. S value extracted is used for the input of Algorithm 4.

Algorithm 2 State of the appliances generation algorithm

Require: Activepwr (P)

Ensure: State of appliance (S)

$L_p \longleftarrow$ length of P

$bin \longleftarrow$ histogram (P)

$threshold \longleftarrow bin$

$OFF\_state \longleftarrow 0$

$ON\_state \longleftarrow 1$

for i in range $(0, L_p)$ **do**

  if $P[i] = 0$ **then**

    $S \longleftarrow S||0$

  else if $P[i] < threshold$ **then**

    $S \longleftarrow S||OFF\_state$

  else

    $S \longleftarrow S||ON\_state$

  end if

end for

return S

Algorithm 3 is to extract new transient signal from active power signal, denoted by T. T value is determined based on calculating prior P data (*pre_data*) and post P data (*post_data*). To assign *pre_data* and *post_data* values, the requirement is to use the window size w of P time series data and time shifting time series data of P is +1 for *post_data* value. The size of w depends on the sample size of P and periodicity of the data. To get smaller smoothing moving the average of time series data, there is an initialization of w value is 5. Hence, confidence intervals for the smoothed values are get. T signal extracted from this algorithm becomes to the input of Algorithm 4. If T value has negative value it means corresponding to P active power has falling signal, and then setting appliance's label is OFF event. In contrast to this, the active power has a rising signal, and then setting appliance's label is ON event.

Algorithm 3 Transient signal extraction algorithm

```
Require: Activepwr (P)
Ensure: Transient signal extraction (T)
L_p ⟵ length of P
w ⟵ 5
T ⟵ [0]
for i in range (w, L_p − w) do
    T_sum ⟵ 0
    for j in range (15) do
        pre_data ⟵ P[i − w + j]
        post_data ⟵ P[i − w + 1 + j]
        T_list ⟵ post_data − pre_data
        T_sum ⟵ Σ(T_list)
    end for
    T ⟵ T||T_sum
end for
return T
```

In labeling task, this paper generates ON/OFF labeling by the proposed Algorithm 4. T signal and state of the appliances are extracted by Algorithms 2 and 3, respectively. They become to the input of this algorithm. The task of Algorithm 4 is to generate *ON_label* or *OFF_label* from T signal. This study defines *threshold* value is -5 which is a maximum threshold value to determine whether event status changing from ON to OFF event. If T value extracted at time t smaller than -5 it means that at that time has occurred event changing from ON to OFF. In case S variable has 0 value, it means that no event operation. If T value is smaller than *threshold*, it is assigned label is *OFF_label*; else it is assigned label is *ON_label*. *OFF_label* value or *ON_label* value of each appliance are different values. For example, there are two appliances including Air-purifier and Fan. This paper defines sets of ON/OFF labels of these appliances as follow. Labels of Air-purifier device are *OFF_label*=1 and *ON_label*=2. And labels of Fan device are *OFF_label*=3 and *ON_label*=4. For setting ON/OFF labeling values of each appliance and twenty-test cases data, this paper mentions the labels for testing data in next Section.

Algorithm 4 Labeling ON/OFF algorithm

```
Require: Transient signal extracted (T), State of appliance (S)
Ensure: ON_OFF_label
L_T ⟵ length of (T)
ON_OFF_label ⟵[]
threshold ⟵ −5
for i in range (0, L_T) do
    if S[i] = 0 then
        ON_OFF_label ⟵ ON_OFF_label||0
    else if T[i] < threshold then
        ON_OFF_label ⟵ ON_OFF_label||OFF_label
    else
        ON_OFF_label ⟵ ON_OFF_label||ON_label
    end if
end for
return ON_OFF_label
```

In the classification task, this paper evaluates the proposed method using the models in both Machine Learning (ML) and Deep Learning (DL) fields. For learning T signal and ON/OFF labeling signal, Decision Tree and Long Short-Term Memory (LSTM) models are selected in ML and DL, respectively.

- Decision Tree is a supervised learning method. It is used in both classification and regression tasks. The input feature of this model is used to infer the output feature by learning simple decision rules. CART [31] (Classification and Regression Trees) is similar to C4.5, however, two different to C4.5 are it supports the regression task in numerical target variable and does not need to compute rule sets. This algorithm builds binary trees based on the feature and threshold with the largest information gain at each node. In a decision tree algorithm, it needs training vectors $x_i \in R^n, i = 1, \ldots, l$ and a vector $y \in R^l$. This algorithm needs to recursively partition such that grouping the same labels in a group together. A feature f and threshold $thre_f$, partition the data into T_{left} and T_{right} subsets are contained in each candidate split $\theta = (j, thre_f)$

$$T_{left}(\theta) = (x, y)|x_f \leq thre_j \tag{7}$$

$$T_{right}(\theta) = T \setminus T_{left}(\theta) \tag{8}$$

  The impurity is the choice of the classification or regression task. The impurity at j is calculated based on an impurity function $H(j)$.

$$G(T, \theta) = \frac{n_{left}}{N_j} H(T_{left}(\theta)) + \frac{n_{right}}{N_j} H(T_{right}(\theta)) \tag{9}$$

  Select the parameters that minimizes the impurity is formulated by θ^* as follows.

$$\theta^* = argmin_\theta G(T, \theta) \tag{10}$$

  Recurse for subsets $T_{left}(\theta^*)$ and $T_{right}(\theta^*)$ reaching the maximum allowable depth, $N_j < min_{samples}$ or $N_j = 1$. In the classification task, the output classification represents a region R_j with N_j observations taking on value $0, 1, \ldots, K-1$, for node j.

$$pr_{jk} = \frac{1}{N_j} \Sigma_{x_i \in R_j} I(y_i = k) \tag{11}$$

  be the proportion of class k observations in node j. $I(.)$ is a spitting criterion that makes used of the impurity measure.

  Common measures of impurity are Gini index which is calculated as follows.

$$H(X_j) = \Sigma_k p_{jk}(1 - pr_{jk}) \tag{12}$$

- LSTM has been designed by Hochreiter and Schmidhuber in 1997 [32]. LSTM is an elegant Recurrent Neural Network (RNN). The LSTM architecture is combined memory cells by replacing the regular units of the neural network. A memory cell consists of three gates: an *input gate*, a *forget gate*, an *output gate* and the state of memory cell is called *cell state* (Figure 9a). In particular, at the *input gate*, it allows incoming signal to alter the *cell state* or block it. On the other hand, the *output gate* can allow the *cell state* to have an effect on other neurons or prevent it. Finally, the *forget gate* can modulate *cell state* of the memory cell. Besides, it allows the cell to remember or forget its previous state as needed.

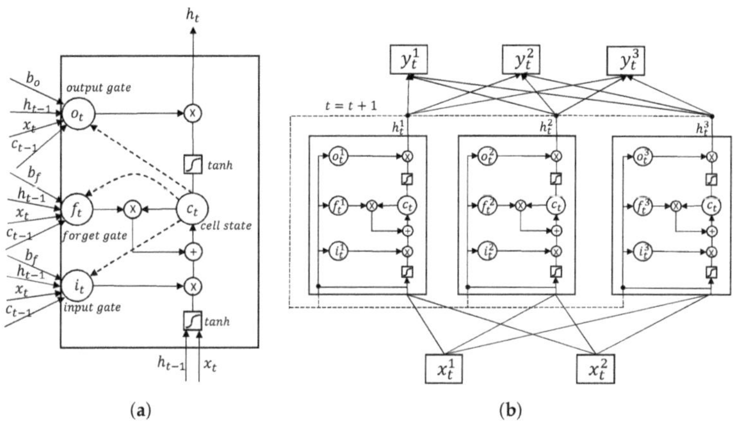

Figure 9. (**a**) LSTM Cell; (**b**) An architecture of LSTM model

The difference between LSTM and the original RNN is hidden units. In LSTM, the hidden units are replaced by LSTM cells. Figure 9b shows an architecture of LSTM having two input units, three LSTM cells as hidden units and three output units. The equations below describe how LSTM processes data. Assumption that $X_t = [x_t^1, x_t^2, x_t^3, ..., x_t^{n_x}]$ is an input vector and $H_t = [h_t^1, h_t^2, h_t^3, ..., h_t^{n_h}]$, $Y_t = [y_t^1, y_t^2, y_t^3, ..., y_t^{n_y}]$, $C_t = [c_t^1, c_t^2, c_t^3, ..., c_t^{n_c}]$ are hidden, output and cell vector, respectively. The elements of each vector are units for layers of LSTM. n_x, n_h, n_c and n_y are a number of each units. σ is the logistic sigmoid function, and i, f and o are the input gate, forget gate and output gate, respectively. The weight matrix superscripts have the obvious meaning. For example, W_t^{hi} is the hidden-input gate weight matrix and W_t^{xo} is the input-output gate weight matrix. b_t^i, b_t^f, b_t^c and b_t^o are bias terms at time t.

First is to compute the value for f_t, the activation of the forget gate. The output range of f_t is from 0 to 1 and the output value will be multiplied by c_{t-1} when calculating c_t. Therefore, f_t means an activation rate of the previous cell state.

$$f_t = \sigma(W_t^{xf} \sum_{i=1}^{n_x} x_t^i + W_t^{hf} \sum_{j=1}^{n_h} h_{t-1}^j + W_t^{cf} \sum_{k=1}^{n_c} c_{t-1}^k + b_f) \quad (13)$$

Second is to compute the value for the input gate i_t. In common with f_t, i_t is the activation ratio of the input value x_t.

$$i_t = \sigma(W_t^{xi} \sum_{i=1}^{n_x} x_t^i + W_t^{hi} \sum_{j=1}^{n_h} h_{t-1}^j + W_t^{ci} \sum_{k=1}^{n_c} c_{t-1}^k + b_i) \quad (14)$$

Third is to compute the value for the cell state c_t. Two factors are combined. The first factor is the previous cell state activated by the forget gate and the second factor is the input value activated by the input gate.

$$c_t = f_t c_{t-1} + i_t tanh(W_t^{xc} \sum_{i=1}^{n_x} x_t^i + W_t^{hc} \sum_{j=1}^{n_h} h_{t-1}^j + b_c) \quad (15)$$

Final is to compute the value of their output gates and use it for the memory block output.

$$o_t = \sigma(W_t^{xo} \sum_{i=1}^{n_x} x_t^i + W_t^{ho} \sum_{j=1}^{n_h} h_{t-1}^j + W_t^{co} \sum_{k=1}^{n_c} c_{t-1}^k + b_o) \quad (16)$$

$$h_t = o_t tanh(c_t) \tag{17}$$

Equations (13)–(17) are processed in one LSTM cell. After all process are done in the hidden layer, it can be calculated for the output units with the hidden vector H_t.

$$y_t = \sigma(W_t^{hy} \sum_{i=1}^{n_h} h_t^i) \tag{18}$$

3.5. Energy Monitoring and Consumption Prediction System

This is a web application to display the energy disaggregation results. This paper uses Node.js, Javascript, and HTML to build the user interface for NILM system. The results of classification are passed to energy monitoring system.

4. Experiments

This section points data preparation for experiment and experiment results in this approach. The results including T signals extracted and classification ON/OFF event. This study sets up the environment for implementation as following Intel ® CoreTM i7-4790 CPU @3.60GHz; GPU: NVIDIA GeForce GTX 750 (NVIDIA Corporation, Santa Clara, CA, USA) ; RAM: 16GB and Operating System (OS) : Windows 10; the language programming in Python.

Besides, this paper uses confusion matrix (CM) to evaluate the approach classification model. CM includes four categories such as True Positive (TP), False Positive (FP), True Negative (TN) and False Negative (FN). This paper assumes that the positive event means an appliance is turned ON and when the appliance is turned OFF that it is the negative event. Hence, Recall, Precision, Accuracy, F1-score are calculated for evaluating the NILM model based on CM.
Recall is a ratio of the number correctly classified to the total number of actual positive samples.

$$\text{Recall} = \frac{\text{TP}}{\text{TP} + \text{FN}} \tag{19}$$

Precision is a ratio of the number of correctly classify to the total number of predicted positive samples.

$$\text{Precision} = \frac{\text{TP}}{\text{TP} + \text{FP}} \tag{20}$$

Accuracy is a ratio of correctly classify to the total test data.

$$\text{Accuracy} = \frac{\text{TP} + \text{TN}}{(\text{TP} + \text{FP}) + (\text{FN} + \text{TN})} \tag{21}$$

F1-score is the harmonic average of Recall and Precision.

$$\text{F1-score} = 2 \times \frac{\text{Precision} \times \text{Recall}}{\text{Precision} + \text{Recall}} \tag{22}$$

Furthermore, this paper measures loss of NILM model using Loss function such as Mean Squared Error (MSE).

$$\text{Loss} = \frac{1}{n} \Sigma_{i=1}^{n} (Y_t - Y_p)^2 \tag{23}$$

where Y_t is the expected output of sample data, Y_p is the predicted output of sample data by NILM approach model.

4.1. Data Preparation for Experiment

Five appliances data individual and twenty test cases are collected and saved in *.csv files. As mentioned in the previous Section, this paper selected active power data for input feature in the experiments. Figure 10 illustrates the active power signals from five active power data of five appliances.

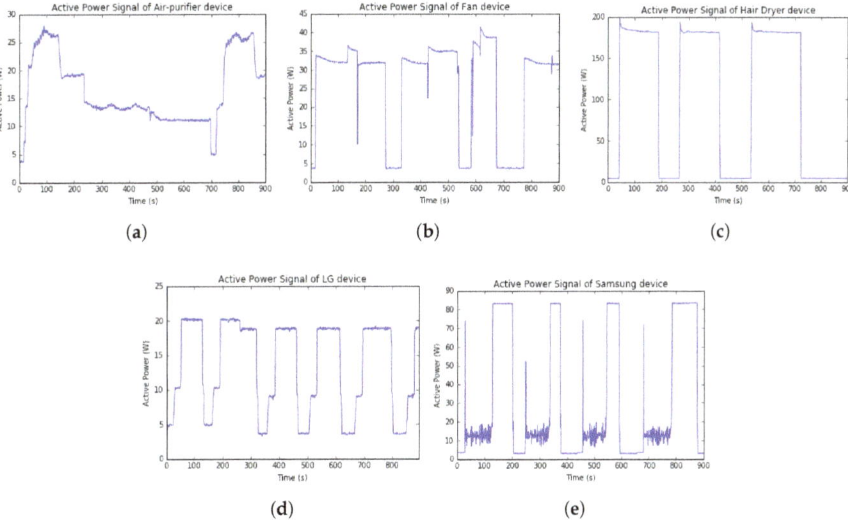

Figure 10. Active power signals: (**a**) Airpurifier; (**b**) Fan appliance; (**c**) Hairdryer appliance; (**d**) LG monitor appliance; (**e**) Samsung monitor appliance.

In testing data, this paper named twenty-test cases data files following in Table 4.

Table 4. Twenty-test cases data file name.

No	Test	File Name	Description
1	Test 1	Airpurifier_Transient_LG_Samsung_Fan_Hairdryer_Steady	
2	Test 2	Airpurifier_Transient_LG_Samsung_Fan_Steady	Air-purifier transient
3	Test 3	Airpurifier_Transient_LG_Samsung_Steady	
4	Test 4	Airpurifier_Transient_LG_Steady	
5	Test 5	Fan_Transient_Airpurifier_Steady	
6	Test 6	Fan_Transient_Samsung_Airpurifier_Steady	Fan transient
7	Test 7	Fan_Transient_Samsung_LG_Airpurifier_Hairdryer_Steady	
8	Test 8	Fan_Transient_Samsung_LG_Airpurifier_Steady	
9	Test 9	Hairdryer_Transient_Airpurifier_LG_Fan_Samsung_Steady	
10	Test 10	Hairdryer_Transient_Airpurifier_LG_Fan_Steady	Hairdryer transient
11	Test 11	Hairdryer_Transient_Airpurifier_LG_Steady	
12	Test 12	Hairdryer_Transient_Airpurifier_Steady	
13	Test 13	LG_Transient_Samsung_Fan_Airpurifier_Hairdryer_Steady	
14	Test 14	LG_Transient_Samsung_Fan_Airpurifier_Steady	
15	Test 15	LG_Transient_Samsung_Fan_Steady	LG monitor transient
16	Test 16	LG_Transient_Samsung_Steady	
17	Test 17	Samsung_Transient_Airpurifier_LG_Haridryer_Fan_Steady	
18	Test 18	Samsung_Transient_Airpurifier_LG_Fan_Steady	Samsung monitor transient
19	Test 19	Samsung_Transient_Airpurifier_LG_Steady	
20	Test 20	Samsung_Transient_Airpurifier_Steady	

Besides, this paper plotted twenty active power signals from twenty test cases in Figure 11.

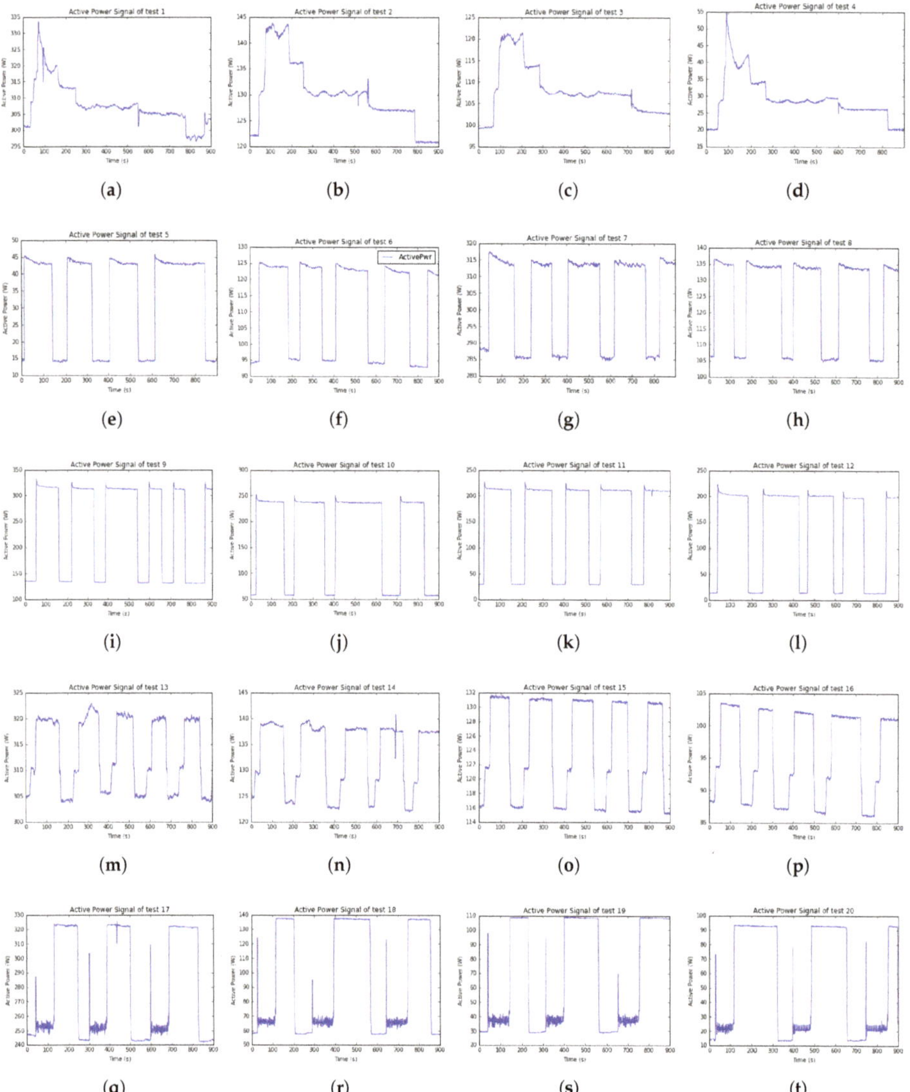

Figure 11. Active Power Signals of twenty test cases data: (**a–d**): test 1 to test 4; (**e–h**): test 5 to test 8; (**i–l**): test 9 to test 12; (**m–p**): test 13 to test 16; (**q–t**): test 17 to test 20.

4.2. Experiment Results

4.2.1. State of Appliance, Transient Signal and ON/OFF Label Results

Firstly, this paper determined and declared ON/OFF label value to use in Algorithm 4. Tables 5 and 6 show ON/OFF labels for five appliances and twenty test cases, respectively.

Table 5. Labeling ON/OFF for each appliance.

Appliance Name	ON/OFF Labeling	
	ON Label	**OFF Label**
Airpurifier	1	2
Fan	3	4
Hairdryer	5	6
LG monitor	7	8
Samsung monitor	9	10

Table 6. Labeling ON/OFF for twenty test cases.

Test Name	ON/OFF Labeling	
	ON Label	**OFF Label**
Test 1, Test 2, Test 3, Test 4	1	2
Test 5, Test 6, Test 7, Test 8	3	4
Test 9, Test 10, Test 11, Test 12	5	6
Test 13, Test 14, Test 15, Test 16	7	8
Test 17, Test 18, Test 19, Test 20	9	10

Secondly, this paper implemented Algorithms 2 and 3 to extract *State* of the appliance and T signals, respectively. After that, the Algorithm 4 are implemented to label ON/OFF based on each T signals from five appliances and twenty test cases data. The left, middle, and right of Figure 12 are *State* S, T and ON/OFF_label signals obtained from five appliances data, respectively. From observing these T signal results in this figure, this paper can determine the number of changing status of event ON/OFF in each appliance. There are two, three, three, five, and four times changing status from ON to OFF corresponding to air-purifier, fan, hairdryer, LG monitor, and Samsung monitor.

In summary, this paper recognizes that when T signal changes the value from positive to negative value and smaller than -5, the event ON/OFF on device will occur. In particular, when the T signal obtains positive values, it means that the appliance is operating (ON). Contrast to this, the appliance switches to OFF.

For visualizing the results of twenty test cases, this paper pointed the S, T, and ON_OFF label results of test 1 to test 4 in Figure 13. Besides, the S, T, and ON_OFF label results of test 5 to test 20 are plotted in Figures A1–A3 at Appendix A, respectively.

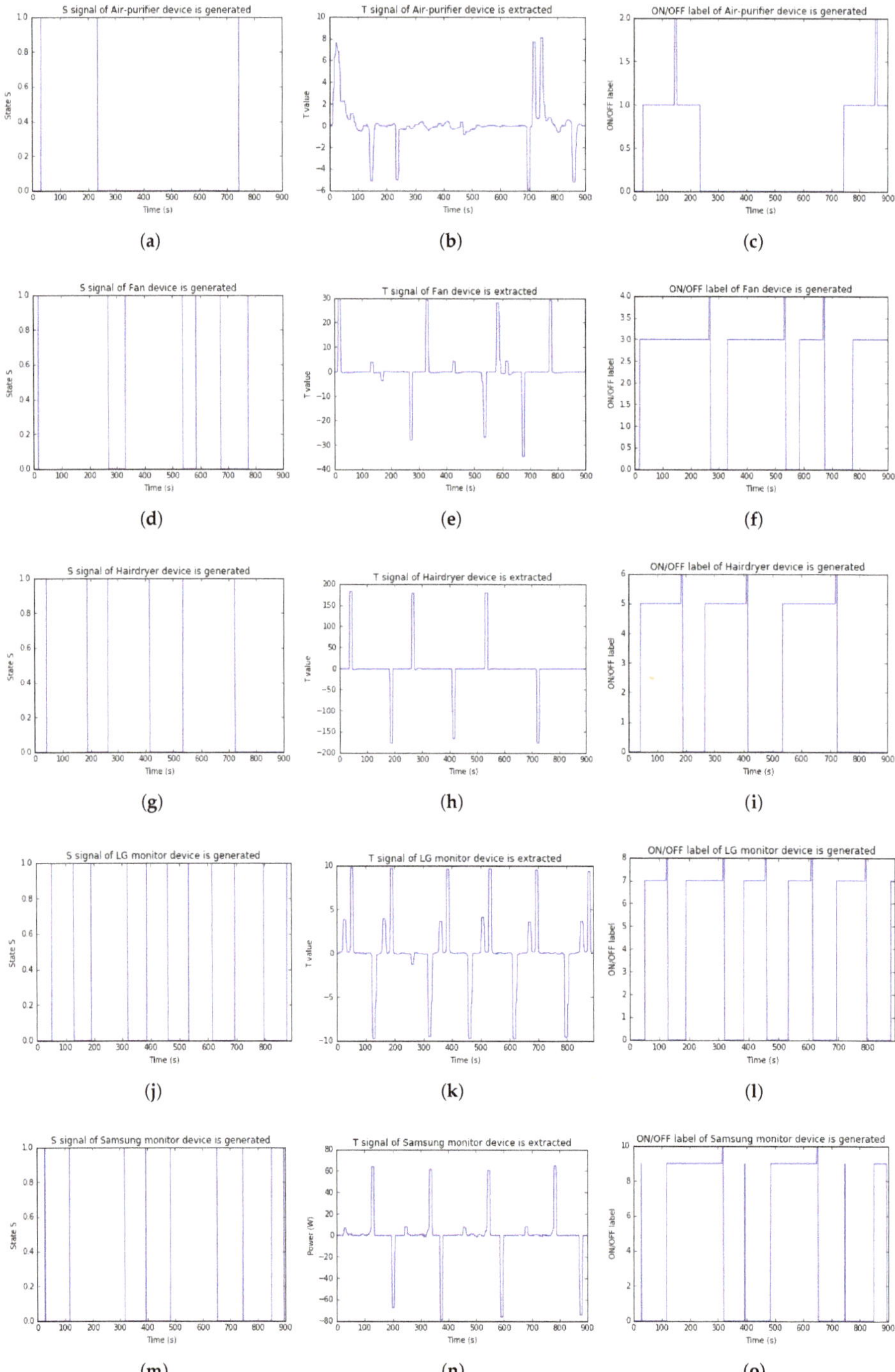

Figure 12. State S, T & ON/OFF label signals of 5 appliances: (**a**–**c**): Air-purifier device; (**d**–**f**): Fan device; (**g**–**i**): Hair Dryer device; (**j**–**l**): LG monitor device; (**m**–**o**): Samsung monitor device.

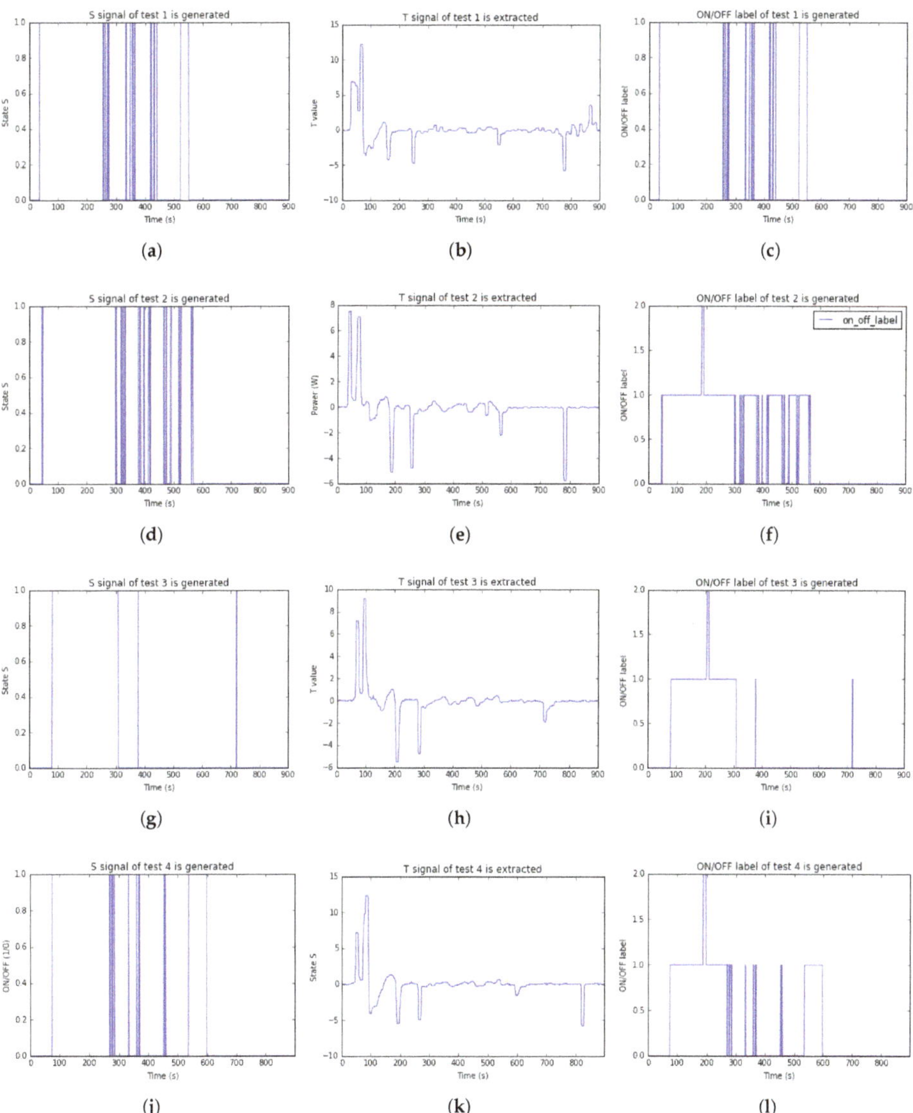

Figure 13. State *S*, *T* & ON/OFF label signals of test 1 to test 4 data: (**a–c**) test 1; (**d–f**) test 2; (**g–i**) test 3; (**j–l**) test 4.

4.2.2. Classification Results

- In ML, this paper applied Decision Tree model for classification ON/OFF event of the household appliance in this study. The reason for selecting Decision Tree model in classification task related to classification accuracy. This study tried to apply other models such as SVM, Random Forest, Multilayer Perceptrons (MLP), etc., however, the result of them obtained not well. This paper used *T* signals for input feature and output feature is ON/OFF labeling for the approached models. Five *T* signals extracted from five appliances are training data. The testing data is twenty *T* signals extracted from twenty tests data. This paper used CM to display the classification result.

The performance classification results for detecting ON/OFF of Decision Tree model are shown in Figure 14. The main diagonal of confusion matrix represents the number of correctly samples are predicted by the applied model.

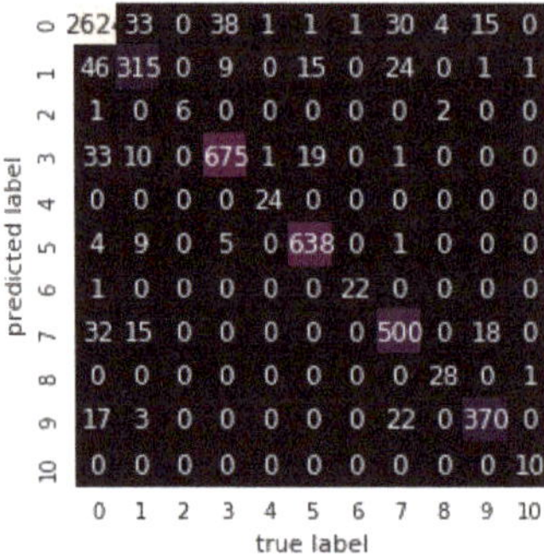

Figure 14. Confusion matrix of Decision Tree model for classification ON/OFF label.

Table 7 is classification report for each event detection of five appliances using other performance metrics such as precision, recall, F1-score. The average accuracy of event detection model on these appliances data obtained 93%.

Table 7. Classification result for event detection ON/OFF of each appliance.

Label	Event Device	Precision	Recall	F1-Score	Support
0	No Event	0.95	0.96	0.95	2747
1	Air-purifier_ON	0.82	0.77	0.79	411
2	Air-purifier_OFF	1.00	0.67	0.80	9
3	Fan_ON	0.93	0.91	0.92	739
4	Fan_OFF	0.92	1.00	0.96	24
5	Hairdryer_ON	0.95	0.97	0.96	657
6	Hairdryer_OFF	0.96	0.96	0.96	23
7	LG-monitor_ON	0.87	0.88	0.87	565
8	LG-monitor_OFF	0.82	0.97	0.89	29
9	Samsung-monitor_ON	0.92	0.90	0.91	412
10	Samsung-monitor_OFF	0.83	1.00	0.91	10
	avg/total	0.93	0.93	0.93	5629

In summary, Table 8 pointed the average performance of the approached model. The performance classification of our approach model achieved 92.64% and loss rate was 1.67.

Table 8. Classification result of Decision Tree model on energy data.

Metric	Performance
Accuracy	0.926413
Precision	0.927533
Recall	0.926413
F1	0.926778
Loss	1.66637

- In DL, LSTM model is used for classification ON/OFF event of appliances. Similar to Decision Tree model, this paper used T signals and ON/OFF labeling for input and output features, respectively. For testing data, this research also used twenty T signals of twenty tests data. Setting up hyper-parameters of LSTM model are described in Table 9 as follows.

Table 9. Hyper-parameter of LSTM model in the approach.

Hyper-Parameter Name	Value
Input	T signals
Hidden size	100
Batch size	893
time step	1
Number of LSTM layers	3
Epochs	5000
Output	On_Off_labeling

To evaluate LSTM model, this paper uses accuracy metric and Root Mean Square Error (RMSE) for measuring performance and loss of the model, respectively. RMSE is the standard deviation of the residuals (prediction errors). The formula to calculate RMSE is as follows.

$$\text{RMSE} = \sqrt{(pr - ex)^2} \tag{24}$$

where pr is predicted output and ex is expected output.

Figure 15 presents the results of LSTM model for classification ON/OFF of five appliances data. Besides, LSTM model's results on test 1 to test 4 data are shown in Figure 16 and the others tests' results in Figure A4 at Appendix A. The average accuracy of LSTM model obtained 96.85% for detecting ON/OFF appliances. The loss of LSTM model (RMSE) obtained 0.60632.

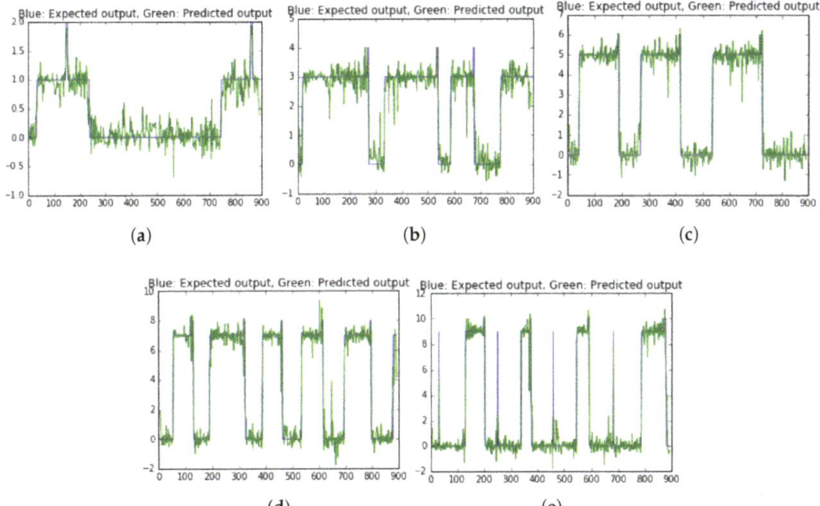

Figure 15. Classification results of LSTM model on 5 appliances data: (**a**) Airpurifier device; (**b**) Fan device; (**c**) Hairdryer device; (**d**) LG monitor device; (**e**) Samsung monitor device.

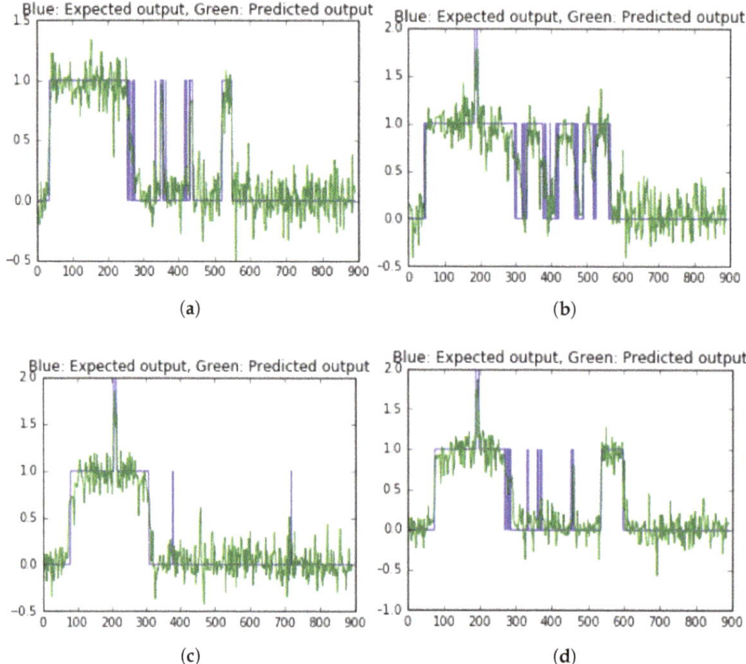

Figure 16. Classification results of LSTM model on test 1 to test 4 data: (**a**) Test 1; (**b**) Test 2; (**c**) Test 3; (**d**) Test 4.

4.2.3. Applying the Proposed Method on a Publicly Available Dataset

There are several NILM available datasets such as Building-Level fUlly-labeled dataset for Electricity Disaggregation (BLUED), UK-DALE, Residential Energy Disaggregation Dataset (REDD), The Almanac of Minutely Power dataset (Version 2) (AMPds2) etc. This paper applied the proposed method on AMPds2 [33] dataset which is low sampling rate data. Although several data file formats are published such as Original file format, Tab-delimited, Rdata format, Variable metadata, etc. The original file format in CSV is used in this paper. In this dataset, there are 19 appliances data in isolation have already collected and tested in electricity data. There are no publicly aggregated data between the loads in this dataset. Therefore, this paper only performed and tested on individual load. Among available loads, 5 appliances data are randomly selected to use in this experiment. These appliances are Master and South Bedroom, Basement Plugs and Lights, Instant Hot Water Unit, Entertainment: TV, PVR, AMP, Kitchen Fridge. Besides, ON/OFF label value are determined and declared from 1 to 10 corresponding to each appliance. The information of these appliances and their ON/OFF labels are described in Table 10 as follows.

Table 10. Five appliances information in AMPds2 dataset selected [34].

CSV File Name	Appliance Name	Watts (W)	ON Label	OFF Label
Electricity_B2E	Master and South Bedroom	175	1	2
Electricity_BME	Basement Plugs and Lights	387	3	4
Electricity_HTE	Instant Hot Water Unit	70	5	6
Electricity_TVE	Entertainment: TV, PVR, AMP	415	7	8
Electricity_FGE	Kitchen Fridge	525	9	10

About processing dataset, in this dataset, there are 11 electric data features in each appliance data. Timestamp (TS) feature is Unix timestamp value in this dataset. Electricity measurement is at one minute intervals. This dataset was collected to total of 1,051,200 readings per meter for 2 years of monitoring from 2012 to 2014. Therefore, these loads data are big dataset. In testing performance of the presently proposed methodology, the range of TS in the first 500 readings/samples value data from 1333263600 to 1333323540 is selected. This range of TS value is converted to Universal Time Coordinated (UTC) corresponding to the date time range from 2012-4-01 7:00 a.m. to 2012-4-01 15:19 p.m. The total of time series data is 500 min. Because real power P is active power data, hence P is selected for the input data in the experiment. Figure 17 visualizes five real power P signals time series data were selected.

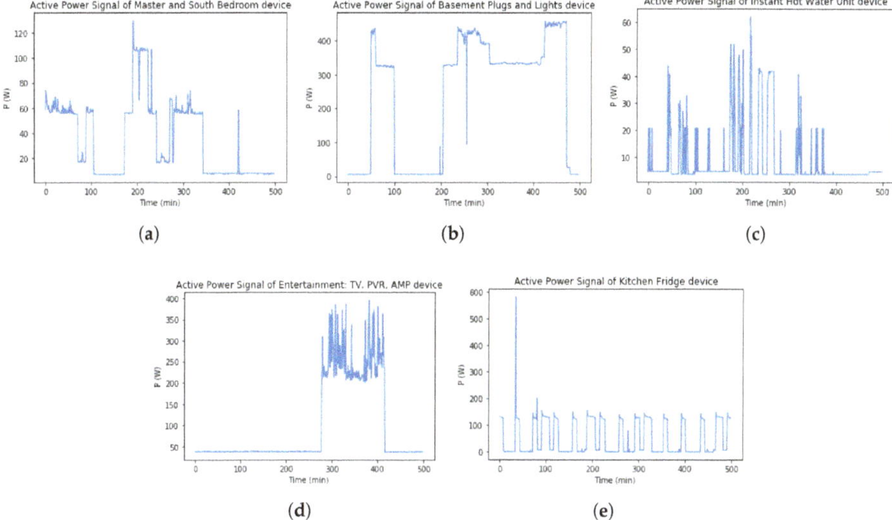

Figure 17. Active power signal of appliance data: (**a**) Master and South Bedroom; (**b**) Basement Plugs and Lights appliance; (**c**) Instant Hot Water Unit appliance; (**d**) Entertainment: TV, PVR, AMP appliance; (**e**) Kitchen Fridge appliance.

About feature extraction result, this paper applied the proposed solution to generate and extract S, T, ON/OFF label signals of each appliance. Figure 18 illustrates these results of each appliance.

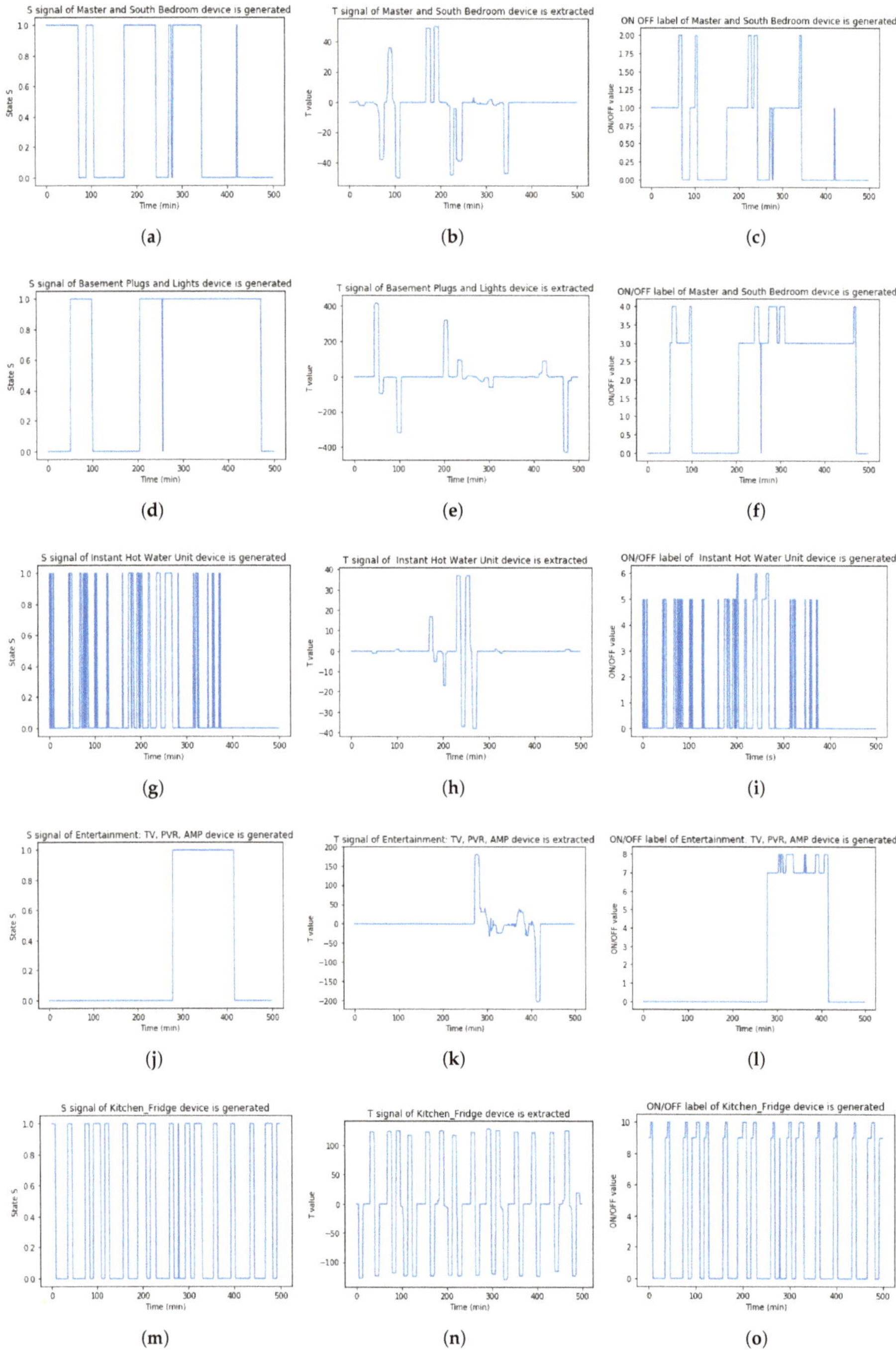

Figure 18. State *S*, *T* & ON/OFF label signals of 5 appliances: (**a–c**) Master and South Bedroom device; (**d–f**) Basement Plugs and Lights device; (**g–i**) Instant Hot Water Unit device; (**j**)–**l**) Entertainment: TV, PVR, AMP device; (**m–o**) Kitchen Fridge device.

About classification result, this paper applied two approached learning models on this selected dataset.

- In ML, the classification result of Decision Tree model is presented in Figure 19. Furthermore, Table 11 presents the event detection results in detail of this confusion matrix result. The average classification performance of Decision Tree model on this dataset achieved 98.6% accuracy and loss rate was 0.488.

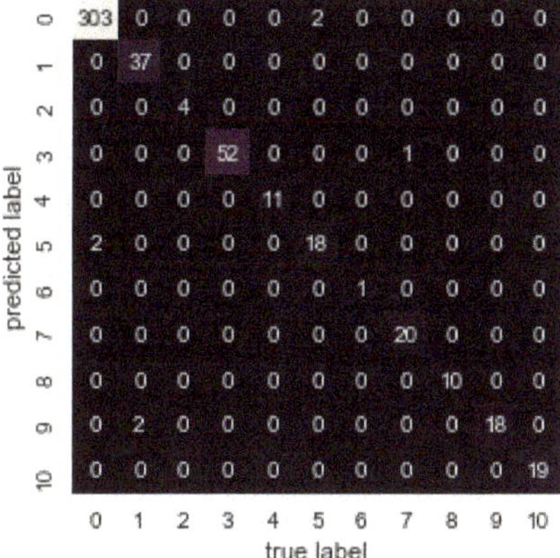

Figure 19. Confusion Matrix of Decision Tree model for classification ON/OFF label on five appliances data in AMPds2.

Table 11. Classification result for event detection ON/OFF of each appliance in AMPds2.

Label	Event Device	Precision	Recall	F1-Score	Support
0	No Event	0.99	0.99	0.99	305
1	Master and South Bedroom_ON	0.95	1.00	0.97	37
2	Master and South Bedroom_OFF	1.00	1.00	1.00	4
3	Basement Plugs and Lights_ON	1.00	0.98	0.99	53
4	Basement Plugs and Lights_OFF	1.00	1.00	1.00	11
5	Instant Hot Water_ON	0.90	0.90	0.90	20
6	Instant Hot Water_OFF	1.00	1.00	1.00	1
7	Entertainment: TV, PVR, AMP_ON	0.95	1.00	0.98	20
8	Entertainment: TV, PVR, AMP_OFF	1.00	1.00	1.00	10
9	Kitchen Fridge_ON	1.00	0.90	0.95	20
10	Kitchen Fridge_OFF	1.00	1.00	0.91	19
	avg/total	0.99	0.99	0.99	500

- In DL, the classification result of LSTM model obtained 96.78% accuracy and RMSE was 0.5124. Figure 20 illustrates the predicted results ON/OFF label of 5 appliances compare to the expected outputs.

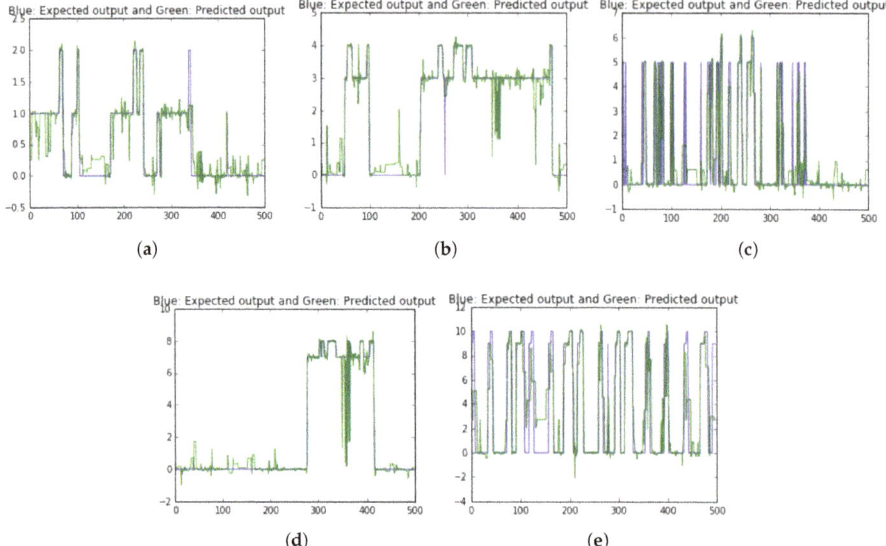

Figure 20. Classification results of LSTM model on 5 appliances data selected in AMPds2: (**a**) Master and South Bedroom; (**b**) Basement Plugs and Lights appliance; (**c**) Instant Hot Water Unit appliance; (**d**) Entertainment: TV, PVR, AMP appliance; (**e**) Kitchen Fridge appliance.

In summary, the presently proposed method achieved high-performance accuracy for load identification on the AMPds2 dataset with over 96% accuracy on both approached learning models.

5. Conclusions and Further Works

This study built a complete NILM framework including data acquisition, appliance feature extraction, classification data and monitoring energy data. First, this paper collected the personal NILM data in low-sampling rate including five household appliances energy data and twenty tests case data in the data acquisition task. Secondly, state of the appliance and transient signals are extracted to generate ON/OFF label on personal data in feature extraction task. The proposed transient signal is a transient signal to detect ON/OFF event of appliances. Thirdly, two models in ML and DL fields are stated to detect event on loads. In particular, the Decision Tree model and LSTM model are applied to perform two classification models with average accuracies achieved 92.64% and 96.85%, respectively. Finally, in monitoring energy data, a website platform system are developed to display the results of load identification. Besides, the proposed method is applied on a publicly available dataset, AMPds2. Both approach models obtained over 96% on this sub-data. In future work, this solution can be applied to extract transient signal on other publicly available datasets to widely apply the proposed methodology.

6. Data Availability

The household appliances energy consumption data set was downloaded from the figshare repository with DOI https://figshare.com/articles/NILM_EnergyData/7269692, and other data generated or analyzed during this study are available from the corresponding author on reasonable request.

Author Contributions: T.-T.-H.L. stated the problems and theirs solution. H.K. managed this project and reviewed the manuscript.

Funding: This research was supported by the MSIT (Ministry of Science and ICT), Korea, under the ITRC (Information Technology Research Center) support program (2014-1-00743) supervised by the IITP (Institute for Information & communications Technology Promotion). This work was supported by the Human Resources Development program (No. 20184030202060) of the Korea Institute of Energy Technology Evaluation and Planning (KETEP) grant funded by the Korea government Ministry of Trade, Industry and Energy.

Conflicts of Interest: The authors declare no conflict of interest.

Abbreviations

NILM	Non-Intrusive Load Monitoring
ILM	Intrusive Load Monitoring
UART	Universal Asynchronous Receiver/Transmitter
EEPROM	Electrically Erasable Programmable Read-Only Memory
RMS	Root Mean Square
AI	Artificial Intelligent
ARM	Acorn RISC Machines
GPU	Graphic Processing Unit
SVM	Support Vector Machine
DNN	Deep Neural Network
MLP	Multilayer Perceptrons
KNN	K-Nearest Neighbour
HMM	Hidden Markov Model
MSE	Mean Squared Error
ML	Machine Learning
DL	Deep Learning
CART	Classification and Regression Trees
LSTM	Long Short Term Memory
RMSE	Root Mean Square Error
ANN	Artificial Neural Network
WTC	Wavelet Transform Coefficient
CM	Confusion Matrix
TP	True Positive
FP	False Positive
TN	True Negative
FN	False Negative
UTC	Universal Time Coordinated
BLUED	Building-Level fUlly-labeled dataset for Electricity Disaggregation
REDD	Residential Energy Disaggregation Dataset
UK-DALE	UK Domestic Appliance-Level Electricity
AMPds2	The Almanac of Minutely Power dataset (Version 2)

Appendix A

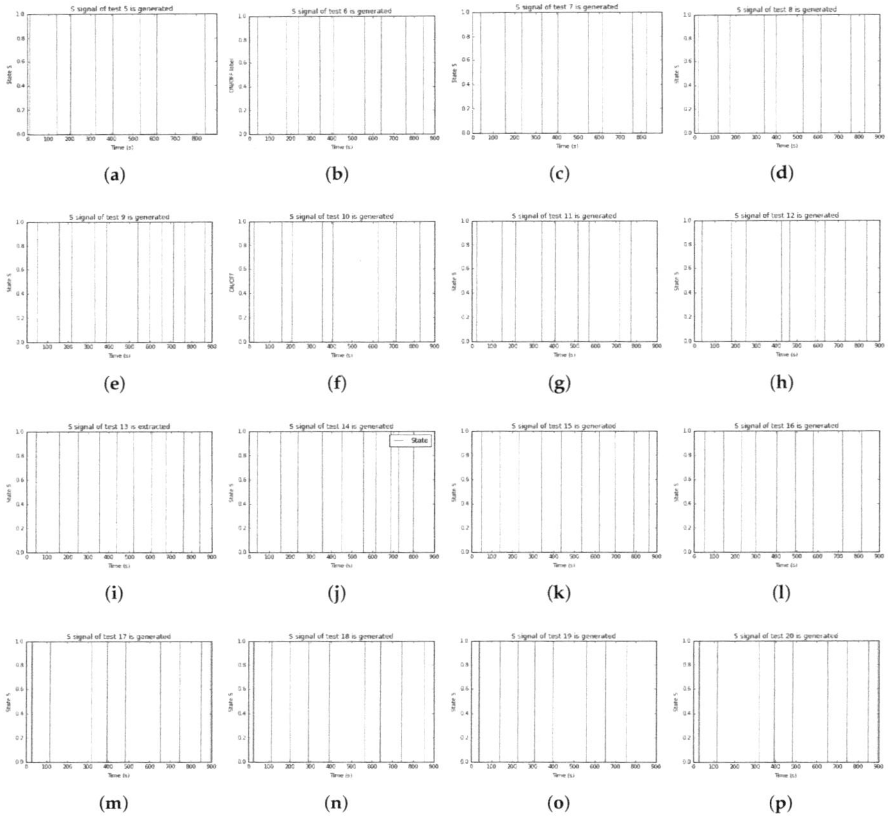

(a) (b) (c) (d)

(e) (f) (g) (h)

(i) (j) (k) (l)

(m) (n) (o) (p)

Figure A1. State *S* signals of test 5 to test 20 data.

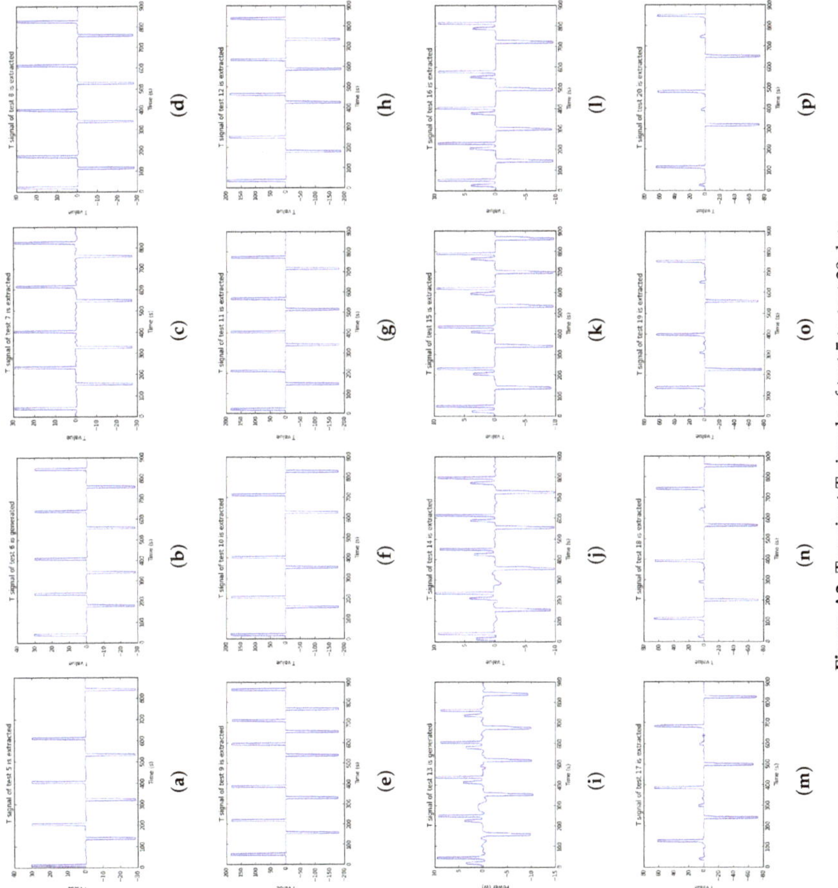

Figure A2. Transient T signals of test 5 to test 20 data.

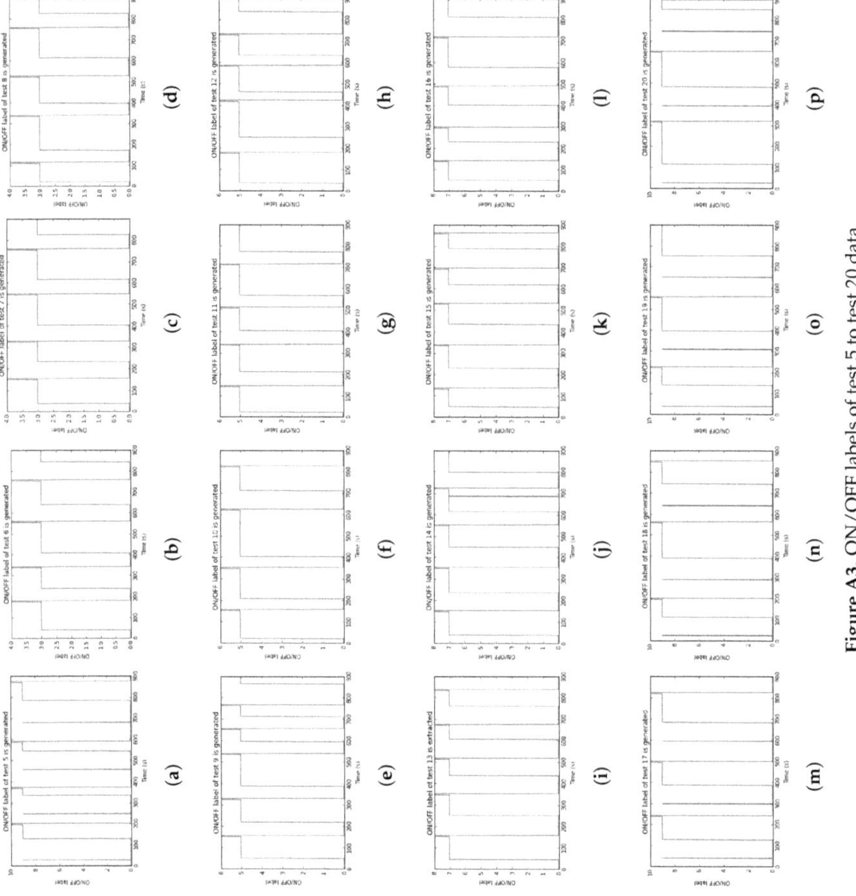

Figure A3. ON/OFF labels of test 5 to test 20 data.

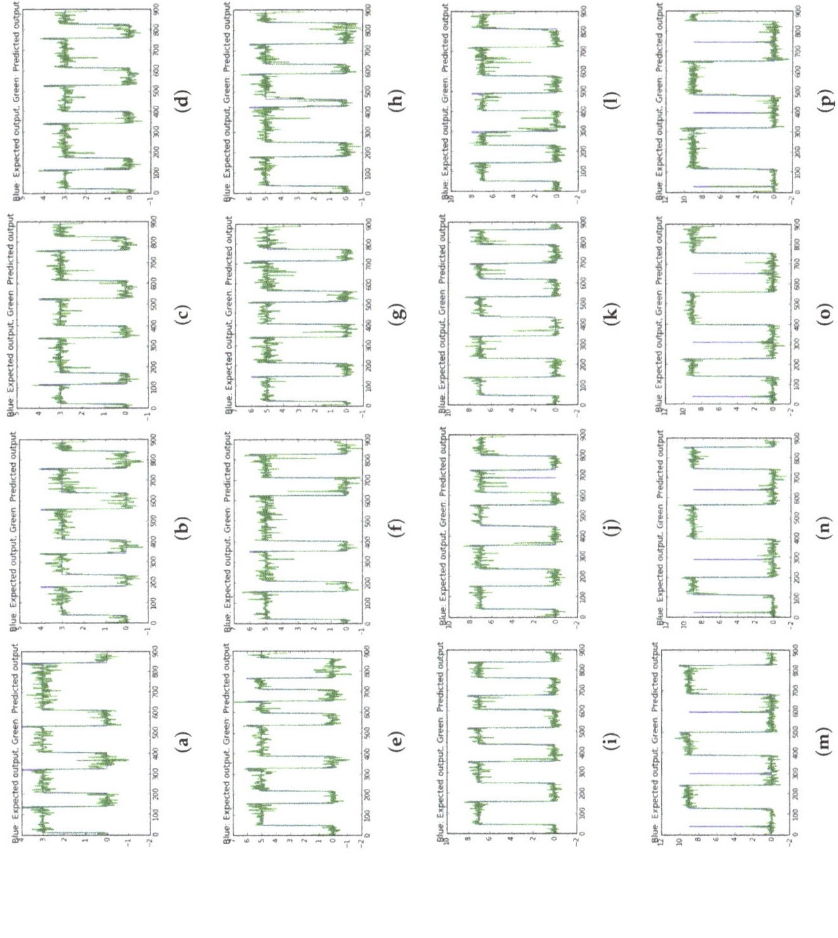

Figure A4. Classification results of LSTM model on test 5 to test 20 data: ON/OFF label is expected output with Blue color; ON/OFF label is predicted output with Green color.

References

1. Lee, J.; Bagheri, B.; Kao, H.A. A cyber-physical systems architecture for industry 4.0-based manufacturing systems. *Manuf. Lett.* **2015**, *11*, 18–23. [CrossRef]
2. Darby, S. *The Effectiveness of Feedback on Energy Consumption a Review for DEFRA of the Literature on Metering, Billing and Direct Displays*; Environmental Change Institute, University of Oxford: Oxford, UK, 2006.
3. Kelly, J.; Knottenbelt, W. Does disaggregated electricity feedback reduce domestic electricity consumption? *arXiv* **2016**, arXiv:1605.00962.
4. Abubakar, I.; Khalid, S.N.; Mustafa, M.W.; Shareef, H.; Mustapha, M. Application of load monitoring in appliances energy management: A review. *Renew. Sustain. Energy Rev.* **2017**, *67*, 235–245. [CrossRef]
5. Hart, G.W. Nonintrusive appliance load monitoring. *IEEE Proc.* **1992**, *80*, 1870–1891. [CrossRef]
6. Roos, J.G.; Lane, I.E.; Botha, E.C.; Hancke, G.P. Using neural networks for non-intrusive monitoring of industrial electrical loads. In Proceedings of the Instrumentation and Measurement Technology Conference, Hamamatsu, Japan 10–12 May 1994; pp. 1115–1118.
7. Sultanem, F. Using appliance signatures for monitoring residential loads at meter panel level. *IEEE Trans. Power Deliv.* **1991**, *6*, 1380–1385. [CrossRef]
8. Laughman, C.; Lee, K.; Cox, R.; Shaw, S.; Leeb, S.; Norford, L.; Armstrong, P. Power signature analysis. *IEEE Power Energy Mag.* **2003**, *1*, 56–63. [CrossRef]
9. Patel, S.N.; Robertson, T.; Kientz, J.A.; Reynolds, M.S.; Abowd, G.D. At the flick of a switch: detecting and classifying unique electrical events on the residential power line. In *International Conference on Ubiquitous Computing*; Springer: Berlin/Heidelberg, Germany, 2007; Volume 4717, pp. 271–288, ISBN 9783540748526.
10. Barsim, K.S.; Streubel, R.; Yang, B. An Approach for Unsupervised NonIntrusive Load Monitoring of Residential Appliances. In Proceedings of the 2nd International Workshop on Non-Intrusive Load Monitoring, Austin, TX, USA, 3 June 2014.
11. Yung, F.W.; Ahmet, Y.S.; Tom, D.; Voon, S.W. Recent approaches to non-intrusive load monitoring techniques in residential settings. In Proceedings of the IEEE Symposium on Computational Intelligence Applications in Smart Grid, Singapore, 16–19 April 2013; pp. 73–79.
12. Hana, A.; Jing, L.; Lina, S.; Vladimir, S. A low-complexity energy disaggregation method: Performance and robustness. In Proceedings of the IEEE Symposium on Computational Intelligence Applications in Smart Grid (CIASG), Orlando, FL, USA, 9–12 December 2014.
13. Hyungsul, K.; Manish, M.; Martin, A.; Geoff, L.; Jiawei, H. Unsupervised Disaggregation of Low Frequency Power Measurements. In Proceedings of the 11th SIAM International Conference on Data Mining, Phoenix, AZ, USA, 28–30 April 2011; pp. 747–758.
14. Zico, K.; Tommi, J.; Kolter, J.Z. Approximate Inference in Additive Factorial HMMs with Application to Energy Disaggregation. In Proceedings of the International Conference on Artificial Intelligence and Statistics, Canary Islands, Spain, 21–23 April 2012; pp. 1472–1482.
15. Oliver, P.; Ghosh, S.; Weal, M.; Rogers, A. Non-intrusive Load Monitoring using Prior Models of General Appliance Types. In Proceedings of the twenty-Sixth AAAI Conference on Artificial Intelligence (AAAI-12), Toronto, ON, Canada, 22–26 July 2012.
16. Stephen, M.; Fred, P.; Ivan, V.B.; Bob, G.; Lyn, B. Exploiting HMM Sparsity to Perform Online Real-Time Nonintrusive Load Monitoring. *IEEE Trans. Smart Grid* **2016**, *7*, 2575–2585.
17. Ahmed, Z.; Alexander, G.; Muhammad, A.I.; Sutharshan, R. Non-intrusive Load Monitoring approaches for disaggregated energy sensing: A survey. *Sensors* **2012**, *12*, 16838–16866.
18. Lukas, M.; Karim, S.B.; Bin, Y. How well can HMM model load signals. In Proceedings of the 3rd International Workshop on NonIntrusive Load Monitoring (NILM 2016), Vancouver, Canada, 14–15 May 2016.
19. Liang, J.; Ng, S.K.; Kendall, G.; Cheng, J.W. Load Signature Study Part I: Basic Concept, Structure, and Methodology. *IEEE Trans. Power Deliv.* **2010**, *25* , 551–560. [CrossRef]
20. Faustine, A.; Mvungi, H.N.; Kaijage, S.; Michael, K. A Survey on Non-Intrusive Load Monitoring Methodies and Techniques for Energy Disaggregation Problem. *arXiv* **2017**, arXiv:1703.00785.
21. Kelly, J.; Knottenbelt, W. Neural NILM: Deep Neural Networks Applied to Energy Disagregation. In Proceedings of the 2nd ACM International Conference on Embedded Systems for Energy-Efficient Built Environments (BuildSys'15), Seoul, Korea, 4–5 November 2015; pp. 55–64.

22. Kim, J.; Le, T.T.; Kim, H. Nonintrusive load monitoring based on advanced deep learning and novel signature. *Comput. Intell. Neurosci.* **2017**, *2017*, 4216281. [CrossRef] [PubMed]
23. Kanghang, H.; Lina, S.; Jing, L.; Vladimir, S. Non-Intrusive Load Disaggregation using Graph Signal Processing. *IEEE Trans. Smart Grid* **2016**, *9*, 1739–1747.
24. Figueiredo, M.B.; De Almeida, A.; Ribeiro, B. *An Experimental Study on Electrical Signature Identification of Non-Intrusive Load Monitoring (NILM) Systems*; Lecture Notes in Computer Science (Including Subseries Lecture Notes in Artificial Intelligence and Lecture Notes in Bioinformatics); Springer: Berlin/Heidelberg, Germany, 2011; Volume 6594, pp. 31–40, ISBN 9783642202667.
25. Chang, H.H.; Chen, K.L.; Tsai, Y.P.; Lee, W.J. A New Measurement Method for Power Signature of Non-intrusive Demoand Monitoring and Load Identification. In Proceedings of the Industry Applications Society Annual Meeting (IAS), Orlando, FL, USA, 9–13 October 2011.
26. Kofi, A.A.; Sekyung, H.; Soohee H. Real-Time Recognition Non-Intrusive Electrical Appliance Monitoring Algorithm for a Residential for a Residential Building Energy Management System. *Energies* **2015**, *8*, 9029–9048. [CrossRef]
27. Chang, H.H.; Lian, K.L.; Su, Y.C.; Lee, W.J. Power-Spectrum-Based Wavelet Transform for Noninstrusive Demand Monitoring and Load Identification. *IEEE Trans. Ind. Appl.* **2014**, *50*, 2081–2089. [CrossRef]
28. Zeifman, M.; Akers, C.; Roth, K. Nonintrusive appliance load monitoring: Review and outlook. *IEEE Trans. Consum. Electron.* **2011**, *57*, 76–84. [CrossRef]
29. Hassan, T.; Javed, F.; Arshad, N. An Empirical Investigation of V-I Trajectory Based Load Signatures for Non-Intrusive Load Monitoring. *IEEE Trans. Smart Grid* **2013**, *5*, 870–878. [CrossRef]
30. Nilson, H.; Kodjo, A.; Suosso, K.; Sayed, S.H.; Michael, F. Power Estimation of Multiple Two-State Loads Using A Probabilistic Non-Intrusive Approach. *Energies* **2018**, *11*, 88. [CrossRef]
31. Breiman, L.; Friedman, J.H.; Olshen, R.A.; Stone, C.J. *Classification and Regression Trees*; Routledge: Wadsworth, OH, USA; Belmont, CA, USA, 1984.
32. Sepp, H.; Jurgen, S. Long short-term memory. *Neural Comput.* **1997** , *9*, 1735–1780.
33. Makonin, S.; Ellert, B.; Bajic, I.V.; Popowic, F. Electricity, water, and natural gas consumption of a residential house in Canada from 2012 to 2014. *Sci. Data* **2016**, *3*, 160037. [CrossRef] [PubMed]
34. Makonin, S. AMPds2: The Almanac of Minutely Power Dataset (Version 2). 2016. Available online: https://doi.org/10.7910/DVN/FIE0S4 (accessed on 30 October 2018).

Article

Concatenate Convolutional Neural Networks for Non-Intrusive Load Monitoring across Complex Background

Qian Wu and Fei Wang *

School of Electronic Information and Communications, Huazhong University of Science and Technology, Wuhan 430074, China; wuqian@hust.edu.cn
* Correspondence: wangfei@hust.edu.cn

Received: 6 April 2019; Accepted: 24 April 2019; Published: 25 April 2019

Abstract: Non-Intrusive Load Monitoring (NILM) provides a way to acquire detailed energy consumption and appliance operation status through a single sensor, which has been proven to save energy. Further, besides load disaggregation, advanced applications (e.g., demand response) need to recognize on/off events of appliances instantly. In order to shorten the time delay for users to acquire the event information, it is necessary to analyze extremely short period electrical signals. However, the features of those signals are easily submerged in complex background loads, especially in cross-user scenarios. Through experiments and observations, it can be found that the feature of background loads is almost stationary in a short time. On the basis of this result, this paper provides a novel model called the concatenate convolutional neural network to separate the feature of the target load from the load mixed with the background. For the cross-user test on the UK Domestic Appliance-Level Electricity dataset (UK-DALE), it turns out that the proposed model remarkably improves accuracy, robustness, and generalization of load recognition. In addition, it also provides significant improvements in energy disaggregation compared with the state-of-the-art.

Keywords: non-intrusive load monitoring; energy disaggregation; deep learning; source separation

---

1. Introduction

Energy consumption has always been a major concern in the world, which can be alleviated with accurate and efficient load monitoring methods. There are two branches of load monitoring, namely intrusive load monitoring (ILM) and non-intrusive load monitoring (NILM). The major difference between them is the number of sensors. The ILM needs to install at least one sensor at each appliance to monitor the load respectively, while the NILM needs to install one sensor on the bus per house merely. The NILM is physically simple at the cost of more complex approaches, thus NILM based approaches are more widely researched.

NILM contains two main objectives: energy disaggregation and load recognition. Conventional energy disaggregation aims to obtain energy consumption for every single appliance, which acts on the entire operation cycle of an appliance, and some research discards event detection and gives a result at the hour level to disaggregate the energy consumption [1]. The approximate power consumption of each appliance in a certain period is its main concern, thus the exact on/off time of appliances is unknown. For online load recognition, appliances are detected from the uncompleted operation cycle, that is the transient-state process of on/off events. In addition, the result of load recognition further benefits energy disaggregation. Some advanced applications in the smart grid need to acquire appliance operation status for remote household control [2], such as demand response. It represents that the power supply side uses induction mechanism (e.g., price changes over time) to improve end users' energy consumption patterns, which demands for the balance of demand and supply in real time [3].

Real-time load recognition requires short sample windows and short execution time. The sample window is usually in seconds, while the execution time is less than a second. Obviously the former has a more significant impact on real time, thus the transient-state process should be captured and recognized. High sampling rate is necessary for sufficient information acquisition from transient-state features. High enough frequency (i.e., 100 kHz or higher) features can easily distinguish different appliances (e.g., electromagnetic interference (EMI) signatures), but EMI features can only be transmitted within a few meters, which is not suitable for household or industrial use. This paper uses 2 kHz features, which introduces interference from background loads, so the emphasis of this paper is to eliminate the influence of complex background loads. There are diverse types of appliances in different houses, and even the same type of appliances in different households vary by brand. It is hard for researchers to collect all combinations of appliances. Briefly, household electrical signals are multi-source single-channel data. Several similar topics are raised at the task of multi-source signals processing, such as the cocktail party problem in speech recognition and blind source separation (BSS) in wireless communications. Multichannel observations are necessary in these areas, whether it is two-channel audio-visual data [4] or two-channel audio observed signal [5], the data source itself gives the possibility of separation. Noiseless samples are used for training and being compared with separated samples in speech recognition methods, and artificial communication signals have the property of time or frequency separability, or they are orthogonal to each other. Unlike these problems, people can hardly obtain individual appliance samples in the single-channel scenario by non-intrusive approaches. To extract the feature of the target appliance from the load mixed with background loads, this paper utilizes the fact that the electrical signal is approximately stationary in a short time, in other words, loads are strongly correlated during this period.

Due to the powerful feature extraction ability of deep learning, this paper proposes an efficient network called the concatenate convolutional neural network. The model combines signal processing and pattern recognition to eliminate the influence of background loads in single-channel data and recognize different appliances. Features in high-dimensional spectrograms can be extracted in virtue of the development of deep neural networks. A series of deep convolutional networks for image classification have been proposed, such as Extreme Inception (Xception) [6], Residual Net (ResNet) [7], Dense Convolutional Network (DenseNet) [8]. In this paper, high-frequency current data is converted to spectrograms by Short Time Fourier Transform (STFT) and set as the model input. Two proven networks are used as the embedding layer to extract spectrogram features. However, it is problematic to apply object recognition methods to the model input without modification. The recognition object covers the background in computer vision, thus the background causes little substantial impact of recognition, whereas the foreground appliance in spectrograms suffers from blurring due to the superimposition of background loads.

The main contributions of this paper are as follows. (1) The model eliminates the impact of background loads to a certain extent, and achieves an F1-score of 89.0% in the classification task. (2) The proposed approach improves the performance of energy disaggregation, especially on multi-state appliances and programmable appliances. It increases an F1-score of 3–73% and reduces mean square error (MAE) of 3.1–24.2 watts. The proposed model is evaluated on the UK Domestic Appliance-Level Electricity dataset (UK-DALE) [9] and the Building-Level fUlly labeled dataset for Electricity Disaggregation (BLUED) [10]. The proposed model is also compared with several works in the aspect of energy disaggregation. In the remainder of this paper, the research status of NILM and the proposed network model will be introduced. Next, the results of the classification and energy disaggregation experiments are presented. Finally, this paper provides conclusions and future work.

2. Related Work

NILM was proposed by Hart more than 20 years ago [11], and numerous related research projects have begun since then. In the demand side management (DSM) program, the home energy management system comprises active appliances and passive appliances [12]. The active appliances consist of energy sources and energy storage systems [13,14]. Like most NILM methods, passive appliances are only considered in this paper. The approaches of improving classification accuracy or energy disaggregation results are closely related to data acquisition and feature selection.

The sampling rate in data acquisition can be simply divided into high-frequency and low-frequency, which directly affects the selection of features and approaches. Low-frequency data is typically used for approaches based on steady-state features, whereas high-frequency data is used for the transient-state analysis. Low-frequency data is easy to acquire and suitable for energy disaggregation [15,16]. The fundamental frequency is either 50 Hz or 60 Hz in most countries, thus sampling at several Hertz or lower is not suitable for spectrum analysis. High-frequency sampling at over 1 kHz captures features of higher harmonics and further recognizes different loads. Prior work [17] proved that the power spectrum computed from 15 kHz transient-state data can identify different loads with the same steady-state active and reactive power.

The higher the sampling frequency, the richer the information obtained. Multiple appliances could be distinguished by high-frequency signatures. Previous research showed that when the sampling rate was as high as 1 MHz, the EMI signatures of almost all appliances were distinguishable on the spectrograms [18], but the signatures could not be transmitted over a long distance, even in tens of square meters house. Low-noise data with no background load has been measured in some datasets [19,20], so there is no need to address these problems with complex networks.

In the field of power, the features are almost directly or indirectly derived from current, voltage, and power. These features in time domain or frequency domain are used in various methods, which mainly focus on some common machine learning algorithms, such as support vector machine (SVM) [21], k-nearest neighbors (kNN) [21], decision tree [22], wavelet design [23], evolutionary algorithm [24], adaptive boost [24], graph signal processing (GSP) [25], hidden Markov model (HMM) [16,26,27], etc. Even though there are methods show the good robustness under noisy conditions [26,27], the noise is different from background loads. The signal noise in these works derives from measurement error and the voltage fluctuation, and prove to have less impact (see Section 3).

The progress of deep learning has a positive impact on feature extraction. Deep neural networks have been exploited to train high dimensional images or sequences. Auto-encoder (AE) [15,28], long short-term memory (LSTM) [15,22], and sequence-to-point (Seq2point) neural networks [29] have been applied to the NILM task to achieve the corresponding targets. The model input of all these methods is low-frequency time series, without considering the influence of background loads. Their target lies in offline energy disaggregation, whereas the proposed approach can eliminate background loads and recognize different appliances. Experiments in Section 4 give a comparison of whether or not the background load has been processed. Some low-frequency NILM approaches perform in real time during the load classification or disaggregation [26,30], and the high-frequency approach in this paper can also achieve short sample windows and short execution time.

3. Concatenate Convolutional Neural Networks

The core idea of this paper is to utilize features extracted by convolutional neural networks (CNNs), then eliminate the influence of background loads by concatenate networks. The interference to recognition is divided into two parts: the voltage fluctuation and multiform background loads. When the voltage fluctuation is regarded as noise, the estimation result on the UK-DALE dataset shows that the signal-to-noise ratio (SNR) of voltage waveforms is 53 dB. For the measurement error, current and voltage waveforms have the SNR of 90 dB [9]. On the other hand, background loads might cause a strong noise on the target appliance. For example, for a 1000 W appliance, the SNR is 10 dB when background loads are only 100 W, and the SNR is 0 dB when background loads are 1000 W. It is obvious that the influence of background loads is much stronger than that of the voltage fluctuation. Therefore, the main purpose of this paper is to eliminate background loads.

3.1. Problem Statement

Originally, the inspiration about this paper comes from the parallel circuit shown in Figure 1. Assume that the main circuit current is I_0, and the branch currents are I_1 and I_2 respectively, and the background load and the target load are Z_1 and Z_2 respectively. When the switch K is turned off, we have the equation $I_{0(off)} = I_{1(off)}$, and $I_{0(off)}$ is known. After the switch K is turned on, we have the equation $I_{0(on)} = I_{1(on)} + I_{2(on)}$ according to Kirchhoff's current law (KCL), and $I_{0(on)}$ is known. In the ideal case, it will be ignored that the effect of Z_2 on Z_1 and the fluctuation of Z_1 branch, thus $I_{1(off)} = I_{1(on)}$, further $I_{2(on)}$ can be expressed by:

$$\begin{aligned} I_{2(on)} &= I_{0(on)} - I_{1(on)} \\ &= I_{0(on)} - I_{1(off)} \\ &= I_{0(on)} - I_{0(off)}. \end{aligned} \tag{1}$$

However, experiments shown in Figure 2 prove that Equation (1) does not strictly hold. Figure 2a shows the microwave spectrogram without background loads, which is measured in the laboratory. Figure 2b shows the microwave spectrogram in the UK-DALE dataset, Figure 2c shows the spectrogram calculated by spectral estimation based on the previous hypothesis of Equation (1). Although there is a strong correlation between Figure 2a and Figure 2c, spectral estimation does not eliminate the background load precisely, there are still intermittent spectral lines before the appliance is turned on. Besides, compared to Figure 2a, some detail components are eliminated in Figure 2c after the appliance is turned on. These phenomena prove that the background load has a certain degree of stationarity, but this does not mean it is exactly unchanged, and $I_{1(off)}$ is not equal to $I_{1(on)}$. Accordingly, the background load needs to be estimated more reasonably.

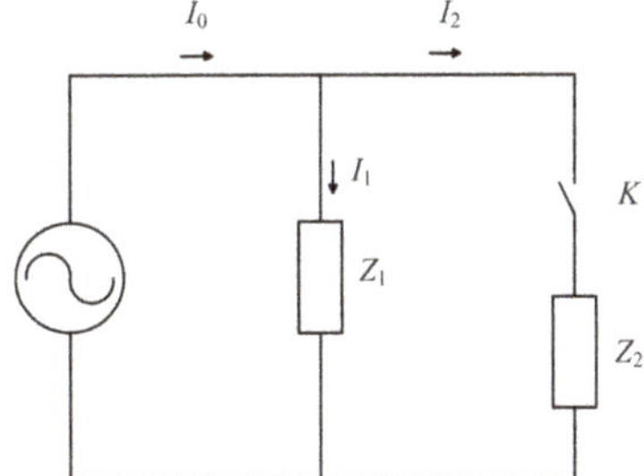

Figure 1. The parallel circuit.

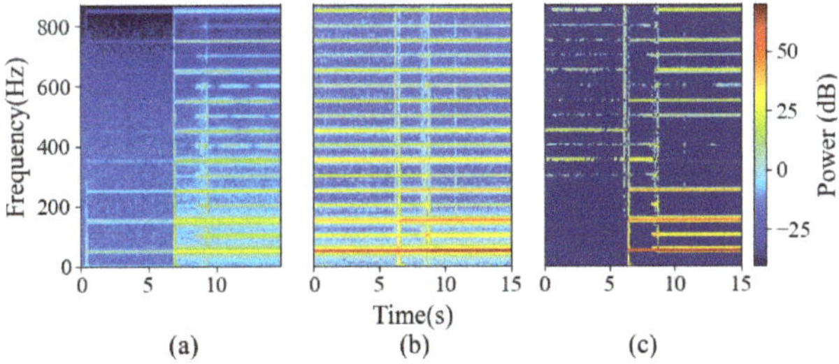

Figure 2. Spectrograms of microwave. (**a**) The spectrogram with no background load. (**b**) The spectrogram with background loads (**c**) The estimated spectrogram.

On this issue, this paper proposes a novel approach, concatenate deep neural network, to estimate the features of $I_{2(on)}$ indirectly, which can be represented as:

$$\begin{aligned} X_{2(on)} &= d_{\phi}(X_{0(on)}, X_{1(on)}) \\ &= d_{\phi}(X_{0(on)}, s_{\varphi}(X_{0(off)}, X_{0(on)})), \end{aligned} \quad (2)$$

where s_{φ} is the function to extract the similar part of two features $\{X_{0(off)}, X_{0(on)}\}$, d_{ϕ} is the function to extract the different part of two features $\{X_{0(on)}, X_{1(on)}\}$, X with different subscripts are features of spectrograms computed by the corresponding current, which can be calculated by:

$$X = f_{\theta}(SI) \quad (3)$$

where SI is the spectrogram. The function f_{θ} is used to extract the features of the mixed load spectrogram and the background spectrogram.

The inspiration of function s_{φ} and d_{ϕ} borrows from Code Division Multiple Access (CDMA), which allows multiple users to transmit independent information within the same bandwidth simultaneously, and the orthogonal spreading code is used to distinguish and extract signals from different users [31].

In the circuit model, the main circuit current $I_0(t)$ is given by:

$$I_0(t) = \sum_{k=1}^{K} I_k(t), \quad (4)$$

where K is the number of branches. The branch current $I_k(t)$ is expressed as:

$$I_k(t) = I_k^{rated} c_k(t) \quad (5)$$

where I_k^{rated} is the rated current of the load on the k-th branch, which is a constant, and the $c_k(t)$ represents the time-varying noise function of the load.

Since appliances are relatively independent in construction and operation, their noise functions are almost uncorrelated with each other and can act as the spreading code in CDMA.

The i-th branch current is recovered by multiplying the noise function:

$$\begin{aligned} z_k(t) &= \int_{t_i}^{t_i+T} I_0(t) c_i(t) dt \\ &= \int_{t_i}^{t_i+T} \sum_{k=1}^{K} I_k(t) c_i(t) dt \\ &= \int_{t_i}^{t_i+T} \sum_{k=1}^{K} I_k^{rated} c_k(t) c_i(t) dt \end{aligned} \quad (6)$$

As the spreading code in CDMA is designed, the noise functions are assumed to be orthogonal in the ideal case, where

$$\int_0^T c_k(t) c_i(t) dt = \begin{cases} T, & \text{for } i = k \\ 0, & \text{for } i \neq k, \end{cases} \quad (7)$$

Therefore the branch current is restored as:

$$z_i(t) = I_i^{rated} T. \quad (8)$$

Unfortunately, unlike CDMA, it is difficult to get the precise "spreading code" for each appliance in practice, and background loads usually consist of multiple loads. Therefore the method of network fitting is used to recover branch current. The first step is to extract the spreading code, and the second step is to reconstruct the feature of the branch current. Then the simplified form of Equation (4) is:

$$\begin{aligned} I_0(t) &= I_B(t) + I_L(t) \\ &= I_B^{rated} c_B(t) + I_L^{rated} c_L(t), \end{aligned} \tag{9}$$

where I_B and I_L represent the branch current of background loads and the load to be recognized (i.e., target load), respectively. $c_B(t)$ and $c_L(t)$ are weakly correlated. In practice, it is easy to get the previous background loads $I_{B'}(t)$ before the target load is turned on. On the hypothesis that background loads are stationary in a short time, the relationship between the noise functions $c_{B'}(t)$, $c_B(t)$ and $c_L(t)$ is stated as:

$$\int_0^T c_i(t) c_{B'}(t) dt \approx \begin{cases} T, & \text{for } i = B \\ 0, & \text{for } i = L \end{cases}, \tag{10}$$

In fact, such estimation is not rigorous, because stationarity does not mean complete equality, and weak correlation does not mean strict independence. Thus the similarity learning module s_φ is equiped to fit Equation (10), and the feature of background loads is obtained by:

$$feature(I_B(t)) = s_\varphi\left(I_0(t), I_{B'}(t)\right). \tag{11}$$

For the branch current, $I_L(t)$ can be calculated by subtraction through Equation (9). For the corresponding feature, the difference learning module d_ϕ is used to fit the subtraction operation and obtain the feature of $I_L(t)$ by:

$$feature(I_L(t)) = d_\phi\left(feature(I_0(t)), feature(I_B(t))\right). \tag{12}$$

In summary, the model consists of an embedding module, a similarity learning module, a difference learning module and a classifier, which realizes the complete process of feature extraction, feature selection, and classification.

3.2. Model Architecture

Based on the previous hypothesis, each spectrogram image is split into two blocks representing the background load and the mixed load as shown in Figure 3. Two blocks are input into the network simultaneously, which can be seen in Figure 4.

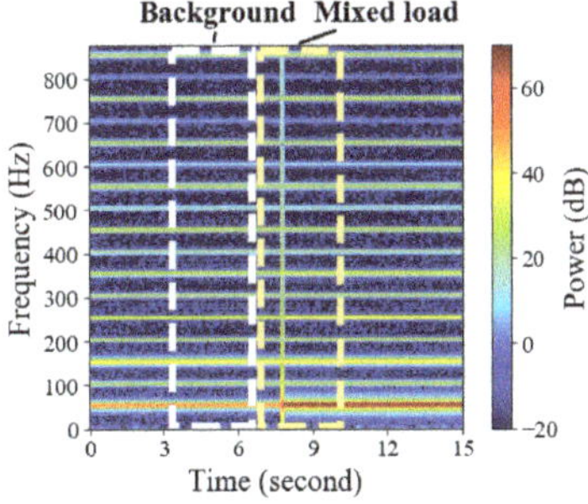

Figure 3. The background load part and the mixed load part of a complete spectrogram.

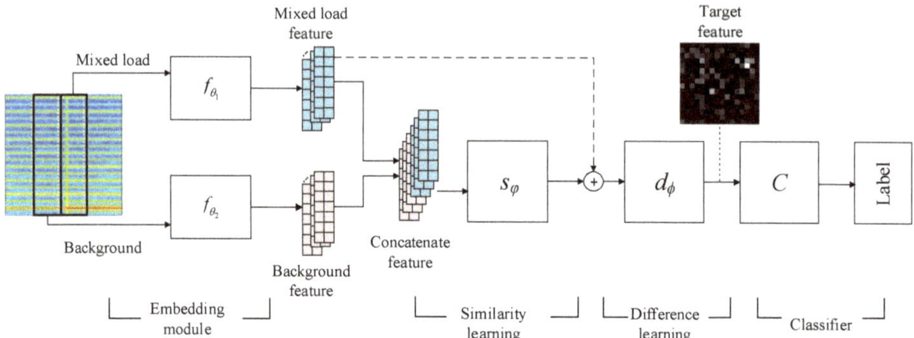

Figure 4. The structure of concatenate convolutional neural networks (CNNs). The concatenation operation is channel-wise.

The CNNs in the embedding module are placed at the front end of the network to convert the image matrix into a vector, which is f_θ in Equation (3), and here the module comprises two networks $\{f_{\theta_1}, f_{\theta_2}\}$ with the same structure and different parameters. The similarity learning module s_φ is used to generate the similar part in the concatenate feature and get the background feature behind the mixed feature, we refer to this as the "implicit background", distinct from the "explicit background" extracted from the background-only load. The concatenation is channel-wise. The difference learning module d_ϕ converts the features of the mixed load and the "implicit background" to the target feature $X_{2(on)}$. The final classifier determines the label of the target load through the target feature maps. The loss function is the cross-entropy function:

$$L = -\left(\sum_{i=1}^{C} y_i \log Z_i + (1 - y_i)\log(1 - Z_i)\right), \tag{13}$$

where y_i is the i-th bit of the one-hot label y of C classes, and Z_i is the i-th element of the network output Z with the softmax activation, which can be represented as:

$$Z = \text{softmax}\left(h_\sigma(X_{2(on)})\right), \tag{14}$$

where h_σ is the classifier to map the obtained target spectrogram feature to the C-dimensional vector, and softmax is the activation.

Figure 5 presents the detail architecture of the proposed network with the embedding module omitted. Similarity learning module seems to be a residual block even though it is not identical. The shortcut connection here aims to convey information about the mixed feature. Each convolutional layer (Conv) comprises a 256-filter convolution of kernel size 7×2 and stride 1. The kernel size is determined by the embedding vector computed by the CNN. In all experiments, the size of the embedding vector (the background feature and the mixed feature) is $7 \times 2 \times 256$. The similarity learning module is followed by an average-pooling layer (AvgPool) with kernel size 7×2 to compress the feature map in height and width. The output size of two fully connected (FC) layers in the difference learning module is 256. For the classifier, two FC layers are 64 and C dimensional, respectively.

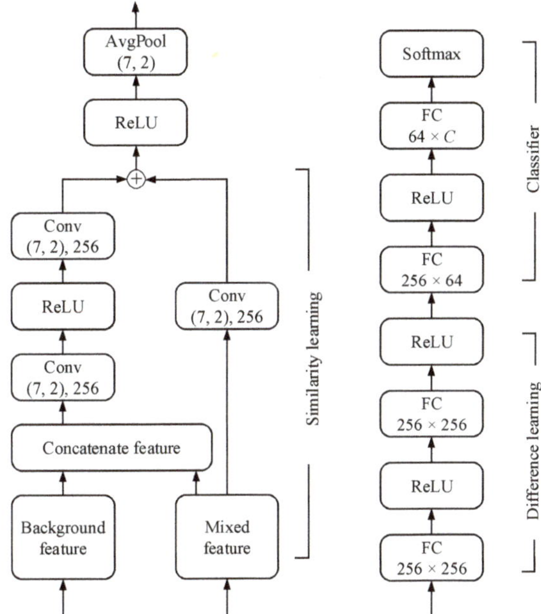

Figure 5. The detail architecture of similarity learning and difference learning module. The addition operation $\oplus$ in similarity learning module is element-wise.

4. Experiments and Discussion

The proposed approach is compared with two baselines and evaluated on the UK-DALE dataset and the BLUED dataset to test its universality. An energy disaggregation result on the UK-DALE dataset will be also provided.

4.1. Dataset and Data Preprocessing

Both the real and imaginary part of spectrograms are taken as model input, in other words, complex spectrograms are computed instead of power spectrograms or magnitude spectrograms. The two-channel data is more expressive not only because it contains phase information, but also due to the additivity of spectrograms in the complex domain. As envisaged, the background load can be removed from the mixed load on this account.

Every switching event contains 7-second current data, that is, 14,000 data points. As shown in Figure 6, plenty of samples are drawn to confirm that background loads are stationary in a 7-second window in most cases. Note that 15-second spectrograms are drawn in this paper in order to give a more complete demonstration, but this does not affect the 7-second spectrograms actually used in experiments. The size of an original 7-second spectrogram is $224 \times 100 \times 2$, the first two dimensions represent frequency and time, and the last dimension represents real and imaginary part. The switching point of each sample is located slightly after the midpoint of the time axis. As shown in Figure 3, two blocks are split from the original image as the proposed network input, and the splitting line is in the midpoint of the time axis. Every block of input image has a size of $224 \times 50 \times 2$.

4.1.1. UK-DALE

Most of our experiments are based on the UK-DALE dataset. The UK-DALE dataset contains electrical data from 5 houses for up to 655 days. A 1/6 Hz aggregate (mains) and individual appliance (submetered) power data were recorded for each house. Houses 1, 2, and 5 are selected as our original data source because only these 3 houses contain 16 kHz aggregate voltage and current data, then 16 kHz data is downsampled to 2 kHz to draw spectrograms. For submetered data, dataset authors did not record 16 kHz data. The switching events are detected with the help of submetered data (6-second time interval). For all experiments on UK-DALE, house 1 and house 5 data are set as the training set, and house 2 data is set as the test set to measure the generalization ability of our model. Seven classes of appliances are used for classification—kettle, fridge, dish washer (DW), microwave (MW), washing machine (WM), laptop or monitor (LCD), and running machine (RM). The number of appliances in each house and appliance spectrograms are shown in Table 1 and Figure 6, respectively.

Table 1. The number of appliance events per house in UK Domestic Appliance-Level Electricity dataset (UK-DALE).

Appliance	House 1	House 2	House 5
Kettle	4430	508	171
Fridge	7525	2172	2453
DW	2797	163	235
MW	536	299	29
WM	12311	785	521
LCD	4022	862	387

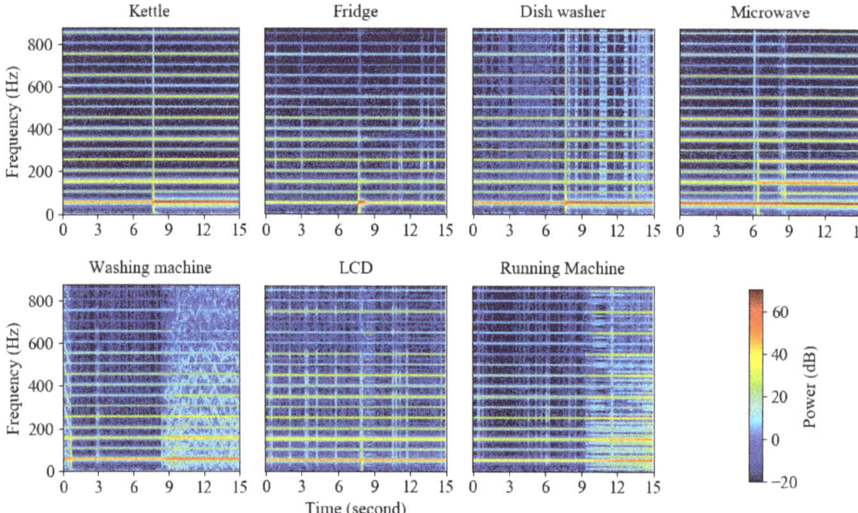

Figure 6. Spectrograms drawn from appliances in UK Domestic Appliance-Level Electricity dataset (UK-DALE).

4.1.2. BLUED

Unlike the UK-DALE dataset, the BLUED dataset collects data from a house for only 8 days, which also has dozens of appliances, so each appliance has very few samples. Table 2 shows the selected appliances to facilitate comparison with the published work based on Fast Shapelets [32]. Raw data (12 kHz) is also downsampled to 2 kHz to draw the same size spectrogram as the one in the UK-DALE dataset. The validation set accounts for 33% and a 3-fold cross-validation is adopted.

Table 2. Appliances used in Building-Level fUlly labeled dataset for Electricity Disaggregation (BLUED).

Appliance	Label	#Events
Fridge	Phase A, 1	282
Lights1 (backyard lights, washroom light, bedroom lights)	Phase A, 2	17
High-Power1 (Hair dryer, air compressor, kitchen aid chopper)	Phase A, 3	18
Lights2 (desktop lamp, basement light, closet lights)	Phase B, 1	38
High-Power2 (printer, iron, garage door)	Phase B, 2	52
Computer and monitor (computer, LCD monitor, DVR)	Phase B, 3	130

4.2. Experimental Infrastructure

All neural networks in classification tasks are implemented using TensorFlow and trained on NVIDIA GeForce GTX 1080 GPUs.

In this paper, all networks are trained using Adam optimizer [33], and the initial learning rate is set to 0.001 for all networks, although some exceptions will be supplemented later.

4.3. The Recognition Result on UK-DALE

Two different prevalent networks are chosen as the embedding module of the model: Xception and DenseNet-121. We name these models Concatenate-CNNs. Meanwhile, two baselines are used to compare with the proposed networks. The model input and description is shown in Table 3.

Table 3. Model input and description.

Model	Model Input	Model Description
CNN (baseline 1)	One $224 \times 100 \times 2$ original spectrogram	CNN with a FC classifier
CNN-SE (baseline 2)	One $224 \times 100 \times 2$ original spectrogram subtracts the estimated background	CNN with a FC classifier
Concatenate-CNN	Two $224 \times 50 \times 2$ spectrograms split from the original spectrograms	CNN with a similarity learning module, a difference learning module and a FC classifier

The first baseline is to classify appliances without the disposal of background loads. The original $224 \times 100 \times 2$ spectrogram is input into the CNN without concatenate features and following networks.

The second baseline is to deal with the background load with spectral estimation (SE). The model input is the estimated target spectrogram which is shown in Figure 2c, the network is the same as the first baseline, and we call this CNN-SE. In this case, the background current is assumed to be invariant during the window. Specifically, the background feature is obtained by calculating the average spectrum of the first 3 seconds in the entire 7-second window, the target spectrogram is obtained by subtraction between the original spectrogram and the calculated background spectrogram.

In Concatenate-DenseNet-121 and its two baselines, the growth rate is set to 32 and the compression rate is 0.5. Other parameters have been illustrated in the last section.

Here are three metrics to evaluate the performance. 'Recall' indicates the proportion of samples of one class are correctly recognized, which is given by:

$$recall = \frac{TP}{TP + FN}. \tag{15}$$

'Precision' represents the proportion of samples recognized as one class and truly belong to that class, which is calculated by:

$$precision = \frac{TP}{TP + FP}. \quad (16)$$

where true positives (TP) are the number of events correctly classified when the appliance was on, false positives (FP) are the number of events classified as on being the appliance off, and false negatives (FN) represent the number of events classified as off being the appliance on.

The F1-score combines recall and precision, here appliance events are classified without estimating the power consumption, thus the F1-score is reported [34], which is given by:

$$F1 - score = 2 \cdot \frac{precision \cdot recall}{precision + recall}. \quad (17)$$

Table 4. Comparison of the proposed model and two baselines on house 2 data of the UK-DALE dataset with recall (%), precision (%), and F1-score (%). Best results are shown in bold.

Model	Metrics	Kettle	Fridge	DW	MW	WM	LCD	RM	Average
Xception	Recall	91.3	98.1	79.1	98.7	96.9	63.5	42.9	81.5
	Precision	91.7	99.9	39.1	98.3	77.6	99.5	81.8	84.0
	F1-score	91.5	99.0	52.3	98.5	86.2	77.5	56.3	80.2
Xception-SE	Recall	97.6	93.9	77.3	98.7	96.3	87.6	33.3	83.5
	Precision	93.0	99.0	35.9	98.7	95.6	98.3	87.5	86.8
	F1-score	95.3	96.4	49.0	**98.7**	**95.9**	**92.6**	48.3	82.3
Concatenate-Xception	Recall	99.6	99.6	84.7	96.7	98.9	71.9	52.4	86.2
	Precision	95.3	99.5	40.7	99.7	93.0	98.4	91.7	88.3
	F1-score	**97.4**	**99.6**	**55.0**	98.1	**95.9**	83.1	**66.7**	**85.1**
DenseNet-121	Recall	100	99.8	71.8	99.7	95.5	81.2	71.4	88.5
	Precision	91.4	99.6	66.9	98.7	85.7	99.4	71.4	87.6
	F1-score	95.5	**99.7**	69.2	99.2	90.3	89.4	71.4	87.8
DenseNet-121-SE	Recall	99.2	86.4	87.7	99.0	97.8	74.2	71.4	88.0
	Precision	95.6	99.9	22.3	99.7	95.5	99.5	78.9	84.5
	F1-score	**97.4**	92.7	35.6	99.3	**96.7**	85.0	**75.0**	83.1
Concatenate-DenseNet-121	Recall	100	99.6	77.9	99.3	92.4	93.0	57.1	88.5
	Precision	92.5	99.2	62.6	100	95.9	98.9	85.7	90.7
	F1-score	96.1	99.4	**69.4**	**99.7**	94.1	**95.9**	68.6	**89.0**

The performance of these models on the UK-DALE dataset is reported in Table 4. The performance of concatenate models is highly depended on the embedding module, and concatenate structure improves the classification result on this basis. For two different CNNs as embedding layers, Concatenate-CNN models perform better than two baselines in average F1-score, as well as recall and precision for most appliances. The proposed model can achieve an equilibrium result among appliances in F1-score. However, the second baseline (spectral estimation) is just slightly better than the first baseline in Xception-SE, and DenseNet-121-SE is even worse than DenseNet-121. Accordingly, Concatenate-CNN models can eliminate background loads better, and the results make clear that background loads are not exact stationary. It is not rigorous to estimate the target spectrogram by directly subtracting the background spectrogram (baseline 2).

In Table 5, six options of network parameters in the similarity learning module and the difference learning module are compared. Small differences among models 1/2/3/4 prove that the proposed model is insensitive to FC output size/Conv filter size. Model 4 is marginally better than model 2, but the huge parameters increase model complexity and inference time. Model 2 is slightly better than model 6 and significantly better than model 5, which indicates that Conv kernel size has an observable effect on the model. To sum up, model 2 is selected for the rest of this paper.

Table 5. F1-score (%) on house 2 data of the UK-DALE dataset. Fully connected (FC) output size/Conv filter size and Conv kernel size are tuned in Concatenate-DenseNet-121.

No.	FC Output Size/Conv Filter Size	Conv Kernel Size	F1-score
1	128	(7,2)	88.1
2	256	(7,2)	89.0
3	512	(7,2)	88.5
4	1024	(7,2)	89.5
5	256	(3,2)	87.1
6	256	(5,2)	88.1

To further visualize the effectiveness of the model, the network's response to two same class samples at different layers is shown in Figure 7, and the mixed feature and the target feature have been indicated in Figure 4. The shown mixed feature is averaged in height and width, retaining channel data. The difference of features is obtained by calculating the absolute value of the difference between the two columns on the left. Within a class, the target feature is more similar than the mixed feature, that is to say, the difference of target features (DF_target) is smaller than the difference of mixed load features (DF_mixed).

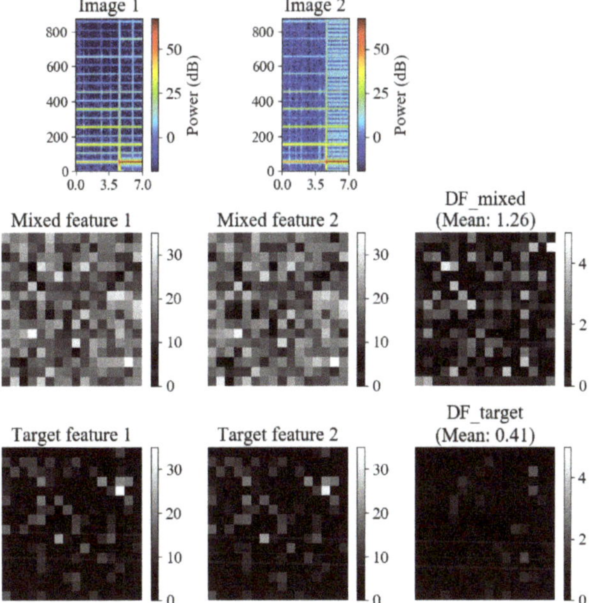

Figure 7. An example of the network Concatenate-DenseNet-121's response at different layers. The model input is two kettle samples with different background loads. For each layer, the output is reshaped to form a 2D image.

4.4. *The Recognition Result on BLUED*

The BLUED dataset contains many types of appliances with few samples, thus the model parameters of Concatenate-DenseNet-121 on UK-DALE are retained and applied to the BLUED dataset to avoid overfitting, then the BLUED data is trained for 30 epochs at a small learning rate of 0.0001.

The result of classification on the BLUED dataset is shown in Table 6. The Concatenate-DenseNet-121 distinguishes 6 classes of appliances, whereas the Fast Shapelets algorithm [32] trains a single

Table 6. The performance of classification on the BLUED dataset.

Label	Recall	Precision	F1-score
A1	98.8	95.5	97.1
A2	50.0	91.7	64.4
A3	71.4	94.4	81.2
B1	82.1	84.2	83.1
B2	94.1	98.0	96.0
B3	87.5	76.4	81.6
Average	**80.7**	**90.0**	**83.9**
Fast Shapelets [32]	77.6	69.7	72.3

classifier on each phase, which only needs to distinguish 3 classes at a time. Nevertheless, the proposed approach improves recall by 3.5%, precision by 19.1%, and F1-score by 11.8%. The result on the BLUED dataset shows the validity of the model on small samples.

4.5. The Energy Disaggregation Result on UK-DALE

The main goal of this paper is to recognize the type of appliances, and energy disaggregation is considered a by-product. Switching events recognized by the previous appliance recognition algorithm determine where the appliance cycle is located, thus the appliance recognition results have a great impact on energy disaggregation. On-events are recognized by the Concatenate-DenseNet-121 network, then off-events are recognized through the power of previous on-events. After determining on/off time, the disaggregated active power data is derived from the translation and smoothing of the aggregate data. The entire process of energy disaggregation is summarized in Algorithm 1 and Figure 8 is an example of energy disaggregation.

Algorithm 1 Work Flow of Energy Disaggregation

Require: The aggregate power data at time t, $P_A(t)$

The on-events recognized by the classification task, $E_{on} = \{E_{on}(1), \ldots, E_{on}(n)\}$

The power of individual appliance l, P_l

Set threshold T_l for appliance l
Detect a falling edge e
if e matches the power of $E_{on}(k) \in E_{on}$ **then**
  Determine the appliance label l and on/off time t_{on}/t_{off} of an operation cycle
  for $i = t_{on}$ to t_{off} **do**
    $P_A(i) \leftarrow P_A(i) - P_A(t_{on})$ # Translation
    if $|P_A(i) - P_l| > T_l$ **then**
      $P_A(i) \leftarrow P_l$ # Smoothing
    end if
  end for
  Delete $E_{on}(k)$ in E_{on}
end if

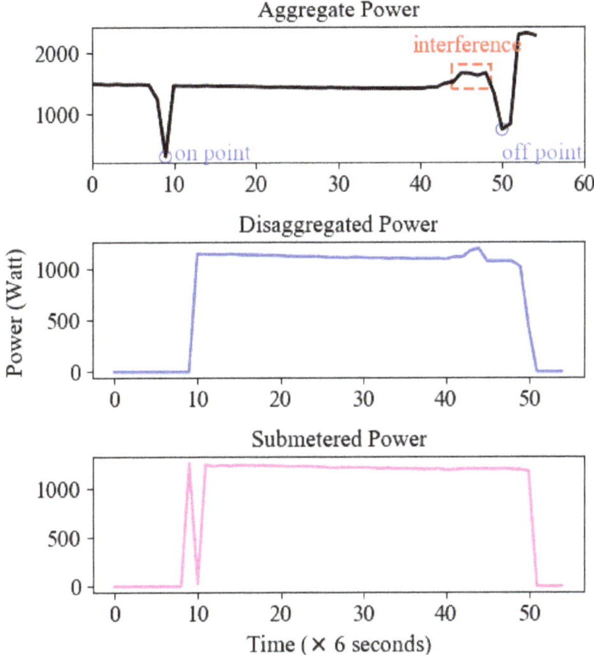

Figure 8. An example of energy disaggregation. The red dashed box stands for interference from other appliances. The blue circle stands for on/off points of an operation cycle.

Five target appliances are selected for energy disaggregation except for LCD and RM in the classification task. The performance is presented on house 2 appliances with the training of house 1 and 5 data. True or false binary judgments no longer depend only on the classification of switching events, but whether each data point matches, thus the F1-score here takes power estimation into account, called the M-fscore in [34]. Mean absolute error (MAE) is introduced to measure the error in every single point [15]. Figure 9 shows the disaggregation result on house 2. The proposed model is compared with the best result in Kelly and Knottenbelt's paper [15] in terms of F1-score and MAE, that is, the "Rectangles" architecture, and the performance is also compared with the result of AE [28] in F1-score and seq2point network [29] in MAE.

The energy disaggregation results of the recognition model outperform the network dedicated to energy disaggregation. With high disaggregation accuracy, concatenate CNNs can be used as an auxiliary tool for energy disaggregation. Compared with the other two methods, all the metrics are greatly improved, especially MAE is reduced in DW, MW, and WM.

The result proves that once the type of switching event is determined, the power waveform does not need a point-to-point reconstruction with neural networks or other approaches. The nuances in waveforms have a small effect, but the influence of incorrect appliance recognition is more critical. The events in the proposed approach are different from those in Kelly and Knottenbelt's paper [15]. The event number of washing machine in this paper (Table 1) far exceeds theirs (less than 600). This is because multiple state transitions in each entire operation cycle are monitored in this paper. The location of every state transition greatly improves the result of energy disaggregation. Multi-state and programmable appliances will be a trend in the future, thus it is difficult to capture and train the entire operating cycle due to numerous combinations of states. It is worth mentioning that the approach in this paper recognizes appliances by high-frequency data, while the other methods in this section used low-frequency data. Obviously, high-frequency data is conducive to locating and distinguishing appliance events, and eliminating the influence of background loads facilitates energy disaggregation.

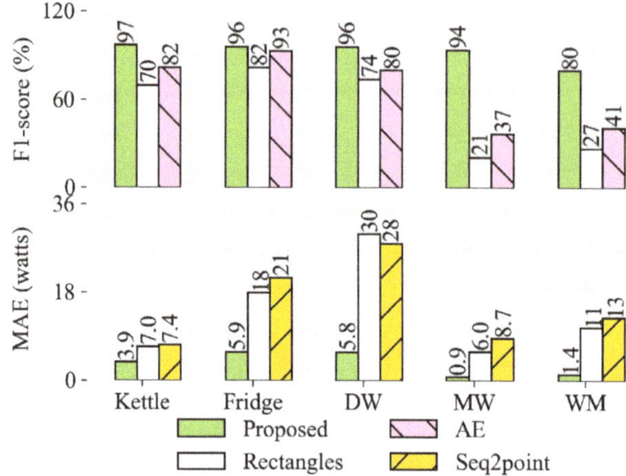

Figure 9. Energy disaggregation results on the UK-DALE dataset.

Table 7 records the average execution time (AET) of the Concatenate-DenseNet-121 model. The CPU option means the model is executed with an Intel Core i3 processor and 8 GB RAM. The GPU option has been stated in Section 4.2. The AET consists of the inference time of the classification task and the running time of the disaggregation task. The AET of the proposed model is short enough to meet the requirement of continuous operation.

Table 7. The average execution time (AET) of the proposed model.

Model	AET (ms)
Concatenate-DenseNet-121 (CPU)	221
Concatenate-DenseNet-121 (GPU)	69

5. Conclusions

The concatenate convolutional neural networks proposed in this paper can apply favorably to appliance recognition and energy disaggregation. The proposed models are evaluated on two real world datasets: UK-DALE and BLUED. Experiment results show the capacity of the proposed model, which can partly resist the interference of the background load on both large and small samples. Besides, the approach presents great generalization ability in appliance recognition and energy disaggregation. The hypothesis of short time stationarity is a premise of the proposed model, which also restricts the window length of samples at the same time. However, there is still the exception that the transient-state process of some appliances is more than three seconds (e.g., running machine). Therefore, we intend to break through the restriction and estimate target features from the longer time mixed load in the future.

Author Contributions: Data curation, formal analysis, software, writing—original draft, Q.W.; resources, F.W.; methodology, writing—review and editing: F.W. and Q.W.

Funding: This research received no external funding.

Conflicts of Interest: The authors declare no conflict of interest.

References

1. Wytock, M.; Kolter, J.Z. Contextually Supervised Source Separation with Application to Energy Disaggregation. In Proceedings of the 28th AAAI Conference on Artificial Intelligence, Québec City, QC, Canada, 27–31 July 2014; pp. 486–492.
2. Yu, L.; Li, H.; Feng, X.; Duan, J. Nonintrusive appliance load monitoring for smart homes: Recent advances and future issues. *IEEE Instrum. Meas. Mag.* **2016**, *19*, 56–62. [CrossRef]
3. Siano, P. Demand response and smart grids—A survey. *Renew. Sustain. Energy Rev.* **2014**, *30*, 461–478. [CrossRef]
4. Ephrat, A.; Mosseri, I.; Lang, O.; Dekel, T.; Wilson, K.; Hassidim, A.; Freeman, W. T.; Rubinstein, M. Looking to listen at the cocktail party: A speaker-independent audio-visual model for speech separation. *ACM Trans. Graph.* **2018**, *37*, 112. [CrossRef]
5. Mogami, S.; Sumino, H.; Kitamura, D.; Takamune, N.; Takamichi, S.; Saruwatari, H.; Ono, N. Independent Deeply Learned Matrix Analysis for Multichannel Audio Source Separation. In Proceedings of the IEEE 26th European Signal Processing Conference (EUSIPCO), Rome, Italy, 3–7 September 2018; pp. 1557–1561.
6. Chollet, F. Xception: Deep learning with depthwise separable convolutions. In Proceedings of the IEEE Conference on Computer Vision and Pattern Recognition (CVPR), Honolulu, HA, USA, 22–25 July 2017; pp. 1251–1258.
7. He, K.; Zhang, X.; Ren, S.; Sun, J. Deep residual learning for image recognition. In Proceedings of the IEEE Conference on Computer Vision and Pattern Recognition (CVPR), Las Vegas, NV, USA, 27–30 June 2016; pp. 770–778.
8. Huang, G.; Liu, Z.; Van Der Maaten, L.; Weinberger, K.Q. Densely connected convolutional networks. In Proceedings of the IEEE Conference on Computer Vision and Pattern Recognition (CVPR), Honolulu, HA, USA, 22–25 July 2017; pp. 2261–2269.
9. Kelly, J.; Knottenbelt, W. The UK-DALE dataset domestic appliance-level electricity demand and whole-house demand from five UK homes. *Sci. Data* **2015**, *2*, 150007. [CrossRef] [PubMed]
10. Anderson, K.; Ocneanu, A.; Benitez, D.; Carlson, D.; Rowe, A.; Berges, M. BLUED: A fully labeled public dataset for event-based non-intrusive load monitoring research. In Proceedings of the 2nd KDD Workshop on Data Mining Applications in Sustainability (SustKDD), Beijing, China, 12–16 August 2012; pp. 12–16.
11. Hart, G.W. Nonintrusive appliance load monitoring. *Proc. IEEE* **1992**, *80*, 1870–1891. [CrossRef]
12. Carli, R.; Dotoli, M. Energy scheduling of a smart home under nonlinear pricing. In Proceedings of the 53rd IEEE Conference on Decision and Control, Los Angeles, CA, USA, 15–17 December 2014; pp. 5648–5653.
13. Sperstad, I. B.; Korpås, M. Energy Storage Scheduling in Distribution Systems Considering Wind and Photovoltaic Generation Uncertainties. *Energies* **2019**, *12*, 1231. [CrossRef]
14. Hosseini, S.M.; Carli, R.; Dotoli, M. Model Predictive Control for Real-Time Residential Energy Scheduling under Uncertainties. In Proceedings of the 2018 IEEE International Conference on Systems, Man, and Cybernetics (SMC), Miyazaki, Japan, 7–10 October 2018; pp. 1386–1391.
15. Kelly, J.; Knottenbelt, W. Neural NILM: Deep neural networks applied to energy disaggregation. In Proceedings of the 2nd ACM International Conference on Embedded Systems for Energy-Efficient Built Environments, Seoul, South Korea, 4–5 November 2015; pp. 55–64.
16. Parson, O.; Ghosh, S.; Weal, M.; Rogers, A. Non-intrusive load monitoring using prior models of general appliance types. In Proceedings of the 26th AAAI Conference on Artificial Intelligence, Toronto, ON, Canada, 22–26 July 2012; pp. 356–362.
17. Chang, H.H.; Chen, K.L.; Tsai, Y.P.; Lee, W.J. A new measurement method for power signatures of nonintrusive demand monitoring and load identification. *IEEE Trans. Ind. Appl.* **2012**, *48*, 764–771. [CrossRef]
18. Froehlich, J.; Larson, E.; Gupta, S.; Cohn, G.; Reynolds, M.; Patel, S. Disaggregated end-use energy sensing for the smart grid. *IEEE Pervasive Comput.* **2011**, *10*, 28–39. [CrossRef]
19. Kahl, M.; Haq, A.U.; Kriechbaumer, T.; Jacobsen, H.A. WHITED—A Worldwide Household and Industry Transient Energy Data Set. In Proceedings of the 3rd International Workshop on Non-Intrusive Load Monitoring, Vancouver, BC, Canada, 14–15 May 2016.
20. Chui, K.; Lytras, M.; Visvizi, A. Energy Sustainability in Smart Cities: Artificial intelligence, smart monitoring, and optimization of energy consumption. *Energies* **2018**, *11*, 2869. [CrossRef]

21. Figueiredo, M.; De Almeida, A.; Ribeiro, B. Home electrical signal disaggregation for non-intrusive load monitoring (NILM) systems. *Neurocomputing* **2012**, *96*, 66–73. [CrossRef]
22. Le, T.T.H.; Kim, H. Non-Intrusive Load Monitoring Based on Novel Transient Signal in Household Appliances with Low Sampling Rate. *Energies* **2018**, *11*, 3409. [CrossRef]
23. Gillis, J.M.; Morsi, W.G. Non-intrusive load monitoring using semi supervised machine learning and wavelet design. *IEEE Trans. Smart Grid* **2016**, *8*, 2648–2655. [CrossRef]
24. Hassan, T.; Javed, F.; Arshad, N. An empirical investigation of V-I trajectory based load signatures for non-intrusive load monitoring. *IEEE Trans. Smart Grid* **2014**, *5*, 870–878. [CrossRef]
25. He, K.; Stankovic, L.; Liao, J.; Stankovic, V. Non-intrusive load disaggregation using graph signal processing. *IEEE Trans. Smart Grid* **2018**, *9*, 1739–1747. [CrossRef]
26. Makonin, S.; Popowich, F.; Bajić, I.V.; Gill, B.; Bartram, L. Exploiting HMM sparsity to perform online real-time nonintrusive load monitoring. *IEEE Trans. Smart Grid* **2016**, *7*, 2575–2585. [CrossRef]
27. Cominola, A.; Giuliani, M.; Piga, D.; Castelletti, A.; Rizzoli, A.E. A hybrid signature-based iterative disaggregation algorithm for non-intrusive load monitoring. *Appl. Energy* **2017**, *185*, 331–344. [CrossRef]
28. Barsim, K.S.; Yang, B. On the Feasibility of Generic Deep Disaggregation for Single-Load Extraction. In Proceedings of the 4th International Workshop on Non-Intrusive Load Monitoring, Austin, TX, USA, 7–8 March 2018; pp. 1–5.
29. Zhang, C.; Zhong, M.; Wang, Z.; Goddard, N.; Sutton, C. Sequence-to-point learning with neural networks for non-intrusive load monitoring. In Proceedings of the 32nd AAAI Conference on Artificial Intelligence, New Orleans, LA, USA, 2–7 February 2018; pp. 2604–2611.
30. Mocanu, E.; Nguyen, P.H.; Gibescu, M. Energy disaggregation for real-time building flexibility detection. In Proceedings of the 2016 IEEE Power and Energy Society General Meeting (PESGM), Boston, MA, USA, 17–21 July 2016; pp. 1–5.
31. Hanzo, L.; Yang, L.L.; Kuan, E.L.; Yen, K. *Single and Multi-Carrier DS-CDMA: Multi-User Detection Space-Time Spreading Synchronisation Networking and Standards*; John Wiley & Sons: New York, NY, USA, 2003; pp. 35–80.
32. Patri, O.P.; Panangadan, A.V.; Chelmis, C.; Prasanna, V.K. Extracting discriminative features for event-based electricity disaggregation. In Proceedings of the 2014 IEEE Conference on Technologies for Sustainability (SusTech), Ogden, UT, USA, 30 July 2014; pp. 232–238.
33. Kinga, D.P.; Ba, J. Adam: A Method for Stochastic Optimization. In Proceedings of the 3rd International Conference for Learning Representations (ICLR), San Diego, CA, USA, 7–9 May 2015.
34. Makonin, S.; Popowich, F. Efficient sparse matrix processing for nonintrusive load monitoring (NILM). *Energy Efficien.* **2014**, *8*, 809–814. [CrossRef]

Article

A Novel Nonintrusive Load Monitoring Approach based on Linear-Chain Conditional Random Fields

Hui He †, Zixuan Liu †, Runhai Jiao * and Guangwei Yan

School of Control and Computer Engineering, North China Electric Power University, Beijing 102206, China; huihe@ncepu.edu.cn (H.H.); zixuanliu@ncepu.edu.cn (Z.L.); yanguangwei@ncepu.edu.cn (G.Y.)
* Correspondence: runhaijiao@ncepu.edu.cn
† These authors contributed equally to this work.

Received: 28 March 2019; Accepted: 7 May 2019; Published: 11 May 2019

Abstract: In a real interactive service system, a smart meter can only read the total amount of energy consumption rather than analyze the internal load components for users. Nonintrusive load monitoring (NILM), as a vital part of smart power utilization techniques, can provide load disaggregation information, which can be further used for optimal energy use. In our paper, we introduce a new method called linear-chain conditional random fields (CRFs) for NILM and combine two promising features: current signals and real power measurements. The proposed method relaxes the independent assumption and avoids the label bias problem. Case studies on two open datasets showed that the proposed method can efficiently identify multistate appliances and detect appliances that are not easily identified by other models.

Keywords: load disaggregation; nonintrusive load monitoring; conditional random fields; feature extraction

---

1. Introduction

1.1. Background

As the core of an interactive service system, smart power utilization is one of the essential components of a smart grid. There are three aspects to the key technologies associated with this: advanced metering infrastructure (AMI) standards, systems, and terminal technologies; intelligent two-way interactive operation mode and supporting techniques; and the interaction between the user's electrical environment and energy consumption patterns. In actual production, we must break through the bottleneck regarding the meter only being able to read the total amount of energy consumption rather than analyzing the internal load components for users. Load monitoring can not only improve the power information collection system and intelligent power system but also support two-way interactive service and smart power utilization. Nonintrusive load monitoring (NILM), which is a vital part of smart power utilization techniques, can achieve fine-grained tracking of energy consumption and provide load disaggregation information without any intrusive device installation. These data can be further applied to optimize energy conservation strategies.

1.2. Literature Review and Motivation

NILM was first proposed by Hart [1], who devised a method for appliance load monitoring by only identifying electrical appliances within the aggregate power consumption data. This method decomposes the aggregated data into the actual power components of each load and avoids cumbersome device installation. Since then, many new methods have been introduced for load disaggregation, such as Bayes [2] and support vector machines (SVMs) [3,4]. Bayes has shown good performance in some experiments. However, it requires the appliances to have stable power measurements, which is almost nonexistent in reality. Comparably, SVM performs better using low-frequency features. Chui [3] proposed a hybrid genetic algorithm support vector machine multiple kernel learning approach (GA-SVM-MKL), which solves the problems that current algorithms are limited by data granularity and consideration of fewer appliances. It has enhanced the performance indicators (sensitivity, specificity, and overall accuracy) up to 21% compared with traditional methods. Lai [4] used a hybrid SVM/GMM classifier that successfully achieved ubiquitous recognition service. In their model, GMM is employed to describe the distribution of the current measurement to find the power similarity, while SVM is applied to identify the appliances. However, an SVM method can be arbitrary given a large dataset and requires tedious training for best kernels and parameters.

The hidden Markov model (HMM) has become a mainstream algorithm, as it can include appliances' state transition in its learning. Specifically, the task of NILM can be considered as assigning label sequences to a set of observation sequences. Thus, for a given set of aggregated load data, HMM-based approaches are naturally suitable for performing tasks such as identifying tags or disintegrating electrical loads. In previous works, HMM and its variants have improved the accuracy of NILM. Zia [5] applied an HMM-based method to identify personal devices and found that it can effectively distinguish the power consumption patterns of the appliances. Kong [6] proposed a hierarchical HMM (HHMM) framework for modeling household appliances, which provides a promising representation of devices with multiple built-in modes and different power consumption profiles. Kim [7] investigated the effectiveness of several unsupervised disaggregation methods and demonstrated that a conditional factorial hidden semi-Markov model performs better than other methods. Kolter [8] adopted factorial HMM and developed a convex formulation of approximate inference to make the inference algorithm computationally efficient and avoid the issues of local optima. Agyeman [9] came up with a variant of the HMM to identify loads and operation states by practicable measurable parameters. Their results show that the method can provide power usage information in a nonintrusive manner and is ideal for participation in the demand response market.

Nevertheless, the above models assume that any observation in the sequences is independent of the other [10]. In other words, the aggregated load data at any given time only depend on the states of loads at that time and have no association with previous ones. That is not appropriate for a realistic environment. The current data, such as the appliance power consumption, are highly relevant to an extended range of previous observations. There is a weakness in HMM-based models called the label bias problem [11]. When one state transitions to another, the Viterbi algorithm may choose the state with fewer outgoing transitions and takes little notice of the observations. Extraordinarily, the algorithm even ignores the observations if a state has a single outgoing transition. In this case, the result is highly relevant to the training set. If one state is slightly more common in the training set, the algorithm will prefer this transition, whatever the next observation may be.

Conditional random fields (CRFs) have also been used by Panikos [12] as an unsupervised model for energy disaggregation. They apply a clustering method and histogram analysis to detect the selected loads for residential users and have obtained higher-accuracy results compared with previous methods. However, they only detect the on/off states of devices and cannot handle multistate appliances. Additionally, CRFs can extract various features for training, but they only use power measurements as a feature, which may fail to make full use of the advantages of the CRF model.

In our paper, we have proposed a method called linear-chain CRFs for load disaggregation, which perfectly solves the above problems. Our linear-chain CRF model defines a log-linear distribution over all of the observation sequences in the aggregate data, which relaxes the requirements for the independence of observation data in HMM. It not only considers the influence of the previous state on the current state but can also incorporate all useful information in the observation, which makes it more viable in reality. Since CRFs define a log-linear distribution over all of the label sequences given in the observations, the transition metric between different state changes and the weights can be traded off. Thus, our linear-chain CRF model avoids the label bias problem. In addition, our model does not require the stable power measurements needed for Bayes as well as the exhausting parameter training needed for SVM. Moreover, by quantizing the power probability density function for each load, we can easily identify multistate appliances. We also employ two promising features: current signals and real power measurements to develop our model. Experimental results on two open datasets demonstrate that the proposed model is feasible for a NILM task.

1.3. Contributions

Our main contributions are as follow:

- We proposed a method called the linear-chain CRF model for load disaggregation and achieved accuracy of 96.04–99.94%. It is demonstrated that this method is effective for the NILM task.
- Because we relaxed the independent assumption required by HMM-based models and avoided the label bias problem, the performance is enhanced by 2.21% compared to existing models.
- We combined two promising features: current signals and real power measurements to build our model, which improved the accuracy of the model significantly.

2. Methodology

Figure 1 shows the goal of our model: breaking down the aggregate data into the actual power consumption of each appliance. Figure 2 illustrates the main framework of our linear-chain CRF model for NILM. First, submeter data of each load was used to create the probability density function for each appliance to acquire the working states. Then, the states of the appliances were grouped to tag and segment the smart meter data. Next, our model extracted features over the training set according to the feature templates. Consequently, the improved iterative scaling algorithm (IIS) was used to train the linear-chain CRF model. Finally, we adopted the Viterbi algorithm to disaggregate the states for each appliance given the aggregate power data.

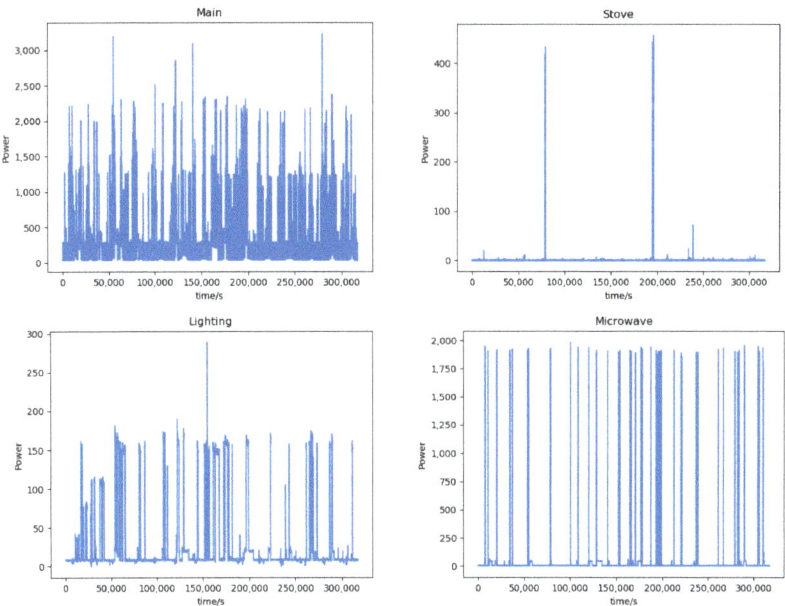

Figure 1. *Cont.*

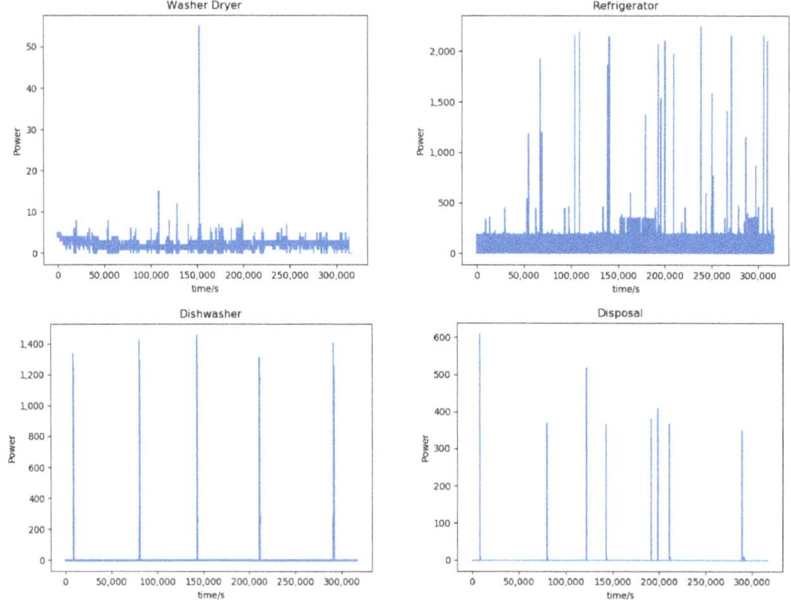

Figure 1. Aggregated load data acquired from smart meters and disaggregating results produced from the model.

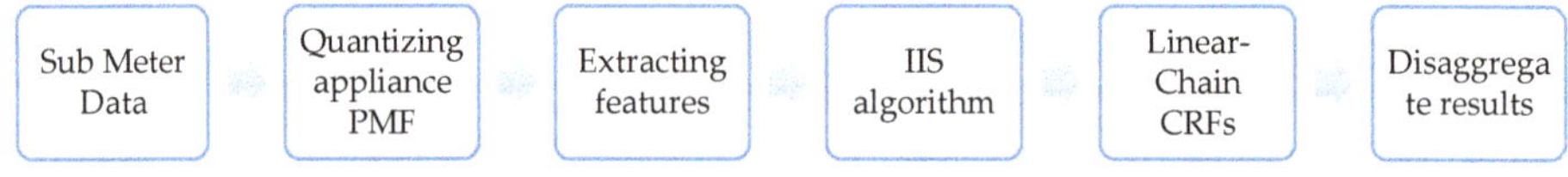

Figure 2. Main framework of our linear-chain CRF model for NILM.

2.1. Probability Mass Functions

Various appliances, such as washing machines, have multiple operating states. The simple on state cannot reflect the real state change when the appliance is working. To identify the different working states of multistate type appliances at a given time, we used the approaches of Stephen [13] to quantize power probability mass function (PMF) for each appliance. We took the PMF as the probability density function (PDF) for their working states. Figures 3 and 4 show the power PDF of some appliances in AMPds2 [14] and REDD house 2 [15]. Compared with low power measurements, most probabilities of high power measurements were excessively low, so we enlarged the y-axis scale appropriately to make it clear.

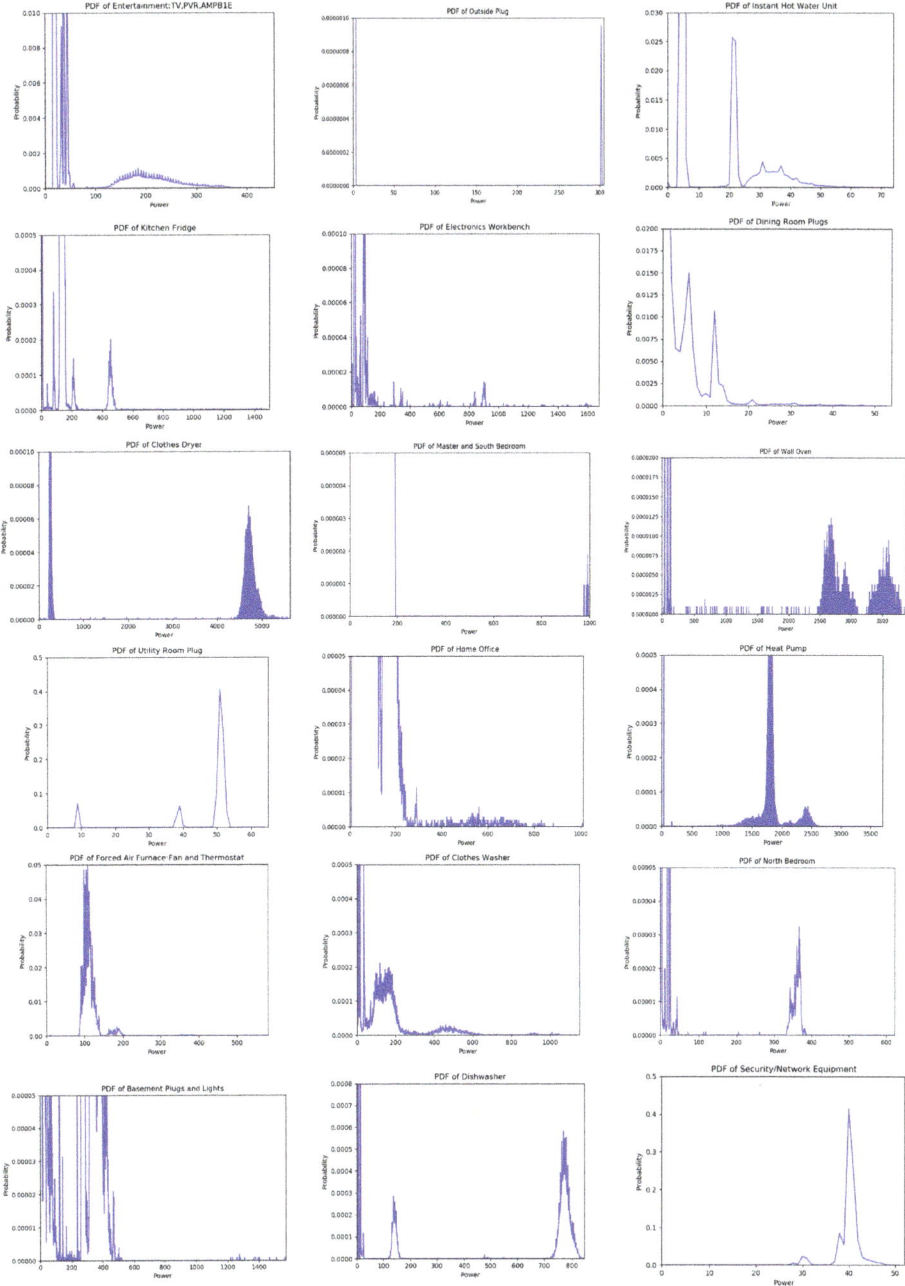

Figure 3. Power probability density function of some appliances in AMPds2.

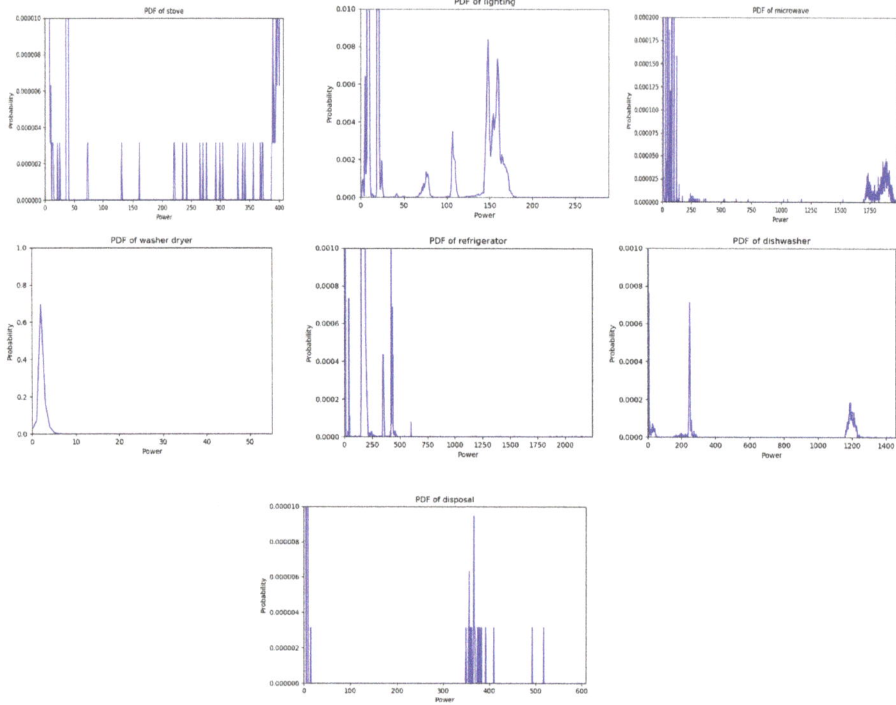

Figure 4. Power probability density function of some appliances in REDD house 2.

When the power measurement of the appliance is distributed in a certain power range, it indicates that the device is in a specific working state. By figuring out the power distribution and finding the power range of the concentrated power distribution, we could analyze the working states for appliances. Let $P(n)$ represent the probability of n, where n is the number of possible observed power measurements. In Stephen's [13] paper, they found the power range by capturing the peak, which is defined as when the slope on the left in the PDF is positive and the slope on the right is negative, which is to say:

$$P(n) - P(n-1) > 0 \tag{1}$$

$$P(n+1) - P(n) \leq 0 \tag{2}$$

$$P(n) > \varepsilon = 0.00021 \tag{3}$$

where ε is used to make sure that the probabilities under this value will not be quantized as a state. However, on the one hand, we considered that the value of ε was hard to generalize since it varied in different datasets and different appliances. Furthermore, this method pays more attention to the peaks with higher probability. However, these peaks are mainly distributed in low power measurements, and most of them are noises rather than states. In fact, some high power measures include some major working states, to which importance should be attached. Therefore, we combined some states with low power measurements and concentrated on the states with high measurements according to the PDF of each appliance. On the other hand, this approach was used to identify a load with a finite number of operating states and that worked worse when the appliances belonged to continuously variable devices. It was apparent to see from the PDF that some appliances, such as dining room plugs and instant hot water units, were not multistate appliances. Thus, it was inapplicable to quantize their

PDFs for working states. We simply determined the on/off states for this type of appliance. More details are discussed in Section 3.

2.2. Segmenting Data

CRFs are a framework for segmenting and labeling sequential data. Let $S = \{s_1, s_2, \ldots, s_n\}$ be the label sequences, and $P = \{p_1, p_2, \ldots, p_n\}$ be the observation sequences. A graphical structure of linear-chain CRFs is shown in Figure 5, which demonstrates that the input of our model is a series of sequences. Our templates then extract features throughout each chain. Therefore, segmenting smart meter data is crucial for feature extraction and model performance. CRFs are adept at dealing with a sentence with no more than 20 tokens. Considering that the working state of an appliance from an hour or 30 min ago has little effect on the current working state, we segmented smart meter data into a sequence for the AMPds2 datasets every 10 min and every minute for the REDD dataset in terms of their different sampling rates (per minute in AMPds2 and per 3 s in REDD). Then, 10 tokens were included in a sequence for the AMPds2 datasets and 20 for the REDD dataset, which made our model perform more efficiently compared with other segmentation methods.

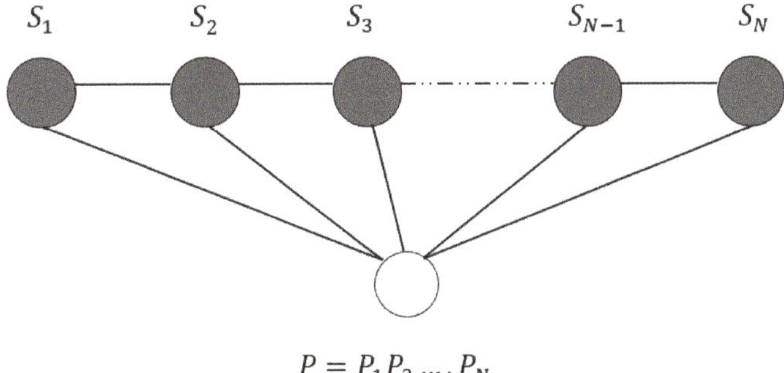

Figure 5. General graphical structure of a linear-chain CRF model.

2.3. Extracting Features

Let $Y = \{y_1, y_2, \ldots, y_n\}$ be the label sequences, $X = \{x_1, x_2, \ldots, x_n\}$ be the observation sequences, $\lambda = \{\lambda_k\} \in R$, $\mu = \{\mu_k\} \in R$ be the parameter vectors, and $P(y|x)$ represent the linear-chain CRFs. Then, define the probability of marking a tag sequence Y on a given observation sequence X as follows [16]:

$$P(y|x) = \frac{1}{Z(X)} \exp\left(\sum_{i,k} \lambda_k t_k(y_{i-1}, y_i, x, i) + \sum_{i,l} \mu_l s_l(y_i, x, i)\right) \tag{4}$$

$$Z(x) = \sum_{y} \exp\left(\sum_{i,k} \lambda_k t_k(y_{i-1}, y_i, x, i) + \sum_{i,l} \mu_l s_l(y_i, x, i)\right) \tag{5}$$

where t_k is a transition feature function depending on the current state i and previous state $i-1$ in the label sequence given the observation sequences; s_l is a state feature function depending on the current state i in the label sequence, which is also viewed as a local feature function; and $\lambda = \{\lambda_k\} \in R$, $\mu = \{\mu_k\} \in R$ are the parameter vectors, which index the weights of the corresponding t_k and s_l function and can be learned by our model.

We defined feature functions t_k and s_l using feature templates. A feature template has the form of a single state S_n or some combination of current states and previous states $S_{n-k} \ldots S_n$. For example, assume that we have a power measurement sequence: 1919, 1918, 1921, 106, 107, 105, 106, 2, 3, 1. The corresponding state sequence is: 2, 2, 2, 1, 1, 1, 1, 0, 0, 0. A single state S_n template refers to series state functions s_{nj}, where n is the position of the current token and j is the number of appliance states.

Let the current token be the fifth one; then, we define s_{51} : if (state = 1 and power measurement = 107) return 1 or else return 0; s_{52} : if (state = 2 and power measurement = 107) return 1 else return 0; s_{53}: if (state = 0 and power measurement = 107) return 1 else return 0. Similarly, templates have the form of $S_{n-k} \ldots S_n$, representing several transition functions t_{nj} where n is the position of the current token, $n-k$ is the position of the previous token, and j is the number of appliance states. Let $k=1$; then, we construct functions as follows: t_{51} : if (state = 1 and power measurements = 106, 107) return 1 else return 0; t_{52} : if (state = 2 and power measurement = 106, 107) return 1 else return 0; t_{53} : if (state = 0 and power measurement = 106, 107) return 1 else return 0. The whole process is shown in Figure 6.

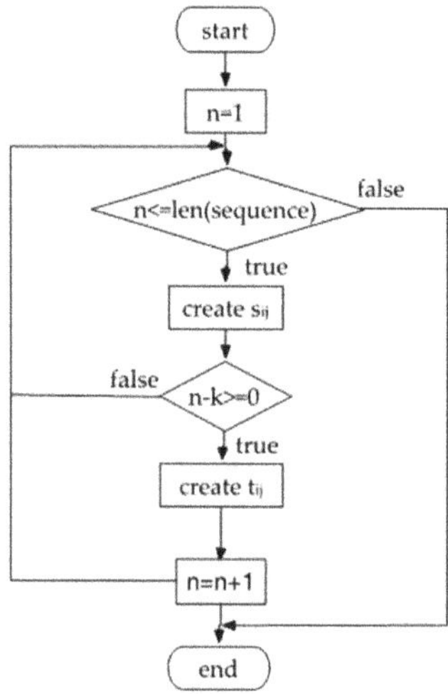

Figure 6. Process of extracting features.

Our model constructs $L * N$ feature functions according to the feature templates designed, where L represents the number of output types, and N represents the number of expanded features. In practice, many feature functions are constructed. For example, in our experiments, 4,704,668 feature functions were produced for five loads (CWE, DWE, FRE, HPE, and WOE) in AMPds2. The excessive feature functions increased the complexity of our model and made it difficult for subsequent training and testing. Actually, some measurements in the dataset were inaccurate or completely noisy, which made those feature functions considering these measurements unnecessary. We found that the frequency of these feature functions' occurrence was much less than normal functions. Therefore, we ignored those functions with fewer than three occurrences, which reduced the complexity greatly.

2.4. Improved Iterative Scaling (IIS) Algorithm

Formulas (4) and (5) define the primary form of linear-chain CRFs. The parameters λ_k and μ_l are the corresponding weights to be estimated from the training set. From Formula (4), we can easily discover that the definition of $P(y|x)$ is similar to a maximum entropy model. Actually, the CRF model is motivated by the principle of maximum entropy. Thus, we could apply the IIS algorithm of the maximum entropy model for parameter learning.

To simplify, let there be M_1 transition feature functions and M_2 state feature functions, $M = M_1 + M_2$, defined as

$$f_k(y,x) = \begin{cases} \sum\limits_{i=1}^{n} t_k(y_{i-1}, y_i, x, i), \ k = 1,2,\ldots M_1 \\ \sum\limits_{i=1}^{n} s_l(y_i, x, i), k = \ M_1 + l; l = 1,2,\ldots,M_2 \end{cases} \tag{6}$$

$$\omega_k = \begin{cases} \lambda_k, & k = 1,2,\ldots M_1 \\ \mu_l, & k = M_1 + l; l = 1,2,\ldots,M_2 \end{cases} \tag{7}$$

$$\omega = (\omega_1, \omega_2, \ldots, \omega_M)^T \tag{8}$$

$$F(y,x) = (f_1(y,x), f_2(y,x), \ldots, f_3(y,x))^T. \tag{9}$$

Then, CRF can be normalized as *a* product of vector ω and $F(y,x)$:

$$P_w(y|x) = \frac{1}{Z_w(x)} \exp(\omega \cdot F(y,x)) \tag{10}$$

$$Z_w(x) = \sum_y \exp(\omega \cdot F(y,x)). \tag{11}$$

Given the empirical distribution $\widetilde{P}(x,y)$, the log-likelihood function $L_{\widetilde{p}}(P_w)$ of conditional probability distribution $P(y|x)$ is defined as:

$$L_{\widetilde{p}}(P_w) = log \prod_{x,y} P(y|x)^{\widetilde{P}(x,y)} = \sum_{x,y} \widetilde{P}(x,y) log P(y|x) \tag{12}$$

When $P(y|x)$ is defined as (10), the log-likelihood function can be derived as followed:

$$\begin{aligned} L_{\widetilde{p}}(P_w) = & \sum_{x,y} \widetilde{P}(x,y) log P(y|x) = \sum_{x,y} \widetilde{P}(x,y) log P_w(y|x) \\ & = \sum_{x,y} [\widetilde{P}(x,y)\omega \cdot F(y,x) - \widetilde{P}(x,y) log Z_w(x)] \\ & = \sum_{x,y} [\widetilde{P}(x,y) \sum_{k=1}^{M} \omega_k f_k(y,x) - \widetilde{P}(x,y) log Z_w(x)] \\ & = \sum_{i=1}^{N} \sum_{k=1}^{M} \omega_k f_k(y_i,x_i) - \sum_{i=1}^{N} Z_w(x_i) \end{aligned} \tag{13}$$

Assuming the current vector $\omega = (\omega_1, \omega_2, \ldots, \omega_M)^T$, the IIS algorithm tries to find the best vector $\omega + \delta = (\omega_1 + \delta_1, \omega_2 + \delta_2, \ldots, \omega_M + \delta_M)^T$, which increases the value of the log-likelihood function. According to Adam [16], the IIS algorithm finds out the increment vector $\delta = (\delta_1, \delta_2, \ldots, \delta_M)^T$ by solving the renewal equation for transition feature Function (14) and state feature Function (15):

$$\sum_{x,y} \widetilde{P}(x) P(y|x) \sum_{i=1}^{n+1} t_k(y_{i-1}, y_i, x, i) exp(\delta_k T(y,x)) = E_{\widetilde{p}}[t_k] \tag{14}$$

where $k = 1,2,\ldots,M_1$; y_i and y_{i-1} refer to the current and previous power measurements; y depends on all states.

$$\sum_{x,y} \widetilde{P}(x) P(y|x) \sum_{i=1}^{n} s_l(y_i, x, i) exp(\delta_k T(y,x)) = E_{\widetilde{p}}[s_l] \tag{15}$$

where $k = M_1 + l; l = 1,2,\ldots,M_2$, $T(y,x)$ is the summation of all feature functions:

$$T(y,x) = \sum_k f_k(y,x). \tag{16}$$

The complete IIS algorithm is shown in Algorithm 1 below.

Algorithm 1. Improved iterative scaling algorithm.

```
for k ∈ (1, M)
    ω_k = 0
repeat
  for k ∈ (1, M)
      if k ∈ (1, M_1)
          δ_k → Σ_{x,y} P̃(x)P(y|x) Σ_{i=1}^{n+1} t_k(y_{i-1}, y_i, x, i) exp(δ_k T(y, x)) = E_p̃[t_k]
      if k ∈ (M_1 + 1, M)
          δ_k → Σ_{x,y} P̃(x)P(y|x) Σ_{i=1}^{n} s_l(y_i, x, i) exp(δ_k T(y, x)) = E_p̃[s_l]
  ω_k ← ω_k + δ_k
until ω_k converge
```

2.5. Viterbi Algorithm

The Viterbi algorithm for CRF prediction is similar to the one for HMM. Assuming that the observation sequences are $\{x\}$, then the task of prophecy is to find the max probability of label sequences $\{y^*\}$:

$$y^* = \underset{y}{argmax}(P_w(y|x)) = \underset{y}{argmax}\frac{1}{Z_w(x)}\exp(\omega \cdot F(y,x)) = \underset{y}{argmax}\exp(\omega \cdot F(y,x)) = \underset{y}{argmax}(\omega \cdot F(y,x)). \quad (17)$$

Therefore, the prediction problem for CRF is converted to $\underset{y}{max}(\omega \cdot F(y,x))$. The Viterbi algorithm is shown in Algorithm 2 below.

Algorithm 2. Viterbi algorithm for CRF prediction.

```
Step 1: initialization
for j ∈ (1, m)
    δ_1(j) = ω · F(y_0 = start, y_1 = j, x)
Step 2: recursion
for i ∈ (2, n)
        δ_i(l) = max_{1≤j≤m} {δ_{i-1}(j) + ω · F(y_{i-1} = j, y_i = l, x)}
        φ_i(l) = argmax_{1≤j≤m} {δ_{i-1}(j) + ω · F(y_{i-1} = j, y_i = l, x)}
        l = 1, 2, ..., m
 Step 3: terminate
        max_y(ω · F(y, x)) = max_{1≤j≤m} δ_n(j)
        y_n = argmax_{1≤j≤m} δ_n(j)
Step 4: traceback
for i ∈ (n − 1, 1)
        y_i = φ_{i+1}(y_{i+1})
```

3. Experiment and Analysis

3.1. Data

The tests were conducted using real monitoring data from AMPds2 [14] and REDD house 2 [15]. The AMPds2 dataset collected the electricity usage of a Canadian family for two years, with a sampling frequency of one reading every minute. It monitored 24 appliances, but only 21 were kept, for they did not detect any data of the removed appliances for the entire measurement time. There were just a few missing data or errors in the dataset, and the algorithm was used to populate the missing data so that the whole dataset was contiguous. This facilitated the division of sequences in subsequent model training. In terms of electricity data, AMpds2 provided 11 measurements: voltage, current, frequency, displacement power factor, apparent power factor, real power, real energy, reactive power, reactive energy, apparent power, and apparent energy, which made it easy to select different features for improving model performance. Developed specifically for load disaggregation, the REDD dataset gathered real power consumption in some homes over several months, with a sampling frequency of approximately 3 s for every reading. In our experiments, we only used the data of house 2 in the REDD dataset, which included 10 types of equipment: lighting, refrigerator, dishwasher, washer-dryer, bathroom GFI, kitchen outlets, oven, microwave, electric heat, and stove.

3.2. Experimental Setup

Firstly, we segmented smart meter data into a sequence for AMPds2 datasets every 10 min and every 1 min for the REDD dataset, as discussed in Section 2.2. We chose the power measurements and current signals in AMPds2 and single power measurements in REDD for disaggregation. Then, we designed several templates for feature extraction. Table 1 shows the list of the feature templates used in our experiments. Among them, Templates 1 and 2 refer to the single power signature in REDD house 2 and AMPds2, respectively, while Templates 3 represent the double signatures: power and current in AMPds2.

Table 1. List of feature templates.

Templates 1	Templates 2	Templates 3	Meaning of the Template
S_n	S_n	S_n	current state
$S_{n-1}S_n$	$S_{n-1}S_n$	$S_{n-1}S_n$	current state and previous state
$S_{n-2}S_{n-1}S_n$	$S_{n-2}S_{n-1}S_n$	$S_{n-2}S_{n-1}S_n$	current state and previous two states
$S_{n-3}...S_n$	/ [1]	$S_{n-3}...S_n$	current state and previous three states
$S_{n-4}...S_n$	/	$S_{n-4}...S_n$	current state and previous four states
$S_{n-5}...S_n$	/	$S_{n-5}...S_n$	current state and previous five states
$S_{n-6}...S_n$	/	/	current state and previous six states
$S_{n-7}...S_n$	/	/	current state and previous seven states
$S_{n-8}...S_n$	/	/	current state and previous eight states

[1] This template does not have this feature.

Only extracting features over a continuous period of time was meaningful, which directly reflected the influence of previous states. If the time interval between two measurements is too large, for example 30 min, then it is not necessary to construct a transition function for these two measurements, because the state of an appliance half an hour ago has little effect on the current state. However, the timestamps in REDD house 2 was not continuous, so we found those intervals and just segmented data through those continuous data. Next, CRF++ [17] was used to build our model. CRF++ is an open-source CRF tool for continuous data annotation and segmentation, which is easy to use and customizable. We removed the features function for which the occurrences were less than three to further reduce the complexity of our model as claimed in Section 2.3. Additionally, a hyper-parameter C need to be selected in CRF++ to trade the balance between overfitting and underfitting. We found that the optimal value is 1.5 after cross-validation. All our work was carried out in Python 3 and C++. We also used 10-fold cross-validation to acquire the best error estimation.

3.3. Evaluation Metrics

In our paper, let *Acc* be the accuracy, *T* be the correct prediction, and *F* be the incorrect prediction. Then, *Acc* is defined as

$$Acc = \frac{T}{T+F} \tag{18}$$

This metric has normally been adopted by many researchers such as Stephen [13] and Kolter [15]. However, we do not think this indicator can properly reflect the performance of the model. Therefore, we adopted a new evaluation indicator: total load accuracy. Let $x_i, i = 1, 2, \ldots, N$ be the appliances monitored in the house, $l_j, j = 1, 2, \ldots, M$ be the observation sequences, and $TAcc$ be the total loads' accuracy. We employed the following notation:

$$f(x_i, l_j, i, j) = \begin{cases} 1 & \text{if the predicted state of } i \text{ appliance at } j \\ & \text{is the same as the real state} \\ 0 & \text{otherwise} \end{cases} \tag{19}$$

Then, the total loads' accuracy $TAcc$ is defined as

$$TAcc = \frac{1}{M} \sum_{j=1}^{M} \prod_{i=1}^{N} f\big(x_i, l_j, i, j\big). \tag{20}$$

Each load has one state at any given time; the total load's accuracy refers to the accuracy that all appliance states are assigned correctly at a given time. We combined this index for estimation because we believed that it could reflect the overall prediction ability of our model for the whole house. However, the accuracy results were generally lower than results in other papers, which only considered a single appliance's on/off accuracy.

3.4. Experiment Results and Analysis

To better test the accuracy of our linear-chain CRF model, we chose seven appliances in REDD house 2: lighting, stove, microwave, washer-dryer, refrigerator, dishwasher, and disposal. Figure 7 illustrates the seven loads' on-duration accuracy in REDD house 2. Obviously, the refrigerator showed the best score, while the disposal scores were very low. The low accuracy results were due to there being less disposal data working in the training sets. We also found that the power measures of the washer-dryer were mainly distributed from 0 to 10 all the time, which was purely low for a normal washer-dryer and similar to other appliances' off state. Thus, our model mostly tagged the washer-dryer working when the measurements varied from 1 to 10, which made the accuracy results higher. We inferred that the washer-dryer in REDD house 2 did not work and the measurements were completely noisy.

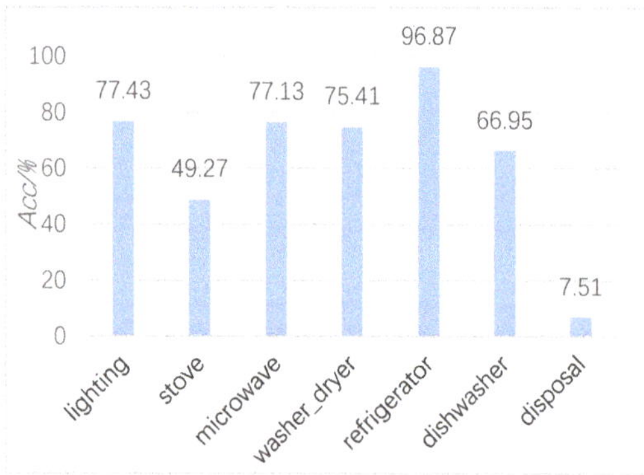

Figure 7. Seven loads' on-duration accuracy in REDD house 2.

We extracted some test sequences, as shown in Figure 8. It illustrates the real state changes of the electrical appliances which were working within a period of 150 s, as well as the inference results of our linear-chain CRF model according to the same data. It is clear that our model worked when different electrical appliances were used at the same time. Nevertheless, errors may have occurred when the power's values of different working states of electrical appliances were similar. For example, during the period from 100 to 150 s, the total power decreased because the refrigerator stopped working. However, our model identified that the light and microwave stopped working while the refrigerator started working.

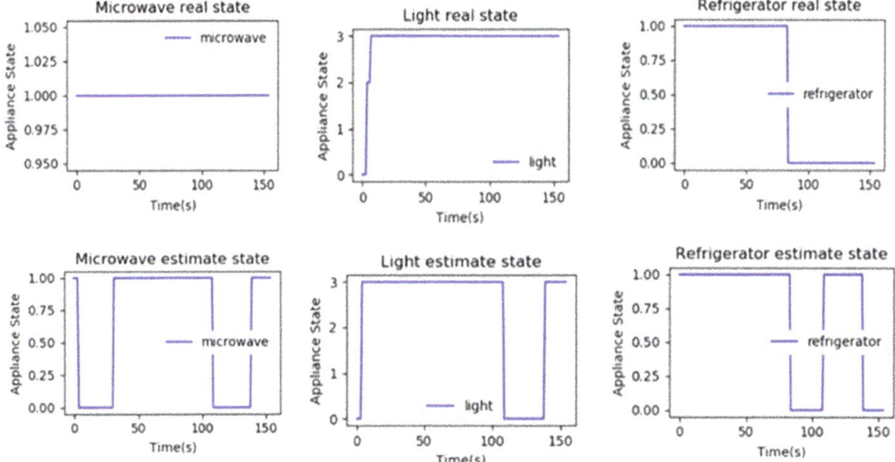

Figure 8. Comparison between appliances' real and estimate states.

Figure 9 shows each test's total loads' accuracy in REDD house 2. Among them, 1 load refers to the refrigerator only; 2 loads mean the refrigerator and microwave; 3 loads stand for the kitchen outlets, microwave, and dishwasher; 4 loads indicate the lighting, microwave, washer-dryer, and refrigerator; 5 loads represent the refrigerator, lighting, dishwasher, microwave, and stove; 6 loads denote the lighting, stove, microwave, refrigerator, dishwasher, and disposal; 7 loads show the lighting, stove, microwave, washer-dryer, refrigerator, dishwasher, and disposal. We can see that the correct rate of accurate prediction of all electrical appliances at each moment was over 88% throughout the test time. This indicates that our model could correctly reflect the working state of all electrical appliances in the house tested at any time, not just for a single device.

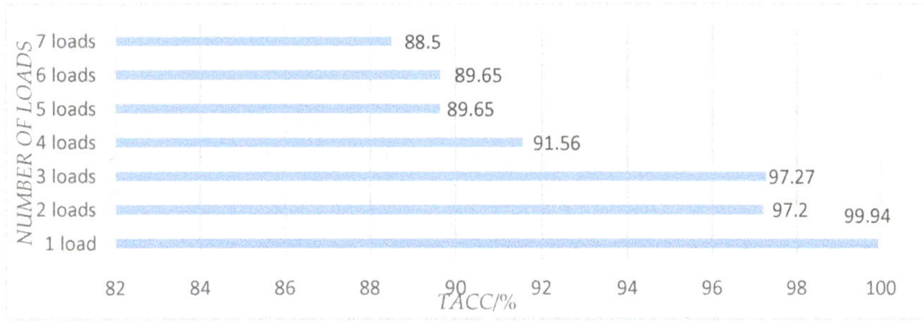

Figure 9. Each test total loads' accuracy in REDD house 2.

As our linear-chain CRF model can combine more than one feature, we chose the current signals together with power measurements for parameter learning to test whether it promotes the performance of our model or not. Further, we hoped to estimate how well our model disaggregates loads in other datasets. We used five loads (including CDE, DWE, FRE, HPE, and WOE) in AMPds2 and verified the accuracy of a single power feature and double features. Figure 10 shows that using dual features can improve the efficiency to some extent. When it is challenging for the classifier to judge the state of the appliance only by the power value, the multiple features can provide an inferential basis by providing other state parameters. For example, our model performed much better in identifying the wall oven using double features, which was better than using a single feature by 32.49%.

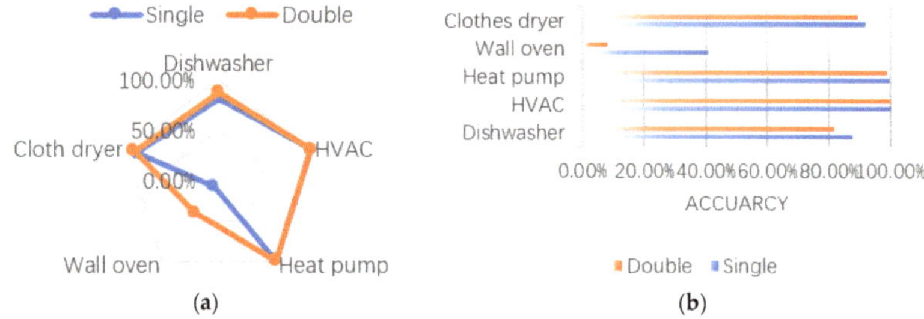

Figure 10. (**a**) Radar charts with five loads (CDE, DWE, FRE, HPE, and WOE) accuracy in AMPds2. (**b**) Histogram charts with five loads accuracy. "Single" means that only power measurements were used, while "Double" means that power measurements and current signals were both included.

In Stephen's [13] paper, they used a sparsity HMM and obtained a perfect result. Thus, an experiment was conducted to assess the performance of the proposed linear-chain CRFs. We used REDD house 2 to test the performance for each model. Stephen divided their tests into three categories: Denoised, Noisy, and Modeled. Our tests belonged to the Noisy configuration, which neither removes the noise in the aggregate observation sequences nor tries to model the noise as a load [13]. Therefore, the Noisy configuration is the most realistic configuration for testing. We found that the use of different datasets and measurement metrics made it nearly impossible to compare different algorithms. Thus, the same datasets and measurement metrics have been used as recommended in Stephen's paper. Firstly, we identified each load working state by quantizing its PMF. In Stephen's [13] paper, they quantized both power and current observations. We just quantized power measurements, because it was enough to describe the working states for each appliance. Tables 2 and 3 show Stephen's and our results for some appliance state quantization in the AMPds2 dataset. '\' refers that the appliance does not have certain working state. In Stephen's results, they classified the low-power operating state of the appliance in detail while grouping all high-power operating states into one state. Hence, the quantization results generated by Stephen are not reasonable. In contrast, we roughly clustered the appliances into a low-power operating state while dividing the high-power operating states in detail. That is more in line with the actual working state of the appliances.

Table 2. Our state quantization results in AMPds2.

Appliance\Max Power	0	1	2	3
B1E	1	6	623	\
BME	10	600	1571	\
DWE	8	300	848	\
EQE	20	34	52	\
FRE	50	300	581	\
HPE	500	2000	3701	\
UTE	0	10	41	65
WOE	0	2300	3200	3896
B2E	9	200	1000	\
CDE	7	1000	5614	\
FGE	8	400	1497	\
OUE	0	305	\ [1]	\

[1] This type of appliance does not have this state.

Table 3. Stephen's state quantization results.

Appliance\Max power	0	1	2	3
B1E	0	1	6	9999
BME	0	5	10	9999
DWE	0	4	8	9999
EQE	0	34	38	9999
FRE	0	100	107	9999
HPE	0	3	39	9999
UTE	0	10	41	9999
WOE	0	2	9999	\
B2E	0	5	9	9999
CDE	0	7	9999	\
FGE	0	3	8	9999
OUE	0	9999 [1]	\ [2]	\

[1] The maximum power value set by Stephen. [2] This type of appliance does not have this state.

Next, we compared some different appliance combinations in REDD house 2, and the results are shown in Table 4. The combinations are the same as in Figure 9. Our disaggregate results were slightly better than Stephen's results, demonstrating that a basic linear-chain CRF model performs better, especially for the case that includes a kitchen outlet (three loads). Most common algorithms cannot deal with kitchen outlets because their power values change irregularly according to the appliances plugged into them. By extracting previous states, our model could improve accuracy to some extent. However, our model scored lower than sparse HMM when it came to four loads involving a washer-dryer. We found that the power value of the washer-dryer in REDD house 2 was excessively low. Thus, compared with HMM, which only extracted the last information, our model was more prone to obtaining errors.

Table 4. Accuracy comparison between the linear-chain CRF model and other algorithms.

Load\Acc (%)	Linear-Chain CRFs	Sparse HMM	SVM-rbf	SVM-Linear	SVM-Sigmoid
1 load	99.94	99.01	99.91	100	94.38
2 loads	99.27	99.00	98.39	96.32	81.82
3 loads	98.80	87.45	81.23	79.81	76.35
4 loads	96.04	98.52	92.40	90.31	88.41
5 loads	96.87	94.69	92.12	88.03	88.85
6 loads	97.40	95.28	93.22	85.83	88.84
7 loads	96.68	95.56	90.90	86.80	87.83

In addition, the proposed method was compared with algorithms which were not based on the probabilistic graph model. We chose the SVM with three different kernels: radial basis function (rbf) kernel, linear kernel, and sigmoid kernel. There were several parameters that had to be determined cautiously to fit for the study, because a higher or lower figure can affect the results considerably and may lead to local maxima or overfitting. "C" is the penalty parameter of all three kernels, and "gamma" is the parameter of the rbf and sigmoid kernels. We employed a grid search to find the best parameters on a small scale of datasets. Then, we employed the best parameters to train the model on all of the training sets and then tested the performance on the test data. The best accuracy rate was obtained when C = 1.0 and gamma = 1.0. The accuracy results are shown in Table 4. It is clear that the rbf kernel was more suitable for identifying appliances in REDD house 2 compared with linear and sigmoid kernels. Moreover, the accuracy rates have a tendency to decrease when there are more appliances, while our model remained reliable. In fact, with the increase in the number of appliances, the total loads' accuracy will decline as shown in Figure 9. However, by extracting a large number of state change characteristics of appliances, the recognition accuracy for most appliances can still be very high.

4. Conclusions

In this paper, we introduced a linear-chain CRF model for load disaggregation and demonstrated that this graphical model is feasible for a NILM task. We combined two features together: the power measurements and current signals. Feature templates were used for constructing feature functions, and the IIS algorithm was applied for parameter learning. Then, the Viterbi algorithm was utilized for decoding and estimated the accuracy results in AMPds2 and REDD house 2. Our accuracy results verified the feasibility and effectiveness of our model.

Author Contributions: Mthodology, validation, formal analysis, investigation, writing—original draft preparation, writing—review and editing, H.H., Z.L.; supervision, funding acquisition, R.J., G.Y.

Funding: This research was funded by the National Key R&D Program of China (2017YFB0202302), the Fundamental Research Funds for the Central Universities (2017MS072), and the College Students' innovation and entrepreneurship training program.

Conflicts of Interest: The authors declare no conflict of interests.

References

1. Hart, G.W. Nonintrusive appliance load monitoring. *Proc. IEEE* **1992**, *80*, 1870–1891. [CrossRef]
2. Lin, G.; Lee, S.; Hsu, J.Y.; Jih, W. Applying power meters for appliance recognition on the electric panel. In Proceedings of the 5th IEEE Conference on Industrial Electronics and Applications, Taichung, Taiwan, 15–17 June 2010; pp. 2254–2259.
3. Chui, K.T.; Lytras, M.D.; Visvizi, A. Energy Sustainability in Smart Cities: Artificial Intelligence, Smart Monitoring, and Optimization of Energy Consumption. *Energies* **2018**, *11*, 2869. [CrossRef]
4. Lai, Y.X.; Lai, C.F.; Huang, Y.M.; Chao, H.C. Multi-appliance recognition system with hybrid SVM/GMM classifier in ubiquitous smart home. *Inf. Sci.* **2013**, *230*, 39–55. [CrossRef]
5. Zia, T.; Bruckner, D.; Zaidi, A. A hidden Markov model-based procedure for identifying household electric loads. In Proceedings of the 37th Annual Conference of the IEEE Industrial Electronics Society, Melbourne, VIC, Australia, 7–10 November 2011; pp. 3218–3223.
6. Kong, W.; Dong, Z.Y.; Hill, D.J.; Ma, J.; Zhao, J.H.; Luo, F.J. A Hierarchical Hidden Markov Model Framework for Home Appliance Modeling. *IEEE Trans. Smart Grid* **2018**, *9*, 3079–3090. [CrossRef]
7. Kim, H.; Marwah, M.; Arlitt, M.; Lyon, G.; Han, J. Unsupervised Disaggregation of Low Frequency Power Measurements. In Proceedings of the 2011 SIAM International Conference on Data Mining, Mesa, AZ, USA, 28–30 April 2011; pp. 747–758.
8. Kolter, J.Z.; Jaakkola, T. Approximate Inference in Additive Factorial HMMs with Application to Energy Disaggregation. *Neural Inf. Process. Syst.* **2010**, *22*, 1472–1782.
9. Agyeman, K.A.; Han, S.; Han, S. Real-Time Recognition Non-Intrusive Electrical Appliance Monitoring Algorithm for a Residential Building Energy Management System. *Energies* **2015**, *8*, 9029–9048. [CrossRef]
10. Wallach, H.M. Conditional Random Fields: An Introduction. *Tech. Rep.* **2004**, 267–272.
11. Lafferty, J.; McCallum, A.; Pereira, F. Conditional Random Fields: Probabilistic Models for Segmenting and Labeling Sequence Data. In Proceedings of the 18th International Conference on Machine Learning, Williams College, Williamstown, MA, USA, 28 June–1 July 2001; pp. 282–289.
12. Heracleous, P.; Angkititrakul, P.; Kitaoka, N.; Takeda, K. Unsupervised energy disaggregation using conditional random fields. In Proceedings of the IEEE PES Innovative Smart Grid Technologies, Europe, Istanbul, Turkey, 12–15 October 2014; pp. 1–5.
13. Makonin, S.; Popowich, F.; Bajic, I.V.; Gill, B.; Bartram, L. Exploiting HMM Sparsity to Perform Online Real-Time Nonintrusive Load Monitoring. *IEEE Trans. Smart Grid* **2016**, *7*, 2575–2585. [CrossRef]
14. Makonin, S.; Popowich, F.; Bartram, L.; Gill, B.; Bajic, I.V. AMPds: A public dataset for load disaggregation and eco-feedback research. In Proceedings of the 2013 IEEE Electrical Power & Energy Conference, Halifax, NS, Canada, 21–23 August 2013; pp. 1–6.
15. Kolter, J.Z.; Johnson, M.J. REDD: A Public Data Set for Energy Disaggregation Research. Available online: http://redd.csail.mit.edu/kolter-kddsust11.pdf (accessed on 10 May 2019).

16. Berger, A.L.; Pietra, V.J.D.; Pietra, S.A.D. A Maximum Entropy Approach to Natural Language Processing. *Comput. Linguist.* **1996**, *22*, 39–71.
17. CRF++. Available online: https://www.findbestopensource.com/product/crfpp (accessed on 21 April 2019).

Article

Towards Effective and Efficient Energy Management of Single Home and a Smart Community Exploiting Heuristic Optimization Algorithms with Critical Peak and Real-Time Pricing Tariffs in Smart Grids †

Muhammad Awais [1], Nadeem Javaid [1,*], Khursheed Aurangzeb [2], Syed Irtaza Haider [2], Zahoor Ali Khan [3] and Danish Mahmood [4]

1 Department of Computer Sciences, COMSATS University Islamabad, Islamabad 44000, Pakistan; amawais@hotmail.com
2 College of Computer and Information Sciences, King Saud University, Riyadh 11543, Saudi Arabia; kaurangzeb@ksu.edu.sa (K.A.); sirtaza@ksu.edu.sa (S.I.H.)
3 Computer Information Science, Higher Colleges of Technology, Fujairah 4114, UAE; zkhan1@hct.ac.ae
4 Shaheed Zulfikar Ali Bhutto Institute of Science and Technology, Islamabad 44000, Pakistan; danishmahmood1@gmail.com
* Correspondence: nadeemjavaidqau@gmail.com or nadeemjavaid@comsats.edu.pk; Tel.: +92-300-579-2728

† This paper is an extended version of paper published in 2018 IEEE 32nd International Conference on Advanced Information Networking and Applications (AINA), Kraków, Poland, 16–18 May 2018.

Received: 14 October 2018; Accepted: 6 November 2018; Published: 12 November 2018

Abstract: Nowadays, automated appliances are exponentially increasing. Therefore, there is a need for a scheme to accomplish the electricity demand of automated appliances. Recently, many Demand Side Management (DSM) schemes have been explored to alleviate Electricity Cost (EC) and Peak to Average Ratio (PAR). In this paper, energy consumption problem in a residential area is considered. To solve this problem, a heuristic based DSM technique is proposed to minimize EC and PAR with affordable user's Waiting Time (WT). In heuristic techniques: Bacterial Foraging Optimization Algorithm (BFOA) and Flower Pollination Algorithm (FPA) are implemented. Furthermore, a novel heuristic algorithm has been proposed by merging the best features of the aforementioned existing algorithms. We test the proposed scheme on single homes and on smart community (involving multiple households). Different Operational Time Intervals (OTIs) are also considered for implementation. We have performed simulations for validating the our scheme. Results clearly demonstrate that the proposed Hybrid Bacterial Flower Pollination Algorithm (HBFPA) shows efficacy for EC and for reduction of PAR with reasonable user WT.

Keywords: scheduling; demand side management; smart grid; home energy management

1. Introduction

Human life has been made easy in many aspects due to the progress in various fields of science. Electricity is one field of science that has made human life easier in various ways. It is generated from different aspects of nature, i.e., from wind stations, nuclear power plants, hydro-power plants and water turbines. Electricity providers are not able to resist load requirements of customers because of the huge increase in the human population, buildings and industries and other infrastructure. The demand for electricity has been increased due to the automation in various sectors. For handling this intense condition, utilities propose to their customers to balance their electricity usage during the day and try to avoid maximum usage in the peak hours by keeping their eye on the highly demanding hours and their extra wastage. The energy that is required by the residential sector is thirty to forty five percent of total energy usage around the globe [1]. It is evidently stated that Energy Demand (ED) is growing endlessly, and it is expected that it will reach up to 56% of the current usage in 2040. Using some intelligent systems and techniques, the user can control this high Load Demand (LD). This demand can be achieved by giving some incentives to the customers, i.e., reduction in Electricity Cost (EC) in low demanding hours. Energy consumption minimization using Demand Side Management (DSM) can also be attained by planning the usage of electricity and by adopting the deployment methods which directly impact the customer's LD. There are different pricing tariffs that are being used by the utility. These pricing tariffs are Critical Peak Pricing (CPP) tariff, Real-Time Pricing (RTP) tariff, Day Ahead pricing (DAP) tariff and the Time of Use (TOU) tariff. Many of these have been implemented to get incentives. The domestic customers consume 18% of this electricity 2011 [2] and it is continuously increasing. Many algorithms are being proposed to optimize the energy demand and many of them are inspired from biology, artificial intelligence and nature.

In our model, we have considered 14 appliances which are categorized by consumer's actions. The Length of Operational Time (LOT) is dissimilar for the different appliances [3]. The considered Parameters are: reduction of EC, PAR and load consumption with affordable Waiting Time (WT). We are optimizing the problem using a novel hybrid approach for single homes as well as multiple homes, i.e., ten, thirty and fifty homes with power consumption patterns and dynamic Power Rating (PR). Some steps we mentioned here are:

- Load shifting strategy from high power demanding hours to low power demanding hours.
- Planning constraints in which different incentives are given to the customers by utility for flexible load shifting because of variation in quality of services provided by them.

This work is an extension of [4]. Motivated from meta-heuristic algorithms, this paper considers the Flower Pollination Algorithm (FPA), Bacterial Foraging Optimization Algorithm (BFOA) and their hybrid algorithm for optimizing the energy consumption in single as well as multiple homes. Our current work presents CPP and RTP pricing tariffs in Home Energy Management (HEM) to decrease the consumer's EC and Peak to Average Ratio (PAR). The PAR is considered to minimize the rebound peaks in off-peak hours.

The rest of the paper is planned as following: the literature review is described in Section 2. Section 3 covers the problem statement and our planned solution with detailed description is explained in Section 4. Our achieved results and the discussions based on them are proved via simulations in Section 5 and what has been learned through this research is summarized and concluded in Section 6 with its future work.

2. Literature Review

Recently, many optimization methods have been proposed to accomplish various objective functions, i.e., lessening the EC and PAR. Various struggles are made for the cost-effective consumption of electricity. In the current section, the previous research explored and described by other researchers on various optimization methods are discussed.

The meta-heuristic methods named: FPA and Harmony Search Algorithm (HSA), which are applied by other researchers to assess the performance in HEM [5]. Tariq et al. have considered one home with various appliances working both automatically and manually. The CPP tariff is used with the proposed scheme to shift the load from on-peak to off-peak hours for EC minimization and User Comfort (UC) maximization. Nonetheless, no consideration has been given to the dynamic PRs of the various appliances.

In [6], the researchers explored an effective DSM model for the residential users. The DSM model uses the Binary Particle Swarm Optimization Algorithm (BPSO), Genetic Algorithm (GA) and Ant Colony Optimization (ACO) Algorithm. All of these algorithms are heuristic algorithms that reduce the EC, PAR and maximize UC. Main objectives of this research are deduced using three basic methods:

GA, BPSO and ACO using TOU and Inclined Block Rate (IBR) pricing tariffs. However, computational complexity is not considered by Rahim et al.

HEM controller is designed by applying heuristic techniques as: GA, BFOA, Wind Driven Optimization Algorithm (WDO), BPSO and Genetic Binary Particle Swarm Optimization (GBPSO) in [7] using RTP pricing tariff. In this paper, main emphasis is on reduction in cost and PAR. The GA accomplishes well in the reduction of PAR and BPSO outperforms in cost minimization. However, a trade-off between EC and delay is existing in the proposed techniques. The simulations showed that the method achieved better results in the specified circumstances. Table 1 shows the brief summarized related work.

In [8], the HEM system is explored which integrate Renewable Energy Resources (RES) using different energy supplements and performs with DSM instantaneously. Therefore, the recommended HEM supports the users in curtailing the consumption and scheduling of appliances in the home to a specific level. When the boundary exceeds a specific level, the appliances will get-off by the utility itself. In this paper, DAP signals and heuristic algorithms are applied to acquire optimum solutions.

For the optimization of optimization problem, Power Scheduling Technique (PST) is explored, where the user can flow the initial and finish times of the appliances and can decrease the power usage. EC in the current exploration is publicized by electric suppliers beforehand [9]. The results validate that the scheduling technique outperformed in terms of low EC highly proficiently.

With the consideration of supplier's services, to stabilize the load for avoiding huge electricity usage is necessary. In addition, the supplier has to generate excess electricity to achieve the load demand. To annihilate the aforementioned issue, a crucial requirement to rise the power consumption agreement with low rates [10] is presented here. The results based on simulations proved that there is a specific limit beyond which the algorithm schedular keep working and pause some appliances for some duration. It supports to continue the LD for low EC and pause the appliances for use at a later time. The TOU tariff is applied for balancing the load appropriately.

By shifting the load, the customer can delimit the high LD. A critical situation for the checker creates, when the shareholder's capacity of increasing energy incomes get inadequate [11]. Generally, peaks can be decreased and valleys can be occupied which helps in balancing the LD.

In the paper [12], the deliberation of the Smart Meter (SM) model is presented, this model comprises Demand Side (DS) production to improve the cost. The cost is used to control the correction variables to improve the solution. This work halts a new pricing tariff for optimum procedures of the Smart Grid (SG), which uses PSO scheme to find the optimum results to loss limits and Demand Response (DR) stability.

The BFOA is hybrid with GA and is applied in [13]. In their work, they used RTP pricing scheme is taken into account to improve the load of the customer, UC and EC. Nevertheless, the dynamic PR for a smart community with varying appliances is not considered.

In this paper [14], DR patterns and procedures are implemented and reviewed by classifying different schemes to decrease power consumption for DR variables which reduce the total power usage by effectively applying DR method, depend on the contribution of customer and contribute to power decrease in the highly demanding hours by consuming Dynamic Pricing Patterns (DPP). The proposed method produce the DPP for predicting methods that reflects the probabilistic performance of the appliances. In [15], the researchers categorized the appliances into three categories, which are base load appliances, non-deferrable and deferrable. The key objectives attained are WT minimization, PAR reduction and cost reduction.

The DSM approach is applied to attain the objective of load balancing for residential, commercial and industrial zone [16]. It reduces the demand of load in peak hours with signification decrease in the bill. The results based on simulations shows an obvious reduction in cost by applying the PSO method.

The researcher in [17] proposed an effective HSA method that is implemented to schedule the DSM. The pricing tariff used by Geem et al. is TOU. Therefore, many simulations are performed and their results show that HSA performs better compared GA.

The hybrid of a fuzzy technique with FPA is applied to accept possibility that uses various methods for accommodating variation in mutually local and global pollination [18]. The FPA outperforms than its hybridized fuzzy version. Results are compared with different mathematical models. However, time computation is ignored in this paper.

The altered FPA is explored in [19], which uses scaling factor for controlling local pollination. The efficiency is intended from various mathematical equations, simulations and four diverse power systems.

In [20], storage devices, i.e., batteries have a limit for charging and discharging to avoid any type of loss, and controllers are installed on them. In the residential area, controller's load is shifted into the slot of low-cost, which means that charge the battery when EC is low and consumes that electricity when the demand is high in order to meet the load demand. At the end, it is concluded that higher battery capacity results in optimal usage of power and EC minimization. The EC and PAR are minimized by the proposed scheme in [21]. However, renewable energy resources and installation cost is completely ignored.

In [22], the cost is reduced by using GA. However, the UC is ignored and PAR is also neglected in this paper. The EC is minimized with the reduction in PAR in [23]. However, UC is neglected completely. A famous approach with its hybrid version is proposed in [24] using a hybrid algorithm of FPA and Tabo Search Algorithm (TSA) to solve the optimization for unconstrained problems. However, UC is ignored. In [25], the authors used FPA to optimize linear antenna arrays. The FPA outperforms other nature-inspired algorithms, including PSO, ACO and cat swarm optimization.

In [26], the authors proposed distributed energy resource aggregator as a new player to manage the energy along with the financial interactions in the day-ahead market. Graditi et al. in [27] propose a novel algorithm based on glowworm swarm particles optimization for optimal management of shiftable load in micro-grids (MGs). The authors also propose an optimal bidding strategy in an MG environment for selling and buying of energy by prosumers [28]. An analog ensemble method is applied for uncertainty caused by the intermittent nature of the renewable energy sources (RESs). The results revealed that the proposed methodology is effective during the hours when electricity price is high and the prosumer is willing to take risks. In [29], the authors use MG, which connects wind turbine and photovoltaic in a grid-connected mode to supply the energy in a smart home. Meta-heuristic algorithms and their hybrids are proposed to reduce the consumer's EC, PAR and WT. The results show that the hybrid schemes outperform the existing algorithms.

SGs have emerged in order to overcome the challenges of energy i.e., aging infrastructure, electricity losses, environmental problems, etc., caused by traditional grids [30]. In [31], the authors use metaheuristic algorithms along with optimal stopping rule to reduce the EC, PAR and WT of appliances. The rebound peak created in the off-peak hours are mitigated through multiple knapsack capacity limits. The authors in [32] define the user comfort by employing time and device based appliances' priorities. An evolutionary accretive comfort algorithm based on GA is considered to achieve the maximum user comfort in three different user budget scenarios. The results reveal that an increase in the budget value increases the user comfort along with cost per unit comfort index value. However, the OTI considered in [31,32] is composed of one hour. In reality, some appliances may take less time to complete their operations.

Table 1. Summarized literature review

Schemes	Achievement	Limitations
GA, BPSO, ACO [6]	Cost, PAR reduction and UC maximization	Computational complexity is not considered
BFOA, BFOA, GA, BPSO, WDO, GBPSO [7]	Reduces the EC and limits PAR	Trade-off between EC and PAR is not considered
GA, PSO, WDO, BFO, HGPSO [8]	Minimizes the electricity bill by scheduling household appliances	EC and PAR reduction are not considered
BFOA [9]	Reduces the EC with affordable UC	Trade-off among EC and UC is not considered
BPSO [10]	Develops efficient scheme to minimize the EC	Privacy of user is not considered
MOEA [11]	EC minimization and reduction in WT	Consumer's threshold limit is not focused
DR programs [13]	Minimize power consumption	Implements the DR program peak demand hours not considered
BPSO [16]	Reduces peak hours demands Reduction in the bill	Peak demand is reduced Electric cost is not considered
HSA [17]	Reduces operational cost	UC is not considered
DSM model is presented using GA [18]	Reduces operational cost, PAR	Time complexity is completely ignored
In-place (PL) generalized algorithm [19]	EC and UC trade-off	Ignores the system complexity
GA current procedural terminology [20]	EC and PAR reduction	System complexity is ignored
GA [21]	EC and PAR is minimized by the proposed scheme	Installation cost is completely ignored
GA [22]	EC is reduced by using GA	UC is ignored and PAR is also neglected
GA [23]	EC is minimized with reduction in PAR	UC is neglected
Hybrid algorithm using FPA and TS [24]	Hybrid version to optimize unconstrained problems	Ignores the optimization problem with multiple constrains
FPA [25]	Side lobe level minimization and null placement	Ignores interferences in undesired direction

3. Problem Statement, Objectives and Mathematical Formulation

In this section, the problem statement of this paper and mathematical formulation are discussed.

3.1. Problem Statement

In the aforementioned literature, complete benefits have not been taken from a smart grid. EC and PAR are minimized by many researchers and some researchers focus on UC and load shifting to off-peak hours from on-peak hours as in [13,33]. However, aforementioned parameters are not catered by any related work simultaneously. OTI of one hour is taken into consideration in aforementioned papers, which is not feasible, i.e., if the OTI is of one hour and the operational time interval of any appliance is 40 min, then the remaining slots will be unused. To tackle this problem, 1, 20, 30 and 60 min Operational Time Intervals (OTIs) are taken into account and their comparative analysis is also done in this paper. Towards efficient energy management positive features of both BFOA and FPA are exploited and a novel proficient heuristic hybrid algorithm is proposed in real time and the CPP tariff to tackle EC and PAR minimization problems with an affordable user's WT simultaneously.

3.2. Objectives

The main objectives of this paper include:

- Scheduling of home appliances,
- EC and PAR reduction,
- Balancing the load,
- Maximizing the UC,
- Trade-off between EC and UC exploited,
- Comparative analysis is also presented.

3.3. Mathematical Modeling

For solving the problem, mathematical equations has been modled. The Total Cost (TC) is determined for four different OTIs using Equations (1)–(4) with CPP and RTP tariffs in cents and power usage of various appliances in kilowatt hours (kWh). The various terms in equations and their symbolizations are provided in Table 2. The fitness function is determined using Equations (5) and (6).

The total load usage of a complete day for four OTIs is determined using Equations (7)–(10) while Equation (11) calculates the load per slot. α shows the 'ON/OFF' status of an appliance that is shown in Equation (12). The main emphasis is on objective functions as stated before. We have to decrease the total EC as in Equation (13) and diminish the PAR achieved using Equation (14) with affordable WT. One of our key objectives is load shifting as assessed in Equation (15). We have used some functions in these equations, which are: mean to find mean value, minimum as 'min' to find minimum value and 'std' is used to find standard deviation.

A day is divided into two scenarios: the first one is on-peak hours representing high electricity cost and the other one is off-peak hours representing a low electricity cost while seeing the 'mean' of given pricing tariff. The proposed Algorithm needs to fulfil our objective functions, and the algorithm will move the load to off-peak hours from the on-peak hours. It helps in reducing the PAR and EC. PAR is calculated using Equation (16), which is a ratio between maximum scheduled load and average unscheduled load. The list of appliances and their PRs are given in Table 3:

Table 2. Terms used in equations and their notations.

Terms	Notations
Electric rate per slot (t)	EP^t_{rate}
Power rating per appliance (ap)	P^{ap}_{rate}
Maximum population size	N_p
Appliance load	L_{oad}
Scheduled load	L^{sch}_{oad}
Unscheduled L	L^{unsch}_{oad}
Domain of electric rate	E_{rate}
Fitness function	E_F
Load per slot (t)	L^t_{oad}
Appliances	app

$$TC = \begin{cases} \sum_{t=1}^{1440} EP^t_{rate} \times P^{ap}_{rate}, & (1) \\ \sum_{t=1}^{72} EP^t_{rate} \times P^{ap}_{rate}, & (2) \\ \sum_{t=1}^{48} EP^t_{rate} \times P^{ap}_{rate}, & (3) \\ \sum_{t=1}^{24} EP^t_{rate} \times P^{ap}_{rate}, & (4) \end{cases}$$

$$E_f = min \begin{cases} l^{i \in N_{Pop}}_{oad} \geq mean(l^{Unsch}_{oad}) \ if \ EP^t_{rate} \leq mean(E_{rate}), & (5) \\ l^{i \in N_{Pop}}_{oad} > std(l^{Unsch}_{oad}) \wedge l^{i \in N_{Pop}}_{oad} < mean(l^{Unsch}_{oad}), \ EP^t_{rate} > mean(E_{rate}), & (6) \end{cases}$$

$$L^{sch}_{oad} = \begin{cases} \sum_{t=1}^{1440} L^t_{oad}, & (7) \\ \sum_{t=1}^{72} L^t_{oad}, & (8) \\ \sum_{t=1}^{48} L^t_{oad}, & (9) \\ \sum_{t=1}^{24} L^t_{oad}, & (10) \end{cases}$$

$$L^t_{oad} = P^{ap}_{rate} \times app, \tag{11}$$

$$\alpha = \begin{cases} 1, & if\ the\ appliance\ is\ ON, \\ 0, & if\ the\ appliance\ is\ OFF, \end{cases} \tag{12}$$

$$Object_1 = min(cost), \tag{13}$$

$$Object_2 = min(PAR), \tag{14}$$

$$Object_3 = L_{oad}, \tag{15}$$

$$PAR = \frac{max(L^{sch}_{oad})}{Average(L^{sch}_{oad})}. \tag{16}$$

Table 3. Power rating and length of operational time for operational time interval 20 min.

Group	Appliances	PR (kWh)	LOTs
Controllable Appliances	Oven	1.30	10.0
	Kettle	2.00	1.00
	Coffee Maker	0.80	4.00
	Rice Cooker	0.85	2.00
	Blender	0.30	2.00
	Frying Pan	1.10	3.00
	Toaster	0.90	1.00
	Fan	0.20	15.0
Shiftable appliances	Washing Machine	0.50	6.00
	Clothes Dryer	1.20	6.00
Non-Shiftable Appliances	Dish Washer	0.70	8.00
	Vacuum Cleaner	0.40	8.00
	Hair Dryer	1.50	2.00
	Iron	1.00	6.00

4. Proposed Methodology

In proposed methodology for HEM, appliances are categorized into three types, which are explained in detail below.

4.1. System Model

In this section, the architecture of the proposed model is discussed in detail as shown in Figure 1. It elaborates both single home and a smart community with multiple homes.

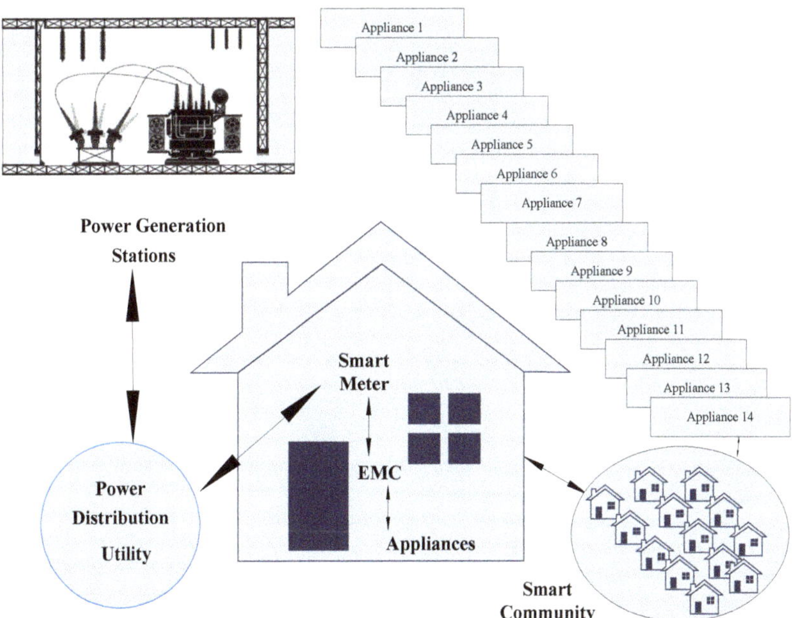

Figure 1. Proposed system model.

In our scenario, single and multiple homes: 10, 30 and 50 homes with four different time slots (1, 20, 30 and 60 min) are considered with dynamic PR appliances. RTP and CPP tariffs are used to calculate the EC. Meanwhile, selection of proper time slots help to achieve aforementioned objective functions and improve the efficacy of proposed model.

4.2. Appliance Classification

In a specific time slot appliances are presented with 'ON/OFF' status using [0, 1] and categories into two types; by determining their behavior, which are given as follows;

- Shiftable Appliances ,
- Controllable Appliances,
- Non-Shiftable Appliances.

4.2.1. Shiftable Appliances

Deferrable appliances are named as shiftable appliances because these appliances can be moved to another slot but without interruption during the working slot. It cannot be halted until its time slot terminates. In our scenario, washing machine and dishwasher are shiftable appliances.

4.2.2. Controllable Appliances

Interruptible appliances are termed as controllable appliances. Operational time of such appliances cannot be altered, i.e., fan.

4.2.3. Non-Shiftable Appliances

Uninterruptible appliances are schedulable but non-shiftable. These appliances cannot be interrupted and their energy feeding configurations and OTI cannot be altered, i.e., dishwasher. All appliances with their PRs are listed in Table 3.

4.3. Pricing Tariff

The price is determined agreeing to utility defined tariffs. Various pricing tariffs are applied to decrease the EC and PAR, which inspires the customers to move the load to off-peak hours from on-peak hours. Different tariffs such as DAP, TOU, CPP, IBR and RTP are available in literature. Among all aforementioned pricing tariffs, CPP and RTP pricing tariffs are considered to conduct simulations. A brief introduction is explained below.

4.3.1. CPP

CPP is an electric service price with respect to time which is implemented on people having usage of electricity cost more than 20 kW and they are equipped with a meter that records its usage after every fifteen minutes. The main purpose of CPP is to provide people with more information so that they can decide when and how they have to use electricity. CPP rates are applied only if your usage is more than 20 kW. If you are not following a CPP tariff, then you have to pay alternative rates. The CPP pricing scheme is shown in Figure 2. In CPP, a communication device is needed that has to communicate in both ways:

- to send how much consumption has taken place from customers to utilities,
- and to send information to customers from utilities

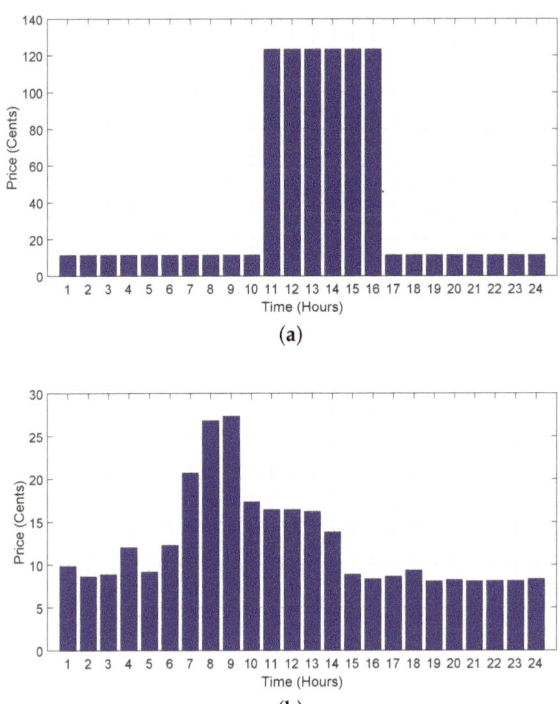

Figure 2. Pricing tariff. (**a**) Critical peak pricing.; (**b**) Real time pricing.

4.3.2. RTP

In the RTP scheme, price depends on hours and it varies hourly. Normally, prices are fixed in RTP, which influence the customer's usage in peak hours. It reflects marginal cost like Wholesale EC, atmospheric conditions and generator faults. Different pricing schemes from utilities impose different retail prices for different hours. RTP is implemented in the interval metering technologies that measure its consumption. It also records separate consumptions for every hour. The RTP scheme is shown in Figure 2.

4.4. *Implemented Techniques*

In this section, implemented techniques (BFOA, FPA, and Hybrid Bacterial Flower Pollination Algorithm (HBFPA)) are explained in detail.

4.4.1. BFOA

A beautiful aspect of nature is that it eliminates the animals that have less foraging behavior. The nature helps those species which possess the good searching behaviors and methods. After few generations, the weak ones are substituted by the healthy ones. Initially, the BFOA method has been proposed by 'Passsino' and 'Kevin' [34] in 2002. The approach in BFOA is that, initially, it allows the cells to group arbitrarily. Three successive phases are of the BFOA are explained below and algorithm for BFOA is discussed in [4] .

BFOA Steps

BFOA steps are explained as follows:

- Chemotaxis: the period of a bacteria's life is measured by the number of these steps, where the fitness J (i) of the bacteria is measured by contiguity to another bacteria's new position O (i), then a tumble besides the measured price surfaces one at a time by adding a unit step scope C (i) and it lies between $[-1, 1]$ in the direction of tumble. We generate a vector A (i) for representation of this random direction called as 'Tumble'.
- Reproduction: where bacteria performed well to move on their generation and the only cells that can perform are those that have done well in their life time.
- Elimination and Dispersal: where cells are discarded and new random cells are inserted having low probability.

The algorithm for BFOA is explained in [4].

4.4.2. FPA

The FPA is a nature inspired method that is stimulated by the pollination procedure of plants [35]. It is predictable that more than one million different classes of plants exist in the world and most of them are from flowering classification. The basic tenacity of flower is to reproduce their offspring via pollination process. The types of pollination are as follows.

Types of Pollination

There are two kinds of flower pollination, namely;

- Biotic pollination,
- Abiotic pollination.

Most of the plants go through biotic pollination (also called as 'local pollination'), which involves transfer of pollens within flowers of the same plant. A few of them perform abiotic pollination. Wind and many other natural processes help the flowers to perform pollination either locally or globally.

Pollinators are basically the pollen vectors that are huge in number. It is estimated that more than twenty thousand types of pollen vector exist in the world. Honeybee is the best example of a pollinator. These pollinators visit certain flowers simultaneously and maintain the consistency of flowers. This consistency involves the advantage of evolutionary purposes, which maximizes the reproduction steps of flowers. This consistency helps the pollinator in different ways, as they require minimum cost investment and more guarantee to intake the pollens.

Types of Different Processes in Pollination

The pollination can be achieved by two processes explained as follows:

- Self-pollination,
- Cross-pollination.

Cross-pollination is done from pollen vectors of different species of flowers while self-pollination occurs in the same flower. Cross pollination occurs at longer distances, which is why it is also called global pollination. One of the most important steps of this procedure is flower consistency, which can be achieved as an incremental stepping process using the difference or similarity between flowers. According to biological evolution, the survival of the best plant is the main objective of flower pollination, which is considered as a plant optimization process of different species.

From the biological evolution point of view, the objective of flower pollination is the survival of the plant and the optimal reproduction of plants in terms of number as well as the fitness. This can be considered as a plant species optimization process. All of the above factors and processes of flower pollination interact with each other to achieve the optimal reproduction. Therefore, this motivates us to design new optimization algorithms.

FPA Steps

In 2012, the FPA algorithm was developed by 'Xin-She' Yang and named as 'FPA for Global Optimization' [36]. For easiness, the following four steps are used.

- Biotic cross-pollination is calculated as a process of global-pollination in which pollen vectors transport pollinators by means of Levy flights.
- Abiotic and self-pollination are used for local-pollination.
- Pollinators sustain flower's uniformity by reproduction probability.
- The transferring of local and global pollination is calculated by a switch probability p, belongs to [0, 1].

The FPA Algorithm is shown in [4].

4.5. Hybridization

The method of hybridization basically contains a combination of two or more meta-heuristic techniques. If a technique maintains its identity while coupling with others, it will be a 'strong coupling' and, if other techniques take charge of its inner work, it will be a 'weak coupling'. During this hybridization, these techniques follow the steps of other techniques and control the strategy of newly proposed hybrid algorithms.

4.5.1. HBFPA

The steps of our explored HBFPA are explained in Algorithm 1. Terms used in HBFPA are explained in Table 4 and probability value is taken as (0.5) in HBFPA. Fitness in HBFPA is calculated using Equations (17) and (18):

$$F = (1 - u(1))^2 + D, \tag{17}$$

where 'D' is

$$D = 100 \times (u(2) - u(1)^2)^2 + 100 \times (u(3) - u(2)^2)^2. \tag{18}$$

Here, 'u' is the appliance's cost.

Table 4. Terms used in hybrid bacterial flower pollination algorithm and their notations.

Terms	Notations
OTIs	t
Total time in hours	T
Upper bound	α
Lower bound	β
Appliances	D
Fitness	E_F
Maximum population size	N_p
Newly generated population	Xnew
Old generated population	X

In order to have a better understanding of HBFPA, a numerical example is illustrated in Figure 3. In Figure 3, a solution represents the status of 14 appliances. The one value states that the appliance is ON in the given time-slot while the 0 value shows the device OFF status. The HBFPA fitness function evaluates each solution in terms of EC and given constraints. Finally, a solution based on minimum EC is selected for that time-slot and the population is updated for the next generation.

Algorithm 1 Algorithm for HBFPA

```
Parameters Initilization
The lower bound (β)
The upper bound (α)
For appliances d∈ D
All OTIs t in T
for Population← 1 to N_p do
    for j ← 1 to D-1 do
        Flowers are generated randomly
    end for
end for
Levy flight
N = Maximum iterations
for j ← 1 to N do
    for Pop← 1 to N_p do
        for i ← 1 to N_s do
            Tumble using Levy flight
            Calculate the E_f using Eq. (5), (6)
            Go to new location
            Calculate the E_f again using Eq. (5), (6)
            if E_f(Xnew) < E_f(X) then
                Update the population using swimming
                Calculate the E_f using Eq. (5), (6)
            else
                Bacteria tumble using Levy flight
                Take a random direction move
                Calculate the E_f using Eq. (5), (6)
            end If
        end for
        Update global population
    end for
end for
Return the best solution with less EC
```

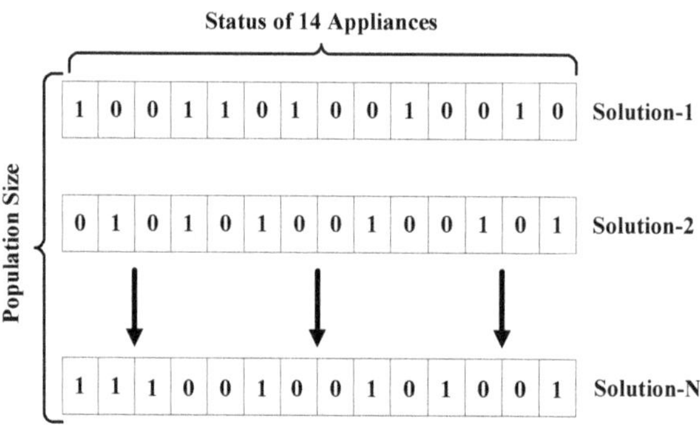

Figure 3. Hybrid bacterial flower pollination algorithm numerical example.

5. Simulation Results and Discussion

In this section, simulations and results with proper justification are described in order to specify the performance of the proposed hybrid algorithm using RTP and CPP price tariffs. Therefore, to judge the productivity of proposed technique and to describe its optimality for single home and a smart community, we have done different simulations for variable time horizon using different OTIs (24, 48, 72 and 1440) for a complete day, starting from 1:00 a.m. to the following 1:00 a.m.

5.1. For Single Homes

In this portion, a single home is considered with 14 appliances and EMC is installed in the home for scheduling of appliances according to price tariff defined by the utility side. Plots for load, EC, PAR and affordable WT using OTI of 20 min are given below:

5.1.1. Load Consumption

Load consumption for single homes using both pricing tariffs are explained as follows.

Load Consumption using CPP

The performance of the proposed hybrid algorithm is evaluated using a CPP price tariff. Our proposed hybrid algorithm outperformed as compared to benchmark schemes. Algorithm is envisioned to evade peak formation in any obvious slots of working hours. Therefore, price reduction happens. Our proposed and implemented technique performed fabulously in the case of different power consumption patterns. Figure 4 shows the behavior of load using CPP with four different OTIs. However, total load should be equal before and after scheduling.

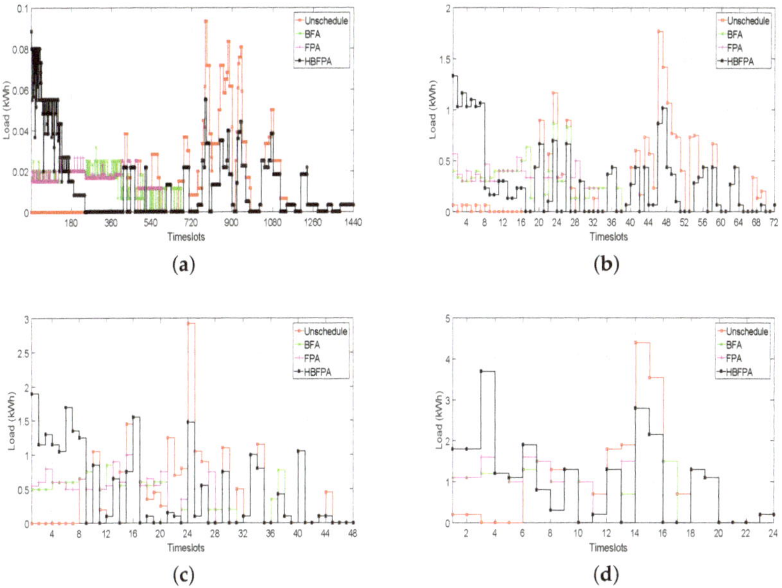

Figure 4. Load for single home using critical peak pricing. (**a**) load using operational time interval of 1 min; (**b**) load using operational time interval of 20 min; (**c**) load using operational time interval of 30 min; (**d**) load using operational time interval of 60 min.

Load Consumption Using RTP

Load consumption patterns using RTP are shown in Figure 5 for four different OTIs. In RTP pricing tariff, the scheduler efficiently manages the load and shifts from high rated hours to low rated hours. This load shifting is performed to reduce the EC. BFOA and FPA reduce the EC to some extent; however, the proposed algorithm outperforms. However, this load shifting affects UC.

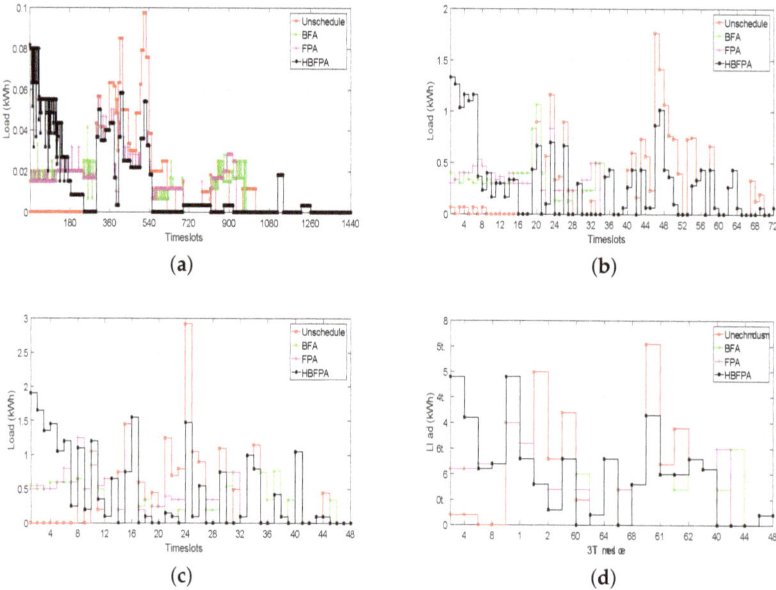

Figure 5. Load for single home using real time pricing. (**a**) load using operational time interval of 1 min; (**b**) load using operational time interval of 20 min; (**c**) load using operational time interval of 30 min; (**d**) load using operational time interval of 60 min.

5.1.2. Electricity Cost

EC for a single home using both pricing tariffs are explained as follows.

Cost Using CPP

The appliance performance in terms of cost are calculated using heuristic optimization techniques and, as a result of this work, the hourly cost is reduced. The proposed technique performed well as compared to FPA and BFOA. CPP remains the same throughout the year except critical peak periods where the price is high. Therefore, the cost pattern is almost similar for all OTIs because, in CPP, peak generation time is the same for all OTIs. Simulation results reveal that the proposed optimization technique reduces the total EC as shown in Figure 6a using CPP. EC values for a single home using CPP are shown in Table 5.

Table 5. Electricity cost for 24 h (for single home).

Techniques	Cost (Cents) Using CPP			
	1 min	20 min	30 min	60 min
Unscheduled	1.5323×10^3	1.1210×10^3	1.2912×10^3	1.1319×10^3
BFOA	848.3800	829.7933	753.6100	952.7100
FPA	785.6600	777.5267	804.0100	1.0423×10^3
HBFPA	785.6600	725.2600	608.0100	762.3100

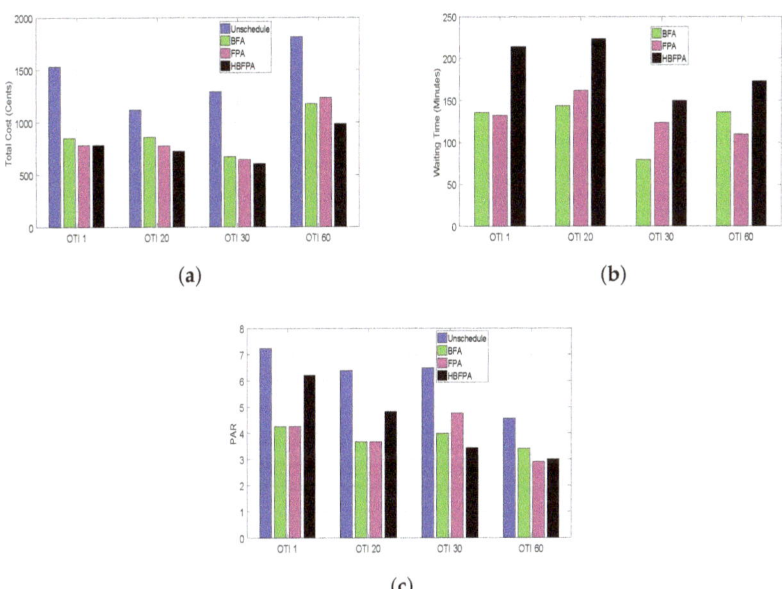

Figure 6. Total cost, waiting time and peak to average ratio for a single home using critical peak pricing. (**a**) Electricity cost using operational time interval of 1, 20, 30, 60 min; (**b**) Waiting time using operational time interval of 1, 20, 30, 60 min; (**c**) Peak to average ratio using operational time interval of 1, 20, 30, 60 min.

Cost Using RTP

Cost minimization is the main objective for which the hybrid heuristic technique is designed to optimize the DSM using the RTP tariff. Figure 7a elucidates EC of all OTIs. The figures clearly demonstrate that benchmark schemes outperformed in terms of EC; however, the proposed hybrid algorithm outperformed by sacrificing UC with affordable WT. EC values for single homes using RTP are shown in Table 6.

Table 6. Electricity cost for 24 h (For single home).

Techniques	Cost (Cents) Using RTP			
	1 min	20 min	30 min	60 min
Unscheduled	362.3626	269.1267	344.2463	333.1345
BFOA	280.0945	265.0310	291.4513	275.6915
FPA	286.3116	267.5580	300.4783	276.2255
HBFPA	267.9894	235.0647	269.3313	275.1495

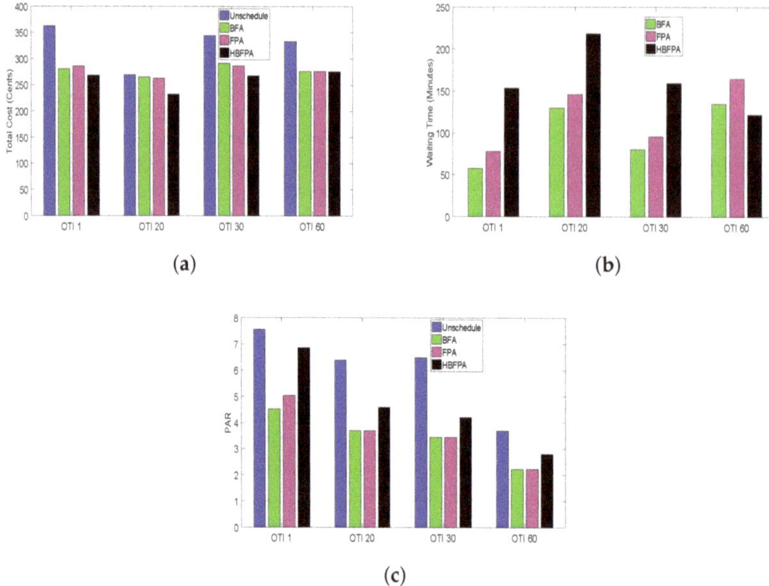

Figure 7. Total cost, waiting time and peak to average ratio for single home using real time pricing. (**a**) Electricity cost using operational time interval of 1, 20, 30, 60 min; (**b**) Waiting time using operational time interval of 1, 20, 30, 60 min; (**c**) Peak to average ratio using operational time interval of 1, 20, 30, 60 min.

5.1.3. Peak-to-Average Ratio

PAR is the value of maximum peak created to the average of total load which a user consumes in a day. The objective is to reduce PAR, which guarantees that created peaks are minimal. PAR for single homes using both pricing tariffs are explained as follows.

PAR Using CPP

When the proposed hybrid algorithm is applied to compute PAR using CPP, the proposed technique outperforms then state of the art schemes. The PAR reduction helps the utility to maintain its constancy and finally the price reduces. The HBFPA schedules the appliances efficiently and turns the appliances 'ON' in off-peak hours to avoid generating extra peaks in on-peak hours as shown in Figure 6c for four different OTIs using CPP tariffs. PAR values for a single home using CPP are shown in Table 7.

Table 7. Peak to average ratio for 24 h (for single homes).

Techniques	PAR Using CPP			
	1 min	**20 min**	**30 min**	**60 min**
Unscheduled	7.241	4.9	3.47	3.3784
BFOA	3.4365	6.48	3.99	3.43
FPA	3.5248	3.68	2.22	2.2289
HBFPA	6.2047	4.4388	3.4657	3.2657

PAR Using RTP

Figure 7c shows PAR using RTP tariffs for different OTIs (1, 20, 30 and 60 min). PAR is calculated by a proposed algorithm that gives good results with maximum PAR reduction. Hence, it can be clearly observed that our hybrid algorithm performs efficiently in the variable OTIs and improves the efficacy of scheduler. PAR values for single homes using RTP are shown in Table 8.

Table 8. Peak to average ratio for 24 h (for single homes).

Techniques	PAR Using RTP			
	1 min	20 min	30 min	60 min
Unscheduled	7.5619	6.3920	6.4850	3.6803
BFOA	4.5242	3.6784	3.8245	3.3175
FPA	5.4937	3.6784	3.4365	2.2289
HBFPA	6.2047	4.3417	4.2125	3.2138

5.1.4. Waiting Time

EC is the important parameter in HEM. There exists a trade-off between EC and UC. Here, UC is basically the WT of appliances to turn them 'ON' (how much a user waits to turn the appliances 'ON').

Waiting Time Using CPP

Figure 6b shows a WT comparison for four different OTIs using CPP price tariffs. If the user turns appliance an 'ON' without considering peak hours, then there will be no WT in that case. By applying the proposed algorithm, consumer have to wait for off-peak hours. The selection of variable OTI affects the WT. It can be easily evaluated that the proposed hybrid algorithm has maximum affordable WT than aforementioned algorithms. WTs for single homes using CPP are shown in Table 9.

Table 9. Waiting time for 24 h (for single homes).

Techniques	Waiting Time Using CPP			
	1 min	20 min	30 min	60 min
BFOA	137.3321	154.1667	86.7857	139.2857
FPA	140.0812	153.3929	102.6786	147.8571
HBFPA	214.3473	227.5595	149.8214	135.00

Waiting Time Using RTP

HBFPA successfully achieves the PAR and cost reduction using RTP tariff. However, there exists a trade-off between EC and UC. Therefore, users are unable to attain much UC. WT of four scenarios using RTP are shown in Figure 7b. It is concluded that affordable WT (delay) with maximum UC is attained. However, for EC reduction, the user's comfort is compromised. WTs for a single home using RTP are shown in Table 10.

Table 10. Waiting time for 24 h (for single homes).

Techniques	Waiting Time Using RTP			
	1 min	20 min	30 min	60 min
BFOA	54.9940	130.9762	80.2500	133.5714
FPA	59.7152	147.9762	90.5357	165.00
HBFPA	153.5777	228.5714	160.1786	130.00

5.2. For Multiple Homes

We optimize the appliances using BFOA, FPA and the proposed hybrid heuristic algorithm for multiple homes (10, 30 and 50 homes). Different homes may have different PRs. Therefore, the proposed algorithm has selected PRs dynamically. Random PRs are mentioned in Table 11.

Table 11. Random power ratings in (kWh).

Appliances	Power Rating 1	Power Rating 2	Power Rating 3	Power Rating 4
Washing-Machine	0.50	0.70	0.90	0.40
Clothes Dryer	10.0	1.20	1.40	1.60
Dish Washer	0.38	0.50	0.70	0.80
Vacuum Cleaner	0.80	1.00	0.20	0.50
Hair Dryer	1.50	1.20	1.40	1.70
Iron	1.00	1.30	1.50	1.20
Oven	1.30	1.50	1.70	1.90
Kettle	2.00	2.15	2.40	2.14
Coffee Maker	0.80	0.40	0.50	0.20
Rice Cooker	0.85	0.89	0.72	0.79
Blender	0.30	0.47	0.40	0.70
Frying Pan	1.10	1.50	1.90	2.00
Toaster	0.90	1.00	0.50	0.70
Fan	0.20	0.50	0.40	0.70

5.2.1. OTI 1 min

Load consumption, EC, PAR and WT for multiple homes using both pricing tariffs with OTI of 1 min are explained as follows.

Load Using CPP

Figure 8 clearly illustrates the load scheduling for multiple homes (10, 30 and 50 homes) using real-time scenarios with random power ratings and power consumption patterns. The notable thing in figures is that BFOA and FPA performed well while shifting load from on-peak to off-peak hours; however, the proposed hybrid algorithm outperformed.

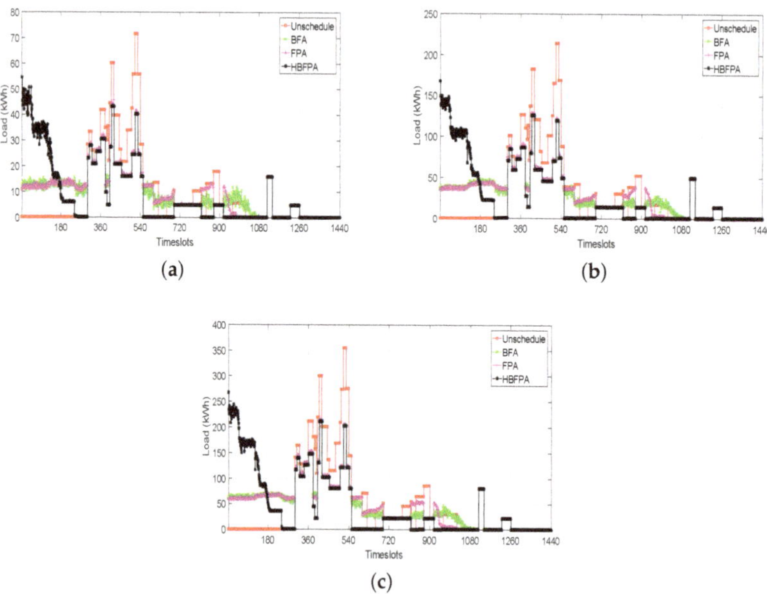

Figure 8. Load for multiple homes: (10, 30 and 50) using critical peak pricing. (**a**) load using operational time interval of 1 min for 10 homes; (**b**) load using operational time interval of 1 min for 30 homes; (**c**) load using operational time interval of 1 min for 50 homes.

EC, PAR and WT Using CPP

We implemented HBFPA for multiple homes to reduce overall cost, reduction in PAR with affordable WT and, as a result of this work, EC per slot is minimized as shown in Figure 9a. Affordable WT is shown in Figure 9b with reduction in PAR as shown in Figure 9c. EC, WT and PAR values for multiple homes are shown in Tables 12–14, respectively.

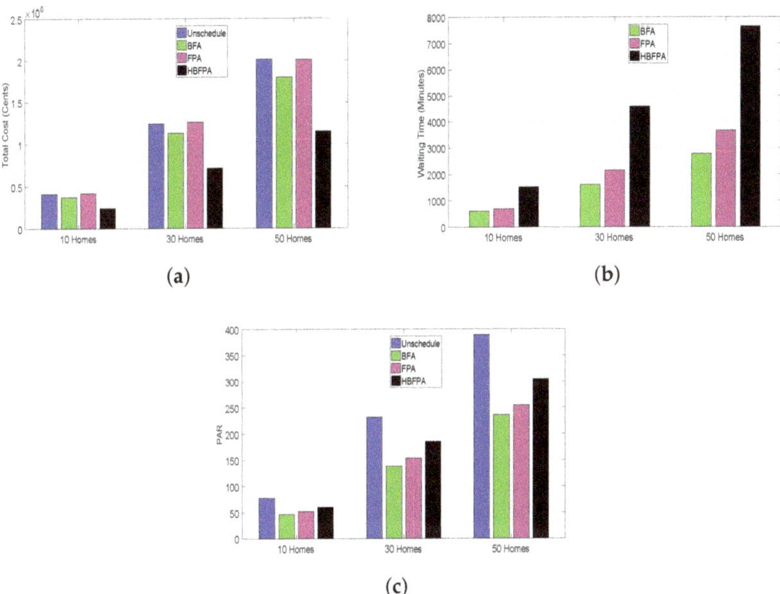

Figure 9. Electricity cost, waiting time and peak to average ratio for multiple homes: (10, 30 and 50) using critical peak pricing. (**a**) Electricity cost using operational time interval of 1 min; (**b**) Waiting time using operational time interval of 1 min; (**c**) Peak to average ratio using operational time interval of 1 min.

Table 12. Electricity cost for 24 h (for multiple homes).

Techniques	Cost (Cents) Using CPP for OTI 1 min		
	10 homes	**30 homes**	**50 homes**
Unscheduled	4.2213×10^5	1.2535×10^6	2.1077×10^6
BFOA	3.8669×10^5	1.1425×10^6	1.9247×10^6
FPA	4.4563×10^5	1.2881×10^6	2.1096×10^6
HBFPA	0.2128×10^5	0.7098×10^5	1.2205×10^6

Table 13. Waiting time for 24 h (for multiple homes).

Techniques	WT Using CPP for OTI 1 min		
	10 homes	**30 homes**	**50 homes**
BFOA	544.0428	1.7259×10^3	2.8516×10^3
FPA	722.7521	2.0134×10^3	3.6380×10^3
HBFPA	1.5302×10^3	4.5947×10^3	7.6534×10^3

Table 14. Peak to average ratio for 24 h (for multiple homes).

Techniques	PAR Using CPP for OTI 1 min		
	10 homes	30 homes	50 homes
Unscheduled	75.9236	233.9259	381.1938
BFOA	46.1947	137.6369	231.5394
FPA	49.2012	150.6242	248.5587
HBFPA	61.5829	180.9010	304.4469

Load Using RTP

Figure 10 shows load shifting for 10, 30 and 50 homes using OTI of 1 min. It is clearly shown in figures that proposed HBFPA outperformed and beat BFOA and FPA very well.

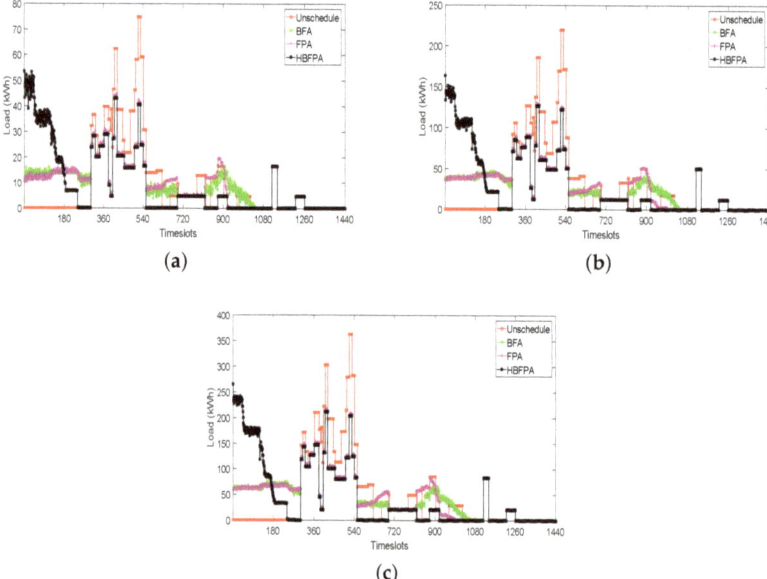

Figure 10. Load for multiple homes: (10, 30 and 50) using Real time pricing. (**a**) load using operational time interval of 1 min for 10 homes; (**b**) load using operational time interval of 1 min for 30 homes; (**c**) load using operational time interval of 1 min for 50 homes.

EC, WT and PAR Using RTP

The proposed algorithm is implemented on multiple homes using the RTP price tariff. The scheduler schedules the appliances for multiple homes and reduces EC by sacrificing UC. Overall cost reduction is shown in Figure 11a while affordable WT is demonstrated in Figure 11b and minimized PAR is shown in Figure 11c. EC, WT and PAR values for multiple homes are shown in Tables 15–17.

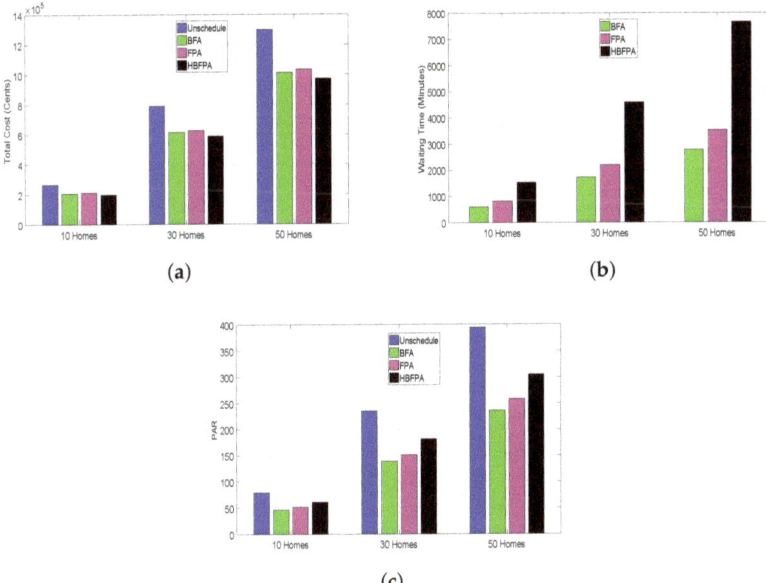

Figure 11. Electricity cost, waiting time and peak to average ratio for multiple homes: (10, 30 and 50) using real time pricing. (**a**) Electricity cost using operational time interval of 1 min; (**b**) Waiting time using operational time interval of 1 min; (**c**) Peak to average ratio using operational time interval of 1 min.

Table 15. Electricity cost for 24 h (for multiple homes).

Techniques	Cost (Cents) Using RTP for OTI 1 min		
	10 homes	**30 homes**	**50 homes**
Unscheduled	2.6912×10^5	7.9680×10^5	1.2990×10^4
BFOA	2.0976×10^5	6.2206×10^5	1.0124×10^6
FPA	2.1458×10^5	6.3795×10^5	1.0328×10^6
HBFPA	2.0176×10^5	5.9593×10^5	9.7160×10^5

Table 16. Waiting time for 24 h (for multiple homes).

Techniques	WT Using RTP for OTI 1 min		
	10 homes	**30 homes**	**50 homes**
BFOA	574.6065	1.6632×10^3	6.7364×10^3
FPA	746.8147	2.1254×10^3	4.5967×10^3
HBFPA	2.7175×10^3	3.6963×10^3	7.6552×10^3

Table 17. Peak to average ratio for 24 h (for multiple homes).

Techniques	PAR Using RTP for OTI 1 min		
	10 homes	30 homes	50 homes
Unscheduled	77.0644	232.1643	394.5380
BFOA	45.6238	138.9392	234.6092
FPA	50.2105	153.5715	253.3993
HBFPA	57.1704	184.7282	302.1077

5.2.2. OTI 60 min

Load consumption, EC, PAR and WT for multiple homes using both pricing tariffs with OTI of 60 min are explained as follows:

Load Using CPP

We have done simulations with OTI of 60 min using CPP tariffs to reduce the creation of peaks in on-peak hours and proposed HBFPA. HBFPA successfully shifts the load from on-peak hours to off-peak hours. Figure 12 clearly elaborates that the proposed scheme is projected to evade peak formations in any obvious slot of occupied hours.

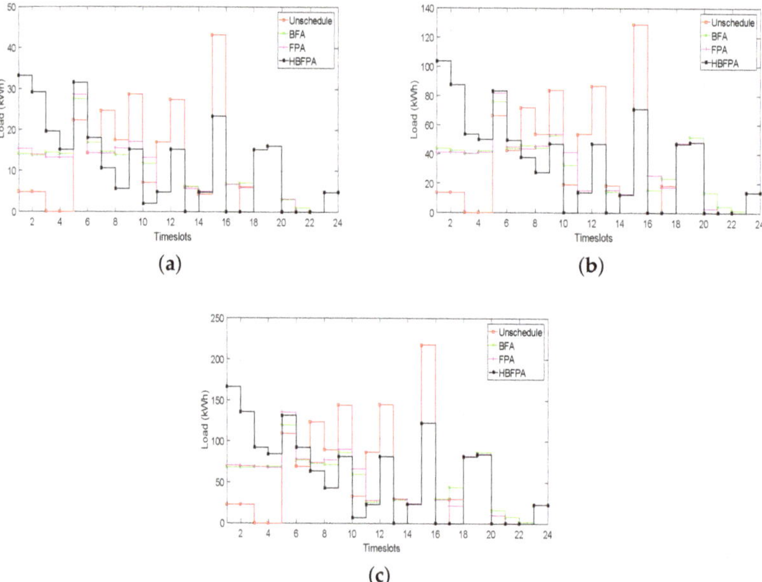

Figure 12. Load for multiple homes: (10, 30 and 50) using critical peak pricing. (**a**) load using operational time interval of 60 min for 10 homes; (**b**) load using operational time interval of 60 min for 30 homes; (**c**) load using operational time interval of 60 min for 50 homes.

EC, WT and PAR Using CPP

The proposed hybrid algorithm optimizes the problem by reducing cost and PAR with affordable WT using a real-time scenario. Thus, overall cost reduction is shown in Figure 13a and UC or WT is demonstrated in Figure 13b and PAR minimization is illustrated in Figure 13c. EC, affordable WT and minimized PAR values are shown in Tables 18–20.

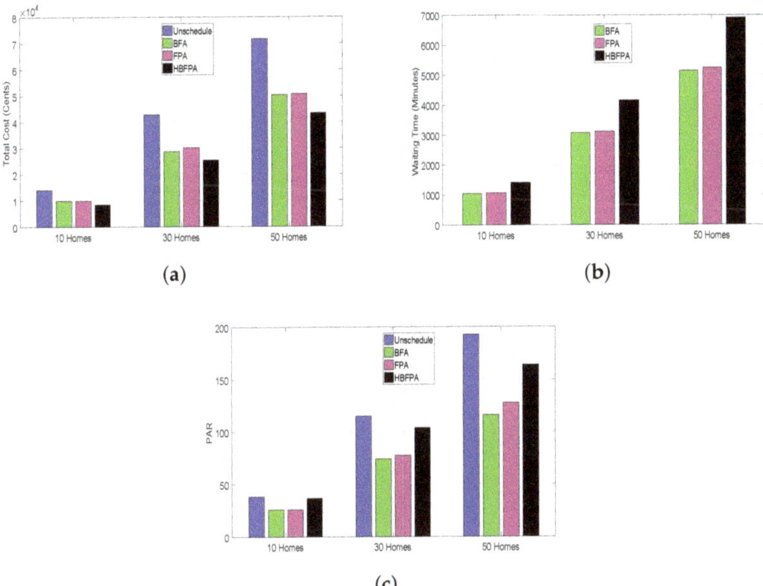

Figure 13. Electricity cost, waiting time and peak to average ratio for multiple homes: (10, 30 and 50) using critical peak pricing. (**a**) Electricity cost using operational time interval of 60 min; (**b**) Waiting time using operational time interval of 60 min; (**c**) Peak to average ratio using operational time interval of 60 min.

Table 18. Electricity cost for 24 h (for multiple homes).

Techniques	Cost (Cents) Using CPP for OTI 60 min		
	10 homes	30 homes	50 homes
Unscheduled	1.5059×10^4	4.2003×10^4	7.1746×10^4
BFOA	1.0666×10^4	2.9568×10^4	4.9553×10^4
FPA	1.0836×10^4	2.9822×10^4	5.1076×10^4
HBFPA	0.83×10^5	2.5075×10^4	4.274×10^4

Table 19. Waiting time for 24 h (for multiple homes).

Techniques	WT Using CPP for OTI 60 min		
	10 homes	30 homes	50 homes
BFOA	1.0043×10^3	3.2093×10^3	5.1507×10^3
FPA	1.0757×10^3	3.2429×10^3	4.1714×10^3
HBFPA	1.3836×10^3	4.1714×10^3	6.8836×10^3

Table 20. Peak to average ratio for 24 h (for multiple homes).

Techniques	PAR Using CPP for OTI 60 min		
	10 homes	30 homes	50 homes
Unscheduled	38.7901	112.7401	189.7837
BFOA	23.5727	78.1047	120.4385
FPA	24.6797	75.6910	129.6416
HBFPA	30.3457	98.0471	167.4858

Load Using RTP

Load plots for multiple homes (10, 30 and 50 homes) using OTI 60 with RTP tariffs are given in Figure 14 for four different OTIs with random PRs and power consumption patterns. HBFPA outperformed in shifting load from on-peak to off-peak hours efficiently for multiple homes (10, 30 and 50 homes) than FPA and BFOA, respectively.

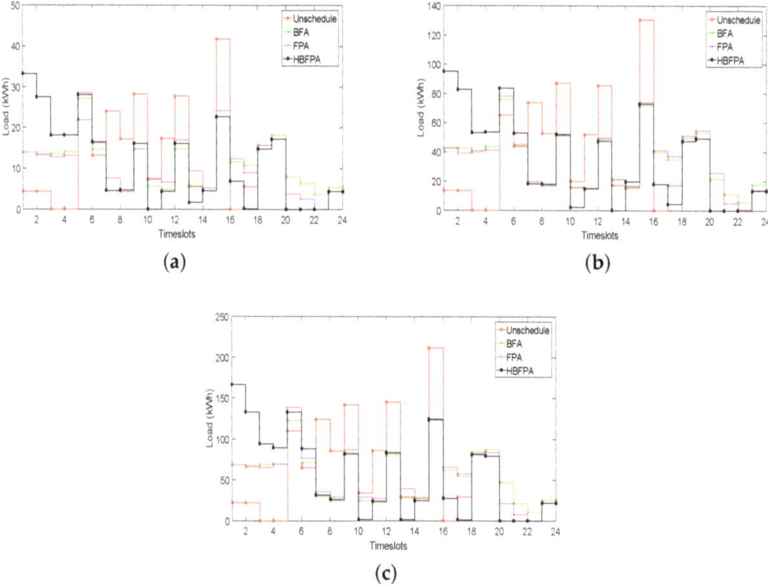

Figure 14. Load for multiple homes: (10, 30 and 50) using real time pricing. (**a**) load using operational time interval of 60 min for 10 homes; (**b**) load using operational time interval of 60 min for 30 homes; (**c**) load using operational time interval of 60 min for 50 homes.

EC, WT and PAR Using RTP

Scheduling using RTP tariffs is discussed in this section. BFA and FPA outperformed in EC minimization for smart community; however, a implemented hybrid algorithm outperformed the aforementioned algorithms. EC minimization is shown in Figure 15a, UC with affordable WT is shown in Figure 15b and PAR reduction is shown in Figure 15c. EC, affordable WT and PAR reduction values are shown in Tables 21–23.

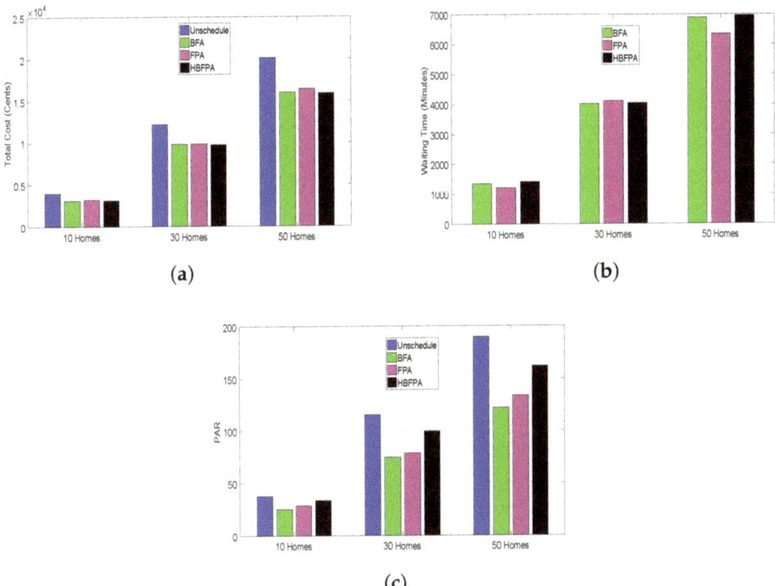

Figure 15. Electricity cost, waiting time and peak to average ratio for multiple homes: (10, 30 and 50) using real time pricing. (**a**) Electricity cost using operational time interval of 60 min; (**b**) Waiting time using operational time interval of 60 min; (**c**) Peak to average ratio using operational time interval of 60 min.

Table 21. Electricity cost for 24 h (for multiple homes).

Techniques	Cost (Cents) Using RTP for OTI 60 min		
	10 homes	30 homes	50 homes
Unscheduled	3.9954×10^3	1.2170×10^4	2.0437×10^4
BFOA	3.1541×10^3	9.6811×10^3	1.6242×10^4
FPA	3.1884×10^3	9.8798×10^3	1.6251×10^4
HBFPA	3.1877×10^3	1.6242×10^3	1.6039×10^4

Table 22. Waiting time for 24 h (for multiple homes).

Techniques	WT Using RTP for OTI 60 min		
	10 homes	30 homes	50 homes
BFOA	1.4314×10^3	4.0729×10^3	6.7364×10^3
FPA	1.4371×10^3	3.8336×10^3	6.869×10^3
HBFPA	1.3964×10^3	4.1579×10^3	6.9236×10^3

Table 23. Peak to average ratio for 24 h (for multiple homes).

Techniques	PAR Using RTP for OTI 60 min		
	10 homes	30 homes	50 homes
Unscheduled	37.3187	115.6367	195.2858
BFOA	25.9190	74.4312	122.2942
FPA	25.8770	80.9041	129.3206
HBFPA	33.8192	102.8952	166.9638

Here, an inclination is made that explains that the more a consumer sacrifices his luxury, low price rates will be given to him by utility. Our proposed technique outperforms in the case of different power consumption patterns and random PRs.

5.3. Feasible Regions

In mathematical optimization, a feasible region is the unique set of nominee solutions for the suggested scheme. Four constraints should be preserved while computing feasible regions:

- Min cost, Min Load,
- Min cost, Max Load,
- Max cost, Min Load,
- Max cost, Max Load.

Point 'P5' is cutting the overall area at point 'P2' in Figure 16 and in Figure 17, which shows the maximum cost in scheduled cases. Therefore, point 'P5' shows feasible regions for our objective functions.

Feasible Region Using CPP

In this paper, feasible regions are formulated using CPP price tariffs. The pointers (P1, P2, P3, P4, P5) have shown the possible feasible regions against different OTIs. The area of feasible regions is shaded with a cyan color. For details, follow Figure 16.

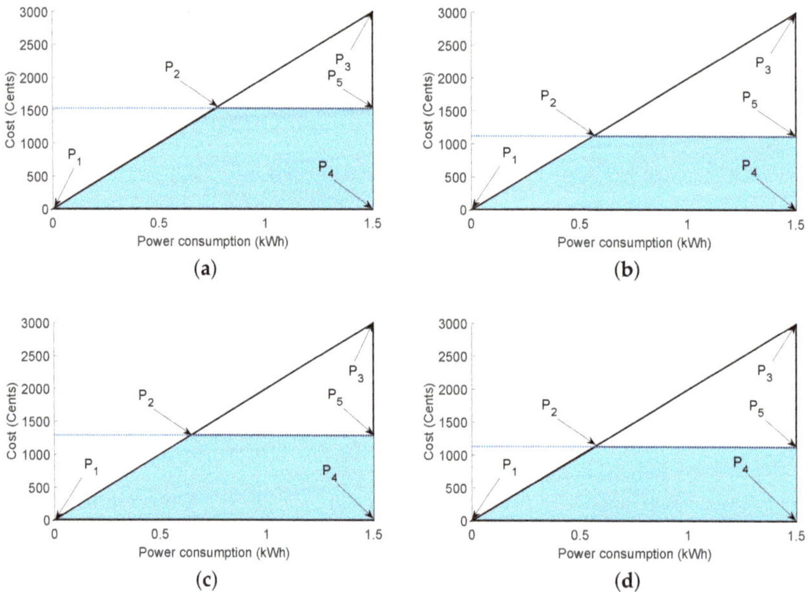

Figure 16. Feasible regions for a single home using critical peak pricing. (**a**) Operational time interval of 1 min; (**b**) Operational time interval of 20 min; (**c**) Operational time interval of 30 min; (**d**) Operational time interval of 60 min.

Feasible Region Using RTP

In this section, feasible regions are formulated using an RTP price tariff. The Pointers (P1, P2, P3, P4, P5) have shown the possible feasible regions against different OTIs using RTP. The area of feasible regions using RTP are shaded with a cyan color. For depth details, follow Figure 17.

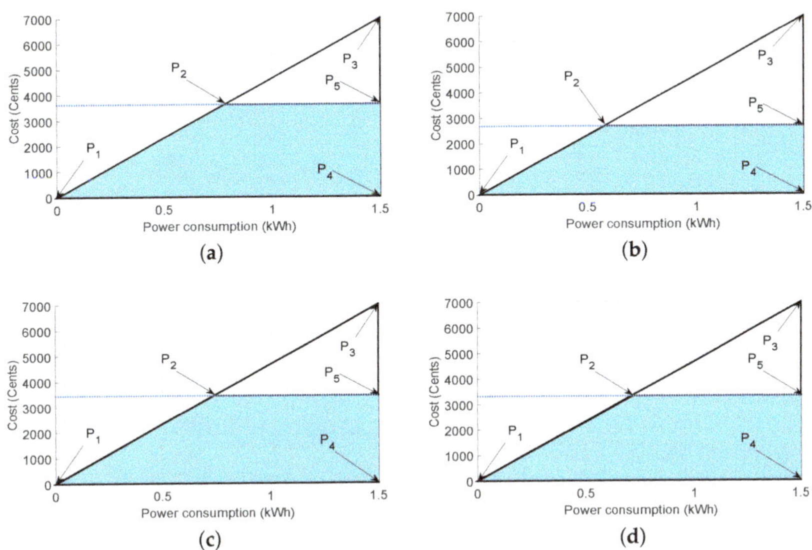

Figure 17. Feasible regions for a single home using real time pricing. (**a**) Operational time interval of 1 min; (**b**) Operational time interval of 20 min; (**c**) Operational time interval of 30 min; (**d**) Operational time interval of 60 min.

5.4. Performance Trade-Off

In this section, the trade-off between two parameters is discussed. Simulation results show the existing trade-off between EC and affordable WT. EC diminishes as the user sacrifices his luxury by deferring his activity. Therefore, to get the benefit in one context, the user has to compromise on some other parameters. The trade-off between EC and UC is demonstrated in the figures above.

6. Conclusions and Future Work

In this research work, DR and DSM are the main factors to maintain a balance between supply and demand of electricity. In this paper, to minimize EC and PAR with an affordable user's WT, a novel heuristic algorithm has been proposed. Our novel algorithm is proposed using two pricing schemes: CPP and RTP and four performance parameters (EC, PAR, load balancing and UC) are considered to evaluate our proposed algorithm with dynamic PR. Results show the efficacy of a novel heuristic hybrid algorithm HBFPA for a single home and smart community with multiple homes. When CPP is used, the HBFPA has achieved an EC value of 762.3100 cents for a single home as compared to unscheduled load EC value 1131.9 cents at 60 min OTI. The BFOA and FPA have achieved EC values of 952.71 cents and 1042.3 cents, respectively. When RTP is used at 60 min of OTI, the ECs achieved for the unscheduled case, BFOA, FPA and HBFPA are 333.1345, 275.6915, 276.2255 and 275.1495 cents, respectively. The HBFPA has achieved the PAR value 3.2657 as compared to unscheduled case 3.3784 for 60 min OTI using CPP. At 60 min OTI, the average WT of HBFPA, BFOA and FPA are 135, 139.2857 and 147.8571 min, respectively. Similarly, EC and PAR are minimized with an affordable user's WT by helping the scheduler to schedule the load from on-peak hours to off-peak hours for multiple home scenarios. From the results, it has been cleared that the implemented scheme outperformed the aforementioned existing algorithms. However, there exists a trade-off between the cost and user's comfort.

In the future, multiple optimization techniques will be integrated with renewable energy resources, dynamic programming and the cloud concept to schedule the home appliances to reduce cost and PAR despite using electricity management controllers. There will be a scenario when some homes may consist of a few number of appliances and there may exist some homes with a greater number of appliances.

Author Contributions: M.A., N.J., and K.A. proposed and implemented the novel schemes; S.I.H., Z.A.K., and D.M. completed the mathematical modeling. All authors together refined the manuscript. Finally, M.A., N.J., responded to the queries of the reviewers.

Funding: The authors extend their appreciation to the Deanship of Scientific Research at King Saud University for funding this work through research group NO (RG-1438-034).

Conflicts of Interest: The authors declare no conflict of interest.

References

1. Mehreen, G.; Patidar, S. Understanding the energy consumption and occupancy of a multi-purpose academic building. *Energy Build.* **2015**, *87*, 155–165.
2. Today in Energy—U.S. Energy Information Administration (EIA). Available online: https://www.eia.gov/todayinenergy/detail.php?id=12251 (accessed on 3 August 2018).
3. Logenthiran, T.; Srinivasan, D.; Shun, T.Z. Demand side management in smart grid using heuristic optimization. *IEEE Trans. Smart Grid* **2012**, *3*, 1244–1252. [CrossRef]
4. Awais, M.; Javaid, N.; Mateen, A.; Khan, N.; Mohiuddin, A.; Rehman, M.H.A. Meta Heuristic and Nature Inspired Hybrid Approach for Home Energy Management Using Flower Pollination Algorithm and Bacterial Foraging Optimization Technique. In Proceedings of the 2018 IEEE 32nd International Conference on Advanced Information Networking and Applications (AINA), Krakow, Poland, 16–18 May 2018; pp. 882–891. [CrossRef]
5. Tariq, M.; Khalid, A.; Ahmad, I.; Khan, M.; Zaheer, B.; Javaid, N. Load Scheduling in Home Energy Management System Using Meta-Heuristic Techniques and Critical Peak Pricing Tariff. In Proceedings of the International Conference on P2P, Parallel, Grid, Cloud and Internet Computing, Barcelona, Spain, 8–10 November 2017; Springer: Cham, Switzerland, 2017; pp. 50–62.
6. Rahim, S.; Javaid, N.; Ahmad, A.; Khan, S.A.; Khan, Z.A.; Alrajeh, N.; Qasim, U. Exploiting heuristic algorithms to efficiently utilize energy management controllers with renewable energy sources. *Energy Build.* **2016**, *129*, 452–470. [CrossRef]
7. Mahmood, D.; Javaid, N.; Alrajeh, N.; Khan, Z.A.; Qasim, U.; Ahmed, I.; Ilahi, M. Realistic scheduling mechanism for smart homes. *Energies* **2016**, *9*, 202. [CrossRef]
8. Ahmad, A.; Khan, A.; Javaid, N.; Hussain, H.M.; Abdul, W.; Almogren, A.; Alamri, A.; Niaz, I.A. An Optimized Home Energy Management System with Integrated Renewable Energy and Storage Resources. *Energies* **2017**, *10*, 549. [CrossRef]
9. Ma, K.; Yao, T.; Yang, J.; Guan, X. Residential power scheduling for demand response in smart grid. *Int. J. Electr. Power Energy Syst.* **2016**, *78*, 320–325. [CrossRef]
10. Muralitharan, R.S.; Shi, Y. Multiobjective optimization technique for demand side management with load balancing approach in smart grid. *Neurocomputing* **2016**, *177*, 110–119. [CrossRef]
11. López, M.A.; Torre, S.; Martín, S.; Aguado, J.A. Demand-side management in smart grid operation considering electric vehicles load shifting and vehicle-to-grid support. *Int. J. Electr. Power Energy Syst.* **2015**, *64*, 689–698. [CrossRef]
12. Chanda, S.; De, A. A multi-objective solution algorithm for optimum utilization of smart grid infrastructure towards social welfare. *Int. J. Electr. Power Energy Syst.* **2014**, *58*, 307–318. [CrossRef]
13. Khalid, A.; Javaid, N.; Mateen, A.; Khalid, B.; Khan, Z.A.; Qasim, U.Demand side management using hybrid bacterial foraging and genetic algorithm optimization techniques. In Proceedings of the 2016 10th International Conference on Complex, Intelligent, and Software Intensive Systems (CISIS), Fukuoka, Japan 6–8 July 2016; IEEE: Piscataway, NJ, USA, 2016; pp. 494–502.
14. Vardakas, J.; Zorba, N.; Verikoukis, C.V. A survey on demand response programs in smart grids: Pricing methods and optimization algorithms. *IEEE Commun. Surv. Tutor.* **2015**, *17*, 152–178. [CrossRef]
15. Aslam, S.; Iqbal, Z.; Javaid, N.; Khan, Z.A.; Aurangzeb, K.; Haider, S.I. Towards efficient energy management of smart buildings exploiting heuristic optimization with real-time and critical peak pricing schemes. *Energies* **2017**, *10*, 2065. [CrossRef]

16. Gupta, I.; Anandini, G.N.; Gupta, M. An hour wise device scheduling approach for demand side management in smart grid using particle swarm optimization. In Proceedings of the 2016 National Power Systems Conference (NPSC), Bhubaneswar, India, 19–21 December 2016; IEEE: Piscataway, NJ, USA, 2016.
17. Geem, Z.W.; Yoon, Y. Harmony search optimization of renewable energy charging with energy storage system. *Int. J. Electr. Power Energy Syst.* **2017**, *86*, 120-126. [CrossRef]
18. Valenzuela, L.; Valdez, F.; Melin, P. Flower pollination algorithm with fuzzy approach for solving optimization problems. In *Nature Inspired Design of Hybrid Intelligent Systems*; Springer: Cham, Switzerland, 2017; pp. 357–369.
19. Dubey, H.M.; Pandit, M.; Panigrahi, B.K. A biologically inspired modified flower pollination algorithm for solving economic dispatch problems in modern power systems. *Cognit. Comput.* **2015**, *7*, 594–608. [CrossRef]
20. Kakran, S.; Chanana, S. Smart operations of smart grids integrated with distributed generation: A review. *Renew. Sustain. Energy Rev.* **2018**, *81*, 524–535. [CrossRef]
21. Mary, G.A.; Rajarajeswari, R. Smart grid cost optimization using genetic algorithm. *Int. J. Res. Eng. Technol.* **2015**, *3*, 282–287.
22. Bharathi, C.; Rekha, D.; Vijayakumar, V. Genetic Algorithm Based Demand Side Management for Smart Grid. *Wirel. Pers. Commun.* **2017**, *93*, 481–502. [CrossRef]
23. Whitley, D. A genetic algorithm tutorial. *Stat. Comput.* **1994**, *4*, 65–85. [CrossRef]
24. Hezam, I.; Abdel-Baset, M.; Hassan, B. A hybrid flower pollination algorithm with tabu search for unconstrained optimization problems. *Inf. Sci. Lett.* **2016**, *5*, 29–34. [CrossRef]
25. Prerna, S.; Kothari, A. Linear antenna array optimization using flower pollination algorithm. *SpringerPlus* **2016**, *5*, 306.
26. Graditi, G.; di Somma, M.; Siano, P. Optimal Bidding Strategy for a DER aggregator in the Day-Ahead Market in the presence of demand flexibility. *IEEE Trans. Ind. Electron.* **2018**, *66*, 1509–1519.
27. Graditi, G.; di Silvestre, M.L.; Gallea, R.; Sanseverino, E.R. Heuristic-based shiftable loads optimal management in smart micro-grids. *IEEE Trans. Ind. Inf.* **2015**, *11*, 271–280. [CrossRef]
28. Ferruzzi, G.; Cervone, G.; Monache, L.D.; Graditi, G.; Jacobone, F. Optimal bidding in a Day-Ahead energy market for Micro Grid under uncertainty in renewable energy production. *Energy* **2016**, *106*, 194–202. [CrossRef]
29. Iqbal, Z.; Javaid, N.; Iqbal, S.; Aslam, S.; Khan, Z.A.; Abdul, W.; Almogren, A.; Alamri, A. A Domestic Microgrid with Optimized Home Energy Management System. *Energies* **2018**, *11*, 1002. [CrossRef]
30. El-Hawary, M.E. The smart grid-state-of-the-art and future trends. *Electr. Power Compon. Syst.* **2014**, *42*, 239–250. [CrossRef]
31. Khan, A.; Javaid, N.; Ahmad, A.; Akbar, M.; Khan, Z.A.; Ilahi, M. A priority-induced demand side management system to mitigate rebound peaks using multiple knapsack. *J. Ambient Intell. Hum. Comput.* **2018**, 1–24. [CrossRef]
32. Khan, A.; Javaid, N.; Khan, M.I. Time and device based priority induced comfort management in smart home within the consumer budget limitation. *Sustain. Cities Soc.* **2018**, *41*, 538–555. [CrossRef]
33. Yang, X.-S. Flower pollination algorithm for global optimization. In Proceedings of the International Conference on Unconventional Computing and Natural Computation, Orléans, France, 3–7 September 2012; Springer: Berlin/Heidelberg, Germany, 2012; pp. 240–249.
34. Passino, K. Biomimicry of bacterial foraging for distributed optimization and control. *IEEE Control Syst.* **2002**, *22*, 52–67.
35. Balasubramani, K.; Marcus, K. A study on flower pollination algorithm and its applications. *Int. J. Appl. Innov. Eng. Manag.* **2015**, *3*, 230–235.
36. Rodrigues, D.; Yang, X.-S.; de Souza, A.N.; Papa, J.P. Binary flower pollination algorithm and its application to feature selection. In *Recent Advances in Swarm Intelligence and Evolutionary Computation*; Springer: Cham, The Netherlands, 2015; pp. 85–100.

Article

Solving the Energy Efficient Coverage Problem in Wireless Sensor Networks: A Distributed Genetic Algorithm Approach with Hierarchical Fitness Evaluation

Zi-Jia Wang [1], Zhi-Hui Zhan [2,*] and Jun Zhang [2]

1 School of Data and Computer Science, Sun Yat-sen University, Guangzhou 510006, China; wangzjia@mail2.sysu.edu.cn

2 Guangdong Provincial Key Laboratory of Computational Intelligence and Cyberspace Information, School of Computer Science and Engineering, South China University of Technology, Guangzhou 510006, China; junzhang@ieee.org

* Correspondence: zhanapollo@163.com; Tel.: +86-138-2608-9486

Received: 30 October 2018; Accepted: 13 December 2018; Published: 18 December 2018

Abstract: This paper proposed a distributed genetic algorithm (DGA) to solve the energy efficient coverage (EEC) problem in the wireless sensor networks (WSN). Due to the fact that the EEC problem is Non-deterministic Polynomial-Complete (NPC) and time-consuming, it is wise to use a nature-inspired meta-heuristic DGA approach to tackle this problem. The novelties and advantages in designing our approach and in modeling the EEC problems are as the following two aspects. Firstly, in the algorithm design, we realized DGA in the multi-processor distributed environment, where a set of processors run distributed to evaluate the fitness values in parallel to reduce the computational cost. Secondly, when we evaluate a chromosome, different from the traditional model of EEC problem in WSN that only calculates the number of disjoint sets, we proposed a hierarchical fitness evaluation and constructed a two-level fitness function to count the number of disjoint sets and the coverage performance of all the disjoint sets. Therefore, not only do we have the innovations in algorithm, but also have the contributions on the model of EEC problem in WSN. The experimental results show that our proposed DGA performs better than other state-of-the-art approaches in maximizing the number of disjoin sets.

Keywords: wireless sensor networks; energy efficient coverage; distributed genetic algorithm

---

1. Introduction

Wireless sensor networks (WSN) have become a hot research topic and have been widely used in numerous real-world applications, such as traffic monitoring [1], mobile computing [2], environmental observation [3], and many others [4–6]. In these sophisticated environments, in order to make full coverage and get more accurate results, many nodes should be randomly deployed in the area, causing a waste of resources. Since the sensor nodes are equipped with limited battery resources and the replacement of the battery is not feasible in many applications, low power consumption has become a critical factor to be considered when designing the WSN. Therefore, research into energy saving to prolong the network lifetime has become one of the most significant issues in WSN. Moreover, the energy saving in WSN is a significant research topic in smart and sustainable energy systems and applications [7–9].

Due to the significance of the energy efficient problem in the WSN [10–12], many efforts have been made to tackle this problem. These proposed techniques are generally divided into two categories, where one is to design efficient protocols to reduce the energy consumption and the other one is to schedule the nodes to work efficiently. For the first category, protocols such as for the medium access control [13], for transmission [14], and for communication [15] have been proposed in the literature for reducing energy consumption. While the second categorized technique concentrates on scheduling the nodes, resulting in energy preservation and longer network lifetime. These two categories both have contributions to the energy efficiency for WSN from different approaches. The focus of this paper belongs to the second category, which is the energy efficient coverage (EEC) problem in the WSN [16,17].

In EEC, the sensor nodes are divided into different disjoint sets with the constraint that each set can guarantee the full coverage of the whole monitored area. In that way, in any time, only the necessary sensor nodes in one set are activated while the other sensor nodes that monitor the same regions can be turned off. These different disjoint sets work one by one and therefore the network lifetime can be prolonged. Moreover, with more disjoint sets that can be formed, longer network lifetime can be obtained because the nodes are scheduled to work energy efficiently [18–20]. Therefore, it is very interesting and promising for the approaches in [21–23] to divide the deployed sensor nodes into maximal disjoint sets and schedule the sets to work in turn. In this paper, we also focus on maximizing the number of disjoint sets.

Even though the above methods have been applied to the EEC problem [21–23], they can easily get trapped into local optima and cannot achieve or guarantee the full coverage because the EEC problem is NPC [21]. Therefore, in this paper, we proposed distributed genetic algorithm (DGA) to solve the EEC problems in WSN to further improve the performance due to the superior adaptation and strong global search ability of GA. Moreover, two major novel designs and advantages of DGA are described as follows:

(1) DGA is realized in the multi-processor distributed environment by the master-slave distributed model, where a set of processors run distributed to evaluate the fitness values in parallel to reduce the computational cost.

(2) When we evaluate a chromosome, different from the traditional model of EEC problem in WSN that only calculates the number of disjoint sets, a hierarchical fitness evaluation mechanism is proposed and a two-level fitness function with biased attention to the sets with larger coverage percentage is designed.

Therefore, not only do we have the innovations in algorithm, but also have the contributions on the model of EEC problem in WSN. The experimental results show that our proposed DGA performs better than other state-of-the-art approaches in maximizing the number of disjoin sets.

The rest of this paper is organized as follows. Section 2 gives the problem description of the EEC in the WSN and reviews some related work. Section 3 proposes our methodology for solving EEC in the WSN by using DGA. Section 4 presents the experimental results between our approach and other state-of-the-art approaches in the literature. Conclusions are given in Section 5.

2. Energy Efficient Coverage Problem in WSN

The EEC problem is a fundamental and significant research topic in the WSN. In this section, we present the formulation of the EEC and review the related work on this problem.

2.1. Problem Formulation

Given an $L \times W$ (Length $\times$ Width) rectangle area A and D randomly deployed sensor nodes. The EEC is to divide the nodes into several disjoint sets and then schedule these sets to work one by one. As the nodes do not have to work all the time, the energy can be significantly preserved. The aim of the EEC is to maximize the number of disjoint sets, with the constraint that each set can provide the full coverage for the monitored area.

In order to know whether the sensor network provides full coverage in the area A, we assume that the location of each sensor node is known in advance [19,24]. Moreover, the area is divided into many grids and the coverage issue can be transformed to check whether each grid is covered by at least one active sensor node [19].

All the D sensors form the sensors set $S = \{s_1, s_2, \ldots, s_D\}$, where each sensor node s_i is with the location (x_i, y_i) and the sensor radius R. For any grid $g(x, y) \in A$ in the monitored area, the relationship between the s_i and the g is defined as:

$$P(s_i, g) = \begin{cases} 1, \text{ if } (x - x_i)^2 + (y - y_i)^2 \leq R^2 \\ 0, \text{ otherwise} \end{cases} \tag{1}$$

where 1 means that the grid g is covered by the sensor s_i while 0 means the sensor s_i does not cover the grid g. Therefore, for any grid g in the monitored area, if there exist at least one sensor $s_i (1 \leq i \leq D)$ that makes $P(s_i, g) = 1$, we say that A is fully covered by the sensor network.

In the EEC, the set S is divided into M subsets S_j $(1 \leq j \leq M)$, and with the objective of maximizing the value of M, as:

$$\begin{aligned} & f = \max K \\ \textit{subject to} \quad & \text{(i) } \cup_{i=1}^{K} S_i \subseteq S \\ & \text{(ii) } S_i \cap S_j = \Phi \\ & \text{(iii) } \left(\oplus_{s_j \in S_i} P(s_j, g) \right) = 1, \quad \forall g \in A \end{aligned} \tag{2}$$

Here, the constraint (i) means that the unitization of all the subsets S_j must belong to the original set S. The (ii) indicates that there is no intersection between any two different subsets S_{j1} and S_{j2}. The (iii) represents that for any grid g in area A, there exists at least one sensor s_i in S_j which can cover the grid g. Obviously, these constraints can guarantee that each subset S_j can fully cover the monitored area.

2.2. Related Work

With the development of evolutionary computations (ECs) like GA [25,26], ant colony optimization (ACO) [27,28], particle swarm optimization (PSO) [29,30], and differential evolution (DE) [31,32], many researchers have applied ECs into solving the EEC problems in WSN, such as PSO-based [33] and ACO-based [34] approaches. Specifically, in [33], Zhan et al. extended the binary PSO (BPSO) to solve the EEC problem by finding a minimal set of nodes again and again to maximize the number of disjoint sets. Lin et al. [34] proposed the ACO-based approach to maximize the number of connected covers, called ACO-MNCC, to maximize the lifetime of heterogeneous WSNs by transforming the search space of the problem into a construction graph. Besides, in [35], Yang et al. proposed a probabilistic model to tackle the EEC problem in WSN, which transforms the area converge into point converge. It greatly reduces the dimension of problem. Lee et al. [36] also tried to solve the EEC problem in the WSN by using the ACO, which designs three pheromones to balance the local exploitation and global exploration. Meanwhile, they also introduced a probabilistic sensor detection model, which makes the algorithms more realistic [37].

Some researchers have also used GA for solving the EEC in the WSN recently. For example, Lai et al. [25] proposed a GA-based method to maximize the disjoint sensor covers called GAMDSC. However, there is still much room for improvement. First, their works only solved the small-scale problems (up to 140 sensors). When the scales of problems increase, their performance often face difficulties. Second, parallel or distributed methods can further improve the performance and save the computational time.

3. Methodology: DGA for the EEC Problem

In this section, the DGA for solving the EEC problem in WSN is implemented and described. The master-slave distributed framework is first given. Then the chromosome representation and the hierarchical fitness evaluation mechanism are described. After that, the genetic operators, including the selection, crossover, and mutation are proposed. At last, the overall algorithm is presented.

3.1. Distributed Framework of DGA

In DGA, the master-slave distributed model is used. The processor whose role is to control the whole evolutionary process including crossover, mutation, and selection operations, is called MASTER. The other processors, which are responsible for chromosomes' evaluation to reduce the computational costs, act as SLAVEs. To give an overview of the DGA, the master-slave distributed structure is depicted in Figure 1. In each generation, the master sends chromosomes to slaves, while the slaves evaluate the chromosomes and return their corresponding fitness values back to the master [38].

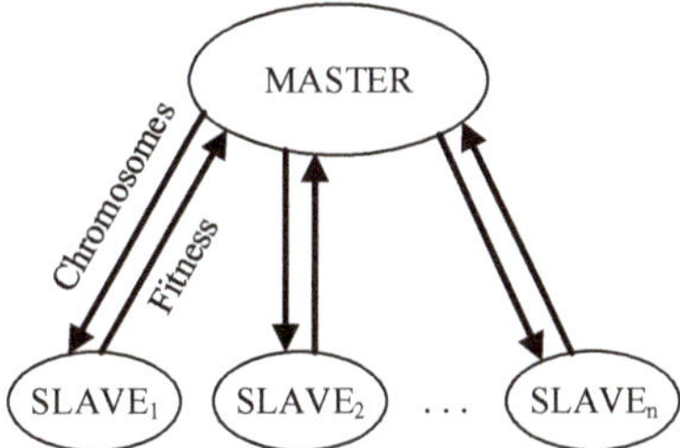

Figure 1. The distributed framework of the distributed genetic algorithm (DGA).

3.2. Chromosome Representation

The aim of the EEC problem is to maximize the number of disjoint sets. Therefore, the first thing is to determine how many sets the nodes can be divided into at most. This is also the upper bound of the disjoint sets number K [23]. Since the area has been divided into many grids, for each grid i, we find out the number of sensors k_i that cover the grid i. Then, the minimal k_i is the upper bound number of disjoint sets, i.e., $K = \min\{k_i\}$.

With the upper bound K determined, we can encode the chromosome in a very intuitive manner that each gene is an integer in the range of [1, K]. Therefore, the chromosome is a string of integers with the length of D, where D is the number of sensor nodes. The representation is

$$X = [x_1, x_2, \ldots, x_D], \text{ where } x_i \in Z \text{ and } 1 \le x_i \le K \tag{3}$$

In (3), the value of x_i denotes which set the ith sensor belongs to. Figure 2 gives two examples of the chromosome representation. In the examples, D is 4 and K is 2. In Figure 2a, the chromosome is X^a = [1, 2, 1, 2], meaning that the first and the third sensors belong to the 1st set while the second and the fourth sensors belong to the 2nd set. In this case, it can be observed that both the 1st and the 2nd sets can provide full coverage for the monitored area. However, in Figure 2b, the chromosome is X^b = [1, 2, 1, 1], meaning that the first, the third, and the fourth sensors belong to the 1st set while only the second sensor belongs to the 2nd set. In this case, only the 1st set can provide full coverage for the monitored area, but the 2nd set cannot.

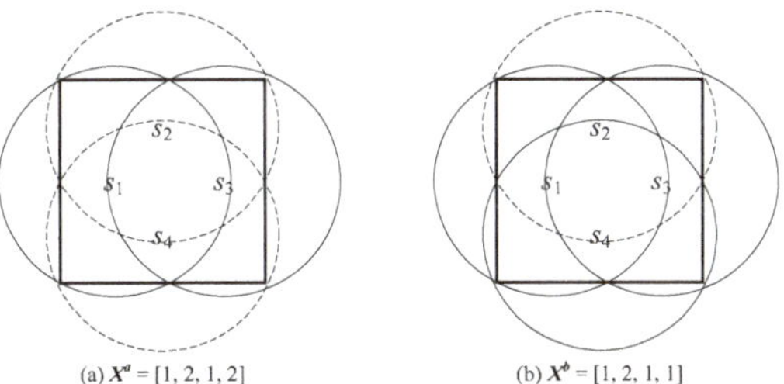

Figure 2. Illustration of the chromosome representation.

3.3. Hierarchical Fitness Evaluation

In order to evaluate the chromosome, herein, a hierarchical fitness evaluation mechanism and a two-level fitness function are proposed.

First, the genes are grouped into different sets according to their values. Then a check procedure is performed on each set to see whether the nodes in the set can provide full coverage. As we divide the monitored area into $L \times W$ grids, for each set S_i, we can count the grids number G that is covered by the sensor nodes in set S_i. Therefore, we can obtain the coverage percentage f_i as:

$$f_i = \frac{G}{L \times W} \tag{4}$$

Note that if f_i is equal to 1, it means that the set S_i can provide full coverage for the monitored area. In this way, we can count and obtain the number of disjoint sets (denoted as M) in the population which can achieve the full coverage. When two chromosomes A and B are compared, A wins B if A has a larger disjoint set number (larger M). The number of disjoint sets M is regarded as the first-level fitness evaluation.

However, to meet the case where the two chromosomes A and B have the same disjoint set number, we further designed a fitness function with biased attention to the sets with larger coverage percentage, so called the second-level fitness evaluation, which is calculated as:

$$F = \sum_{i=1}^{K} w_i \times f_i \tag{5}$$

where K is the upper bound of the disjoint sets and w_i is the weight for the set S_i. Herein, we give the biased attention to the sets with larger coverage percentage. That is because the set with a higher coverage percentage has a higher probability to achieve the full coverage. Therefore, we first sort the sets according to their coverage percentages in descending order, then the weight w_i for the set S_i is set as w_i = 10,000.0/r_i, where r_i is the rank of the set S_i in the sort of coverage percentage.

In that way, if two chromosomes A and B have the same disjoint set number, then the one with a larger fitness value is preferred (larger F).

3.4. Genetic Operators

The genetic operators in DGA consist of the selection, crossover, and mutation. In this subsection, we will briefly describe the implementations of these operators.

3.4.1. Selection

In the selection implementation, we use the tournament selection strategy. Specifically, in each selection, partial of the chromosomes are randomly selected to compete for survival. The winner enters into the next generation. Repeat the selection operator until a population size of chromosomes have been selected.

3.4.2. Crossover

After the selection, the survived chromosomes recombine to create offspring. First, for each chromosome, a random value in range [0, 1] is generated. If the value is smaller than the crossover probability p_c, then the chromosome is used as one of the parents. After all the parents have been determined, every two parents are randomly mated to create two offspring. A crossover position k is randomly generated and the genes from the kth position of the two parents are exchanged, as shown in Figure 3a. In this way, two new chromosomes, so called offspring are generated and enter the population.

3.4.3. Mutation

The chromosomes will perform a mutation operation after the crossover. For each gene in a chromosome, a random value in range [0, 1] is generated and compared with the mutation probability p_m. If the random value is smaller than p_m, then the mutation occurs. As shown in Figure 3b, when the mutation occurs, the gene is set as a random integer in the range [1, K].

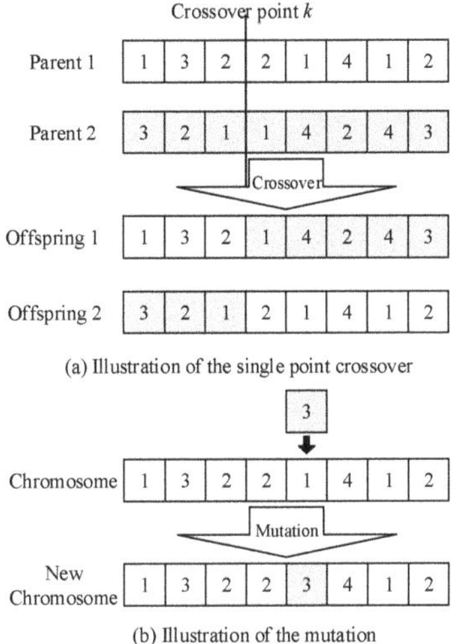

Figure 3. The illustration of the crossover and mutation operators.

3.5. Complete Algorithm

With the designs of the chromosome representation, fitness function, and the genetic operators, the DGA for solving the EEC problem is shown in Figure 4 and is described as the following seven steps.

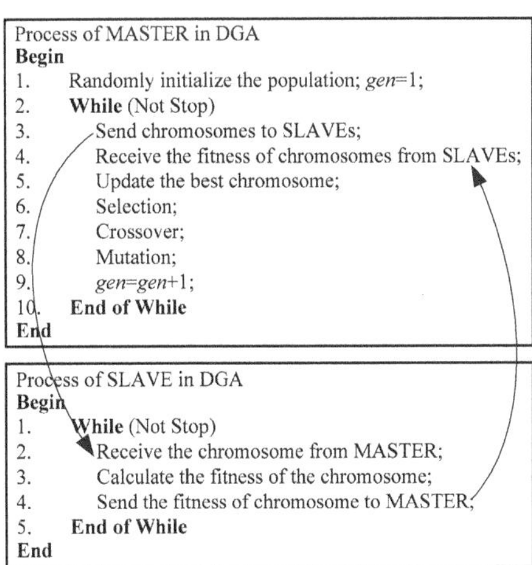

Figure 4. The flowchart of the DGA for solving the energy efficient coverage (EEC) problem in the wireless sensor networks (WSN).

Step 1: Initialization. The N chromosomes in the population are initialized at master. Each gene in the chromosome is set as a random integer value in the range [1, K] as (3).

Step 2: Evaluation. Each chromosome is evaluated as (5) at the slave and returns its fitness to the master. Then the best chromosome with the highest fitness can be determined. This chromosome is compared with the historical best one and will replace the historical best one if it is better. Otherwise, the historical best chromosome will replace the worst chromosome in the current generation.

Step 3: Selection. The selection operation has been discussed in Section 3.4.1.

Step 4: Crossover. The crossover operation has been discussed in Section 3.4.2.

Step 5: Mutation. The mutation operation has been discussed in Section 3.4.3.

Step 6: Termination check. If the termination criteria is met, go to Step 7. Herein, the termination criteria include the obtaining of the upper bound K or reaching the maximal generation number. Otherwise, the algorithm goes to Step 2 and continues the next generation evolution.

Step 7: Terminate. Output the best chromosome as the final solution.

3.6. Computational Complexity

Herein, we denote the population size and the dimension of problem (the number of sensors) as N and D, respectively. First, in the initialization, the computational complexity of DGA is $O(N \times D)$, which is obtained by step 1 in MASTER process of Figure 4. Then, as for the chromosomes evaluation, the computational complexity is $O(N)$, as obtained by steps 3–4 in the MASTER process and steps 2–4 in the SLAVE process of Figure 4. After that, in the chromosome evolution, since we used the tournament selection, the computational complexity is $O(N^2 \times D)$, as obtained by steps 6–8 in the MASTER process of Figure 4. Therefore, the overall computational complexity of DGA is $O(N^2 \times D)$.

4. Experiments and Comparisons

4.1. Algorithms Configurations

The parameter configurations of the DGA are listed as Table 1 and are described as follows.

Table 1. The Parameters Configurations of DGA.

Parameter	Configuration
N	40
Generation	200
p_c	0.8
p_m	0.01

The population size is 40. Several processors are used here to evaluate the fitness of chromosomes. In each tournament selection round, 20% of the chromosomes are randomly chosen and compete for survival. The crossover probability p_c and mutation probability p_m are 0.8 and 0.01, respectively. The algorithm terminates at the maximal generations of 200 or it obtains the upper bound K of the disjoint sets number.

4.2. Experimental Results and Comparisons

In our experiments, we choose several EA-based algorithms, including a PSO-based method (BPSO [33]) and a GA-based method (GAMDSC [25]) to compare with our proposed DGA. To have a reliable and fair comparison, the parameter configurations of all the competitor algorithms are set the same as suggested in their original papers.

Herein, the network topology configurations used in [19] are adopted. That is, with the monitored area 50 m by 50 m, the sensing range R spans 8 m, 10 m, and 12 m, whilst the deployed nodes number N can be 100, 150, 200, 250, and 300. For each case with N original deployed nodes, we randomly deploy the D nodes in the network to form the topology. Note that the D nodes must provide the full coverage. Otherwise, the nodes are randomly deployed again until the network is fully covered. On each case, we generate three different random topologies. Therefore, there are totally $3 \times 3 \times 5 = 45$ cases in this experiment. Due to the stochastic characteristic of the algorithms, we simulate each case on each topology 10 times and the mean results are recorded for statistics. All the approaches are dealing with the same test environments, resulting in a fair comparison.

The simulation results are presented and compared in Table 2. For clarity, the best results are highlighted in boldface. The results show that the DGA outperforms the other approaches on most of the cases. Also, DGA can maximize the number of disjoin sets and obtain the upper bound K in most of the cases, except cases 34, 42–44, while other algorithms can only find few disjoin sets. Moreover, with the increase in the number of nodes, especially in cases 37–45, the superiority of DGA is increasingly obvious, which further indicates the dominance of DGA when solving the complicated EEC problem in WSN.

Table 2. Experimental Results of Comparisons.

Case No.	Topology		Upper Bound K	All Combinations	Mean Disjoint Set Number M			*Error*		
	Node	Range			GAMDSC	BPSO	DGA	GAMDSC	BPSO	DGA
1			2	2^100	**2.000**	**2.000**	**2.000**	**0.000**	**0.000**	**0.000**
2	100	8	2	2^100	**2.000**	**2.000**	**2.000**	**0.000**	**0.000**	**0.000**
3			2	2^100	**2.000**	**2.000**	**2.000**	**0.000**	**0.000**	**0.000**
4			2	2^100	**2.000**	**2.000**	**2.000**	**0.000**	**0.000**	**0.000**
5	100	10	2	2^100	**2.000**	**2.000**	**2.000**	**0.000**	**0.000**	**0.000**
6			2	2^100	**2.000**	**2.000**	**2.000**	**0.000**	**0.000**	**0.000**
7			2	2^100	**2.000**	**2.000**	**2.000**	**0.000**	**0.000**	**0.000**
8	100	12	2	2^100	**2.000**	**2.000**	**2.000**	**0.000**	**0.000**	**0.000**
9			4	4^100	3.900	3.600	**4.000**	0.025	0.100	**0.000**
10			2	2^150	**2.000**	**2.000**	**2.000**	**0.000**	**0.000**	**0.000**
11	150	8	2	2^150	**2.000**	**2.000**	**2.000**	**0.000**	**0.000**	**0.000**
12			2	2^150	**2.000**	**2.000**	**2.000**	**0.000**	**0.000**	**0.000**
13			3	3^150	**3.000**	**3.000**	**3.000**	**0.000**	**0.000**	**0.000**
14	150	10	4	4^150	3.500	3.200	**4.000**	0.125	0.200	**0.000**
15			3	3^150	**3.000**	**3.000**	**3.000**	**0.000**	**0.000**	**0.000**
16			5	5^150	4.200	**5.000**	**5.000**	**0.000**	**0.000**	**0.000**
17	150	12	3	3^150	**3.000**	**3.000**	**3.000**	**0.000**	**0.000**	**0.000**
18			4	4^150	3.800	**4.000**	**4.000**	0.050	**0.000**	**0.000**
19			2	2^200	**2.000**	**2.000**	**2.000**	**0.000**	**0.000**	**0.000**
20	200	8	2	2^200	**2.000**	**2.000**	**2.000**	**0.000**	**0.000**	**0.000**
21			3	3^200	**3.000**	**3.000**	**3.000**	**0.000**	**0.000**	**0.000**
22			3	3^200	**3.000**	**3.000**	**3.000**	**0.000**	**0.000**	**0.000**
23	200	10	4	4^200	3.500	3.300	**4.000**	0.125	0.175	**0.000**
24			5	5^200	**5.000**	4.600	**5.000**	**0.000**	0.080	**0.000**
25			7	7^200	5.900	6.200	**7.000**	0.157	0.114	**0.000**
26	200	12	8	8^200	7.200	6.600	**8.000**	0.100	0.175	**0.000**
27			9	9^200	7.300	7.500	**9.000**	0.189	0.166	**0.000**
28			3	3^250	**3.000**	**3.000**	**3.000**	**0.000**	**0.000**	**0.000**
29	250	8	3	3^250	**3.000**	**3.000**	**3.000**	**0.000**	**0.000**	**0.000**
30			5	5^250	4.900	4.100	**5.000**	0.020	0.180	**0.000**
31			5	5^250	**5.000**	**5.000**	**5.000**	**0.000**	**0.000**	**0.000**
32	250	10	7	7^250	6.300	6.600	**7.000**	0.100	0.057	**0.000**
33			6	6^250	5.700	5.900	**6.000**	0.050	0.016	**0.000**
34			11	11^250	9.900	9.700	10.00	0.100	0.118	0.091
35	250	12	9	9^250	8.600	8.700	**9.000**	0.044	0.033	**0.000**
36			8	8^250	**8.000**	**8.000**	**8.000**	**0.000**	**0.000**	**0.000**
37			6	6^300	**6.000**	**6.000**	**6.000**	**0.000**	**0.000**	**0.000**
38	300	8	3	3^300	**3.000**	**3.000**	**3.000**	**0.000**	**0.000**	**0.000**
39			5	5^300	**5.000**	**5.000**	**5.000**	**0.000**	**0.000**	**0.000**
40			7	7^300	**7.000**	6.600	**7.000**	**0.000**	0.057	**0.000**
41	300	10	7	7^300	**7.000**	6.800	**7.000**	**0.000**	0.029	**0.000**
42			9	9^300	8.300	7.400	**8.400**	0.100	0.178	**0.067**
43			11	11^300	**10.60**	10.00	10.10	**0.036**	0.091	0.082
44	300	12	12	12^300	10.80	10.90	**11.30**	0.100	0.092	**0.058**
45			9	9^300	8.800	8.600	**9.000**	0.022	0.044	**0.000**

We also compare the *Error* values of these approaches when dealing with the EEC problem. The *Error* is calculated as

$$Error = (K - M)/K \quad (6)$$

where K is the upper bound in each case and M is the mean disjoint set number obtained by the approach. The *Error* values of DGA are smaller than 0.1 in all the cases and most of the *Error* values are 0, indicating the DGA is very promising in producing high quality solutions.

4.3. Effects of Parameters

The DGA involves two parameters, the crossover probability p_c and the mutation probability p_m. In this part, we investigate the two parameters based on the same topologies as in Section 4.2.

First, the p_c is set as 0.8 and the p_m varies from 0.01 to 0.09. The mean *Error* values of the 45 cases with different p_m are plotted in Figure 5. As we can see, the small value of p_m generally performs better than the large value of p_m. That may because the larger p_m destroys the convergence ability, which makes the algorithm act like a random search and cannot converge to a promising region. Therefore, the investigated results indicate that the DGA obtains the best performance when the p_m is set as 0.01.

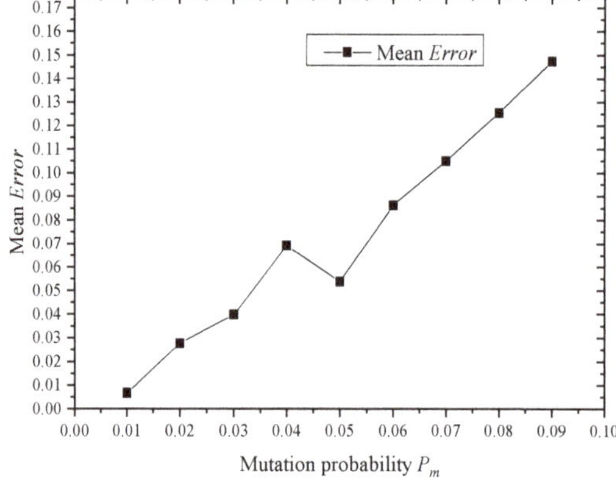

Figure 5. The mean Error values obtained by the DGA with different p_m.

Then, the p_c is investigated, with the p_m set as 0.01. The value of p_c spans from 0.1 to 0.9, with the step 0.1. The mean *Error* values of the 45 cases with different p_c are plotted in the Figure 6. It can be observed that the large value of p_c may be generally better than the small value of p_c. This may be due to large p_c can making the good information of the chromosome spread fast in the population. This is very helpful for the algorithm to use good information to search in the promising region, and therefore results in better performance. However, a too-large p_c value will sometimes make the algorithm converge too fast and be trapped into local optima. For example, when $p_c = 0.9$, the mean *Error* values are worse than that when $p_c = 0.8$. In general, $p_c = 0.8$ is promising for the DGA to obtain satisfying results.

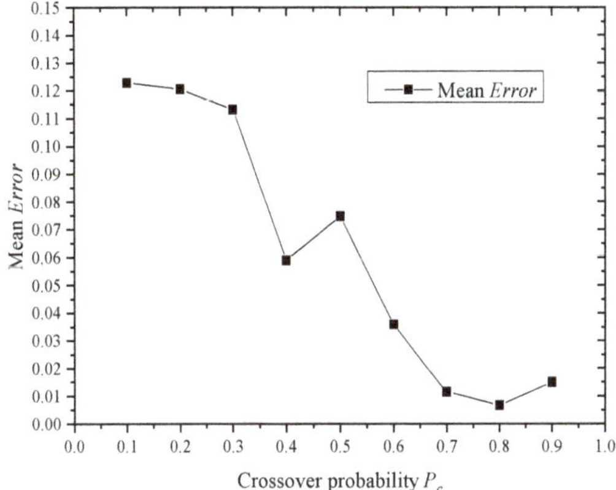

Figure 6. The mean *Error* values obtained by the DGA with different p_c.

4.4. Speedup Ratio

Another advantage of DGA is the parallel implementation, where several processors in DGA are used to evaluate the fitness of chromosomes independently. Table 3 shows the computational times of DGA on all the 45 cases with different processors.

Speedup ratio is the metric to measure the performance of a distributed or parallel algorithm, and it varies with the numbers of processors and the computation cost of the case. Herein, we use the small-scale case 1 and the large-scale case 45 as two representative instances. We test the speedup of DGA on these two cases and draw the speedup curve with the increasing processors, from 2 to 10, shown in Figure 7.

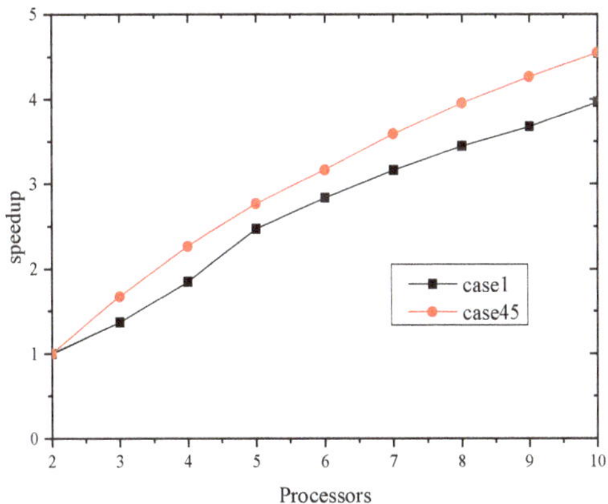

Figure 7. Speedup ratios achieved by DGA on two typical cases.

As we can see, the speedup ratio consistently grows with the increase of the number of processors on both of these two cases. However, the increase of speedup ratio slows down when the number of processors increases from 6 to 10. That may due to the huge communication cost among many processors. Even so, the speedup ratio still keeps growing with the increasing of the number of processors. Moreover, the speedup ratio of case 45 is larger the case 1. That may because case 45 is more complicated than case 1, which will cause more computation cost than case 1. Therefore, under a similar communication cost, the higher computation cost will directly lead to a higher speedup ratio in case 45.

Table 3. Computational time of DGA (s).

Case No. \ Processors	2	3	4	5	6	7	8	9	10
1	101.53	74.26	54.91	41.07	35.78	32.09	29.45	27.62	25.59
2	123.24	90.50	71.44	57.29	53.83	50.38	46.36	42.50	39.85
3	133.24	96.64	77.30	69.96	62.33	59.80	54.87	50.64	48.59
4	116.77	85.19	70.76	63.55	56.58	52.39	48.84	43.78	39.49
5	147.74	100.36	87.53	73.46	65.86	59.73	53.92	50.05	48.74
6	132.90	95.19	83.47	71.41	64.11	58.53	53.29	49.39	47.98
7	158.20	107.27	88.06	76.62	68.54	63.16	56.23	52.28	49.77
8	144.34	103.50	86.77	77.06	68.96	62.38	57.05	52.48	49.21
9	258.93	169.97	130.72	101.29	89.81	80.90	72.86	65.29	61.41
10	188.84	120.19	93.42	82.38	71.84	64.46	56.94	50.00	47.80
11	178.36	118.36	93.21	81.42	70.64	63.89	56.37	49.08	46.32
12	206.96	128.08	98.24	84.18	73.99	65.19	58.03	53.99	50.02
13	253.12	142.81	110.06	94.95	84.88	77.37	73.10	69.00	67.81
14	321.11	205.79	158.70	139.89	126.43	113.42	100.45	90.67	84.10
15	272.37	164.89	142.60	123.77	105.18	90.94	78.09	69.07	63.82
16	386.60	267.58	201.73	175.17	139.95	123.07	110.80	100.43	94.91
17	284.46	214.86	151.65	123.60	104.80	90.44	88.16	79.07	73.94
18	308.81	223.46	176.00	143.88	122.89	110.73	97.94	89.17	82.92
19	267.02	178.39	145.59	123.05	101.54	90.55	81.90	74.43	69.95
20	284.36	193.04	152.46	125.60	104.22	95.95	86.24	79.70	74.18
21	349.97	249.11	185.50	145.96	120.79	108.76	96.98	88.63	83.98
22	351.04	246.52	189.91	156.78	136.28	122.45	109.88	102.45	96.40
23	445.03	323.38	236.05	196.38	167.02	147.60	129.79	120.12	115.23
24	533.63	354.34	286.87	242.26	219.79	189.75	162.89	147.38	139.96
25	627.89	457.94	346.32	271.15	224.81	198.03	176.26	166.13	157.15
26	682.48	475.31	363.82	288.01	236.70	209.82	189.36	175.76	170.80
27	744.13	501.08	389.95	305.20	248.85	228.48	216.66	206.95	199.67
28	436.20	281.56	233.98	198.82	165.39	145.43	134.42	119.83	112.49
29	412.85	265.29	215.59	194.15	169.05	148.05	133.49	120.88	112.55
30	563.58	378.05	295.38	245.90	202.31	171.01	156.44	148.80	141.01
31	591.87	380.09	308.38	254.81	214.68	189.08	169.17	161.44	156.99
32	680.79	485.09	354.86	278.55	236.14	201.17	188.06	179.64	173.99
33	625.32	454.75	336.59	257.97	224.26	200.25	184.99	171.09	166.34
34	1687.20	1153.09	916.81	799.34	684.47	585.99	509.48	451.04	424.53
35	871.04	578.18	449.48	388.74	343.93	309.37	264.36	241.62	225.94
36	778.21	560.31	401.16	372.50	327.59	290.98	262.23	236.73	220.14
37	1014.52	697.61	580.35	502.69	439.96	384.46	321.52	288.86	272.87
38	836.59	580.41	446.17	385.78	332.98	294.35	251.36	222.99	210.33
39	958.63	635.55	528.04	418.04	350.24	303.71	264.91	244.66	233.68
40	1302.70	825.53	671.78	598.59	522.32	451.68	400.26	363.22	342.84
41	1412.63	915.99	732.56	603.58	508.87	449.00	414.48	382.68	367.28
42	2297.81	1542.50	1208.43	1048.97	915.04	801.93	713.63	638.74	586.86
43	2416.68	1651.56	1217.74	1082.13	951.69	837.25	732.56	651.62	606.25
44	2530.80	1754.12	1317.12	1181.00	1008.09	886.37	782.98	696.21	634.14
45	1916.47	1144.10	845.52	692.63	605.70	534.16	484.21	449.11	421.29

5. Conclusions

In this paper, DGA is proposed to solve the EEC problem in the WSN to achieve or guarantee the full converge. In the implementation, DGA is realized in the multi-processor distributed environment, where a set of processors run distributed to evaluate the fitness values in parallel to reduce the computational cost. Moreover, when we evaluate a chromosome, different from the traditional model of EEC problem in WSN that only calculates the number of disjoint sets, hierarchical fitness evaluation mechanism is proposed and a two-level fitness function with biased attention to the sets with larger coverage percentage is further designed. Therefore, not only do we have the innovations in the algorithm, but we also have the contributions to the model of the EEC problem in WSN. Simulations have been conducted and the experimental results confirm the effectiveness and efficiency of the proposed approach when compared with the state-of-the-art approaches.

For future work, we wish to further improve the performance of DGA by considering the potential uncertainties, so as to make DGA more applicable in the real applications. Moreover, we wish to apply the DGA into other more complicated optimization problems, like data routing [39], cloud resource scheduling [40], supply chain managing, even in a dynamic and multi-objective environment [41,42], not only in the WSN.

Author Contributions: Z.-J.W. conducted the experiments, performed the experiments, and wrote the draft of this paper. The idea of this paper was proposed by Z.-H.Z., who also made contributions to help organize and write the paper, together with J.Z.

Funding: This work was partially supported by the Outstanding Youth Science Foundation with No. 61822602, the National Natural Science Foundations of China (NSFC) with No. 61772207 and 61332002, the Natural Science Foundations of Guangdong Province for Distinguished Young Scholars with No. 2014A030306038, the Project for Pearl River New Star in Science and Technology with No. 201506010047, the GDUPS (2016), and the Fundamental Research Funds for the Central Universities.

Acknowledgments: The authors would like to thank the assistant editor and reviewers for their valuable comments and suggestions that improved the paper's quality.

Conflicts of Interest: The authors declare no conflict of interest.

Nomenclature

DGA	distributed genetic algorithm
EEC	energy efficient coverage
WSN	wireless sensor networks
D	the number of sensor nodes
M	the number of disjoint subsets
K	the upper bound of the disjoint sets number
N	the number of chromosomes
p_c	crossover probability
p_m	mutation probability

References

1. Guevara, J.; Barrero, F.; Vargas, E.; Becerra, J.; Toral, S. Environmental wireless sensor network for road traffic applications. *IET Intell. Transp. Syst.* **2012**, *6*, 177–186. [CrossRef]
2. Wang, C.; Guo, S.; Yang, Y. An optimization framework for mobile data collection in energy-harvesting wireless sensor networks. *IEEE Trans. Parallel Distrib. Syst.* **2016**, *15*, 2969–2986. [CrossRef]
3. Martinez, K.; Hart, J.K.; Ong, R. Environmental sensor networks. *IEEE Comput. Soc.* **2004**, *37*, 50–56. [CrossRef]
4. Yahya, A.; Islam, S.; Akhunzada, A.; Ahmed, G.; Shamshirband, S.; Lloret, J. Towards efficient sink mobility in underwater wireless sensor networks. *Energies* **2018**, *11*, 1471. [CrossRef]
5. Al-Jaoufi, M.A.A.; Liu, Y.; Zhang, Z.J. An active defense model with low power consumption and deviation for wireless sensor networks utilizing evolutionary game theory. *Energies* **2018**, *11*, 1281. [CrossRef]
6. Shehadeh, H.A.; Idris, M.Y.I.; Ahmedy, I.; Ramli, R.; Noor, N.M. The multi-objective optimization algorithm based on sperm fertilization procedure (MOSFP) method for solving wireless sensor networks optimization problems in smart grid applications. *Energies* **2018**, *11*, 97. [CrossRef]
7. Marzband, M.; Azarinejadian, F.; Savaghebi, M.; Pouresmaeil, E.; Guerrero, J.M.; Lightbody, G. Smart transactive energy framework in grid-connected multiple home microgrids under independent and coalition operations. *Renew. Energy* **2018**, *126*, 95–106. [CrossRef]
8. Marzband, M.; Fouladfar, M.H.; Akorede, M.F.; Lightbody, G.; Pouresmaeil, E. Framework for smart transactive energy in home-microgrids considering coalition formation and demand side management. *Sustain. Cities Soc.* **2018**, *40*, 136–154. [CrossRef]
9. Tavakoli, M.; Shokridehaki, F.; Akorede, M.F.; Marzband, M.; Vechiu, I.; Pouresmaeil, E. CVaR-based energy management scheme for optimal resilience and operational cost in commercial building microgrids. *Int. J. Electr. Power Energy Syst.* **2018**, *100*, 1–9. [CrossRef]

10. Marzband, M.; Javadi, M.; Pourmousavi, S.A.; Lightbody, G. An advanced retail electricity market for active distribution systems and home microgrid interoperability based on game theory. *Electr. Power Syst. Res.* **2018**, *157*, 187–199. [CrossRef]
11. Tavakoli, M.; Shokridehaki, F.; Marzband, M.; Godina, R.; Pouresmaeil, E. A two stage hierarchical control approach for the optimal energy management in commercial building microgrids based on local wind power and PEVs. *Sustain. Cities Soc.* **2018**, *41*, 332–340. [CrossRef]
12. Marzband, M.; Sumper, A.; Domínguez-García, J.L.; Gumara-Ferreta, R. Experimental validation of a real time energy management system for microgrids in islanded mode using a local day-ahead electricity market and MINLP. *Energy Convers. Manag.* **2013**, *76*, 314–322. [CrossRef]
13. Rajendran, V.; Obraczka, K.; Garcia-Luna-Aceves, J.J. Energy-efficient collision-free medium access control for wireless sensor networks. *Wirel. Netw.* **2006**, *12*, 63–78. [CrossRef]
14. Chlamtac, I.; Petrioli, C.; Redi, J. Energy-conserving access protocols for identification networks. *IEEE/ACM Trans. Netw.* **1999**, *7*, 51–59. [CrossRef]
15. Stine, J.; Veciana, G. Improving energy efficiency of centrally controlled wireless data networks. *Wirel. Netw.* **2002**, *8*, 681–700. [CrossRef]
16. Huang, C.F.; Tseng, Y.C. A survey of solutions to the coverage problems in wireless sensor networks. *J. Internet Technol.* **2005**, *6*, 1–8.
17. Ammari, H.M. Investigating the energy sink-hole problem in connected-covered wireless sensor networks. *IEEE Trans. Comput.* **2014**, *63*, 2729–2742. [CrossRef]
18. Cardei, M.; Wu, J. Energy-efficient coverage problems in wireless ad-hoc sensor networks. *Comput. Commun.* **2006**, *29*, 413–420. [CrossRef]
19. Tian, D.; Georganas, N.D. A node scheduling scheme for energy conservation in large wireless sensor networks. *Wirel. Commun. Mob. Comput.* **2003**, *3*, 271–290. [CrossRef]
20. Zhan, Z.H.; Zhang, J.; Fan, Z. Solving the optimal coverage problem in wireless sensor networks using evolutionary computation algorithms. In Proceedings of the International Conference on Simulated Evolution and Learning, Kanpur, India, 1–4 December 2010; pp. 166–176.
21. Slijepcevic, S.; Potkonjak, M. Power efficient organization of wireless sensor networks. In Proceedings of the IEEE International Conference on Communications, Helsinki, Finland, 11–14 June 2001.
22. Cardei, M.; MacCallum, D.; Cheng, X.; Min, M.; Jia, X.; Li, D.; Du, D.Z. Wireless sensor networks with energy efficient organization. *J. Interconnect. Netw.* **2002**, *3*, 213–229. [CrossRef]
23. Zhang, X.Y.; Zhang, J.; Gong, Y.J.; Zhan, Z.H.; Chen, W.N.; Li, Y. Kuhn-munkres parallel genetic algorithm for the set cover problem and its application to large-scale wireless sensor networks. *IEEE Trans. Evol. Comput.* **2016**, *20*, 695–710. [CrossRef]
24. Li, X.Y.; Wan, P.J.; Frieder, O. Coverage in wireless and ad-hoc sensor networks. *IEEE Trans. Comput.* **2003**, *52*, 753–762. [CrossRef]
25. Lai, C.C.; Ting, C.K.; Ko, R.S. An effective genetic algorithm to improve wireless sensor network lifetime for large-scale surveillance applications. In Proceedings of the IEEE Congress on Evolutionary Computation, Singapore, 25–28 September 2007.
26. Kampelis, N.; Tsekeri, E.; Kolokotsa, D.; Kalaitzakis, K.; Isidori, D.; Cristina, C. 3 Development of demand response energy management optimization at building and district levels using genetic algorithm and artificial neural network modelling power predictions. *Energies* **2018**, *11*, 3012. [CrossRef]
27. Liu, X.F.; Zhan, Z.H.; Deng, D.; Li, Y.; Gu, T.L.; Zhang, J. An energy efficient ant colony system for virtual machine placement in cloud computing. *IEEE Trans. Evol. Comput.* **2018**, *22*, 113–128. [CrossRef]
28. Chen, Z.G.; Zhan, Z.H.; Lin, Y.; Gong, Y.J.; Gu, T.L.; Zhao, F.; Yuan, H.Q.; Chen, X.; Li, Q.; Zhang, J. Multiobjective cloud workflow scheduling: A multiple populations ant colony system approach. *IEEE Trans. Cybern.* **2018**. [CrossRef] [PubMed]
29. Li, Y.H.; Zhan, Z.H.; Lin, S.J.; Zhang, J.; Luo, X.N. Competitive and cooperative particle swarm optimization with information sharing mechanism for global optimization problems. *Inf. Sci.* **2015**, *293*, 370–382. [CrossRef]
30. Liu, X.F.; Zhan, Z.H.; Gao, Y.; Zhang, J.; Kwong, S.; Zhang, J. Coevolutionary particle swarm optimization with bottleneck objective learning strategy for many-objective optimization. *IEEE Trans. Evol. Comput.* **2018**. [CrossRef]

31. Wang, Z.J.; Zhan, Z.H.; Lin, Y.; Yu, W.J.; Yuan, H.Q.; Gu, T.L.; Kwong, S.; Zhang, J. Dual-strategy differential evolution with affinity propagation clustering for multimodal optimization problems. *IEEE Trans. Evol. Comput.* **2018**, *22*, 894–908. [CrossRef]
32. Liu, X.F.; Zhan, Z.H.; Lin, Y.; Chen, W.N.; Gong, Y.J.; Gu, T.L.; Yuan, H.Q.; Zhang, J. Historical and heuristic based adaptive differential evolution. *IEEE Trans. Syst. Man Cybern. Syst.* **2018**. [CrossRef]
33. Zhan, Z.H.; Zhang, J.; Du, K.J.; Xiao, J. Extended binary particle swarm optimization approach for disjoint set covers problem in wireless sensor networks. In Proceedings of the IEEE Conference on Technologies and Applications of Artificial Intelligence, Tainan, Taiwan, 16–18 November 2012.
34. Lin, Y.; Zhang, J.; Chung, H.S.H.; Ip, W.H.; Li, Y.; Shi, Y.H. An ant colony optimization approach for maximizing the lifetime of heterogeneous wireless sensor networks. *IEEE Trans. Syst. Man Cybern. Part C (Appl. Rev.)* **2012**, *42*, 408–420. [CrossRef]
35. Yang, Q.Q.; He, S.B.; Li, J.K.; Chen, J.M.; Sun, Y.X. Energy-efficient probabilistic area coverage in wireless sensor networks. *IEEE Trans. Veh. Technol.* **2015**, *64*, 367–377. [CrossRef]
36. Lee, J.W.; Choi, B.S.; Lee, J.J. Energy-efficient coverage of wireless sensor networks using ant colony optimization with three types of pheromones. *IEEE Trans. Ind. Informat.* **2011**, *7*, 419–427. [CrossRef]
37. Lee, J.W.; Lee, J.Y.; Lee, J.J. Jenga-inspired optimization algorithm for energy-efficient coverage of unstructured WSNs. *IEEE Wirel. Commun. Lett.* **2013**, *2*, 34–37. [CrossRef]
38. Zhan, Z.H.; Liu, X.F.; Zhang, H.; Yu, Z.; Weng, J.; Li, Y.; Gu, T.; Zhang, J. Cloudde: A heterogeneous differential evolution algorithm and its distributed cloud version. *IEEE Trans. Parallel Distrib. Syst.* **2017**, *28*, 704–716. [CrossRef]
39. Shen, M.; Zhan, Z.H.; Chen, W.N.; Gong, Y.J.; Zhang, J.; Li, Y. Bi-velocity discrete particle swarm optimization and its application to multicast routing problem in communication networks. *IEEE Trans. Ind. Electron.* **2014**, *61*, 7141–7151. [CrossRef]
40. Zhan, Z.H.; Liu, X.F.; Gong, Y.J.; Zhang, J.; Chung, H.S.H.; Li, Y. Cloud computing resource scheduling and a survey of its evolutionary approaches. *ACM Comput. Surv.* **2015**, *47*, 1–33. [CrossRef]
41. Liu, X.F.; Zhan, Z.H.; Zhang, J. Neural network for change direction prediction in dynamic optimization. *IEEE Access* **2018**, *6*, 72649–72662. [CrossRef]
42. Zhan, Z.H.; Li, J.; Cao, J.; Zhang, J.; Chung, H.; Shi, Y.H. Multiple populations for multiple objectives: A coevolutionary technique for solving multiobjective optimization problems. *IEEE Trans. Cybern.* **2013**, *43*, 445–463. [CrossRef] [PubMed]

Article

Short-Term Electric Load and Price Forecasting Using Enhanced Extreme Learning Machine Optimization in Smart Grids

Aqdas Naz [1], Muhammad Umar Javed [1], Nadeem Javaid [1,*], Tanzila Saba [2], Musaed Alhussein [3] and Khursheed Aurangzeb [3,*]

1 Department of Computer Science, COMSATS University Islamabad, Islamabad 44000, Pakistan; aqdasmalik17@gmail.com (A.N.); umarkhokhar1091@gmail.com (M.U.J.)
2 College of Computer and Information Systems, Al Yamamah University, Riyadh 11512, Saudi Arabia; tsaba@psu.edu.sa
3 Computer Engineering Department, College of Computer and Information Sciences, King Saud University, Riyadh 11543, Saudi Arabia; musaed@ksu.edu.sa
* Correspondence: nadeemjavaid@comsats.edu.pk (N.J.); kaurangzeb@ksu.edu.sa (K.A.)

Received: 1 February 2019; Accepted: 22 February 2019; Published: 5 March 2019

Abstract: A Smart Grid (SG) is a modernized grid to provide efficient, reliable and economic energy to the consumers. Energy is the most important resource in the world. An efficient energy distribution is required as smart devices are increasing dramatically. The forecasting of electricity consumption is supposed to be a major constituent to enhance the performance of SG. Various learning algorithms have been proposed to solve the forecasting problem. The sole purpose of this work is to predict the price and load efficiently. The first technique is Enhanced Logistic Regression (ELR) and the second technique is Enhanced Recurrent Extreme Learning Machine (ERELM). ELR is an enhanced form of Logistic Regression (LR), whereas, ERELM optimizes weights and biases using a Grey Wolf Optimizer (GWO). Classification and Regression Tree (CART), Relief-F and Recursive Feature Elimination (RFE) are used for feature selection and extraction. On the basis of selected features, classification is performed using ELR. Cross validation is done for ERELM using Monte Carlo and *K*-Fold methods. The simulations are performed on two different datasets. The first dataset, i.e., UMass Electric Dataset is multi-variate while the second dataset, i.e., UCI Dataset is uni-variate. The first proposed model performed better with UMass Electric Dataset than UCI Dataset and the accuracy of second model is better with UCI than UMass. The prediction accuracy is analyzed on the basis of four different performance metrics: Mean Absolute Percentage Error (MAPE), Mean Absolute Error (MAE), Mean Square Error (MSE) and Root Mean Square Error (RMSE). The proposed techniques are then compared with four benchmark schemes. The comparison is done to verify the adaptivity of the proposed techniques. The simulation results show that the proposed techniques outperformed benchmark schemes. The proposed techniques efficiently increased the prediction accuracy of load and price. However, the computational time is increased in both scenarios. ELR achieved almost 5% better results than Convolutional Neural Network (CNN) and almost 3% than LR. While, ERELM achieved almost 6% better results than ELM and almost 5% than RELM. However, the computational time is almost 20% increased with ELR and 50% with ERELM. Scalability is also addressed for the proposed techniques using half-yearly and yearly datasets. Simulation results show that ELR gives 5% better results while, ERELM gives 6% better results when used for yearly dataset.

Keywords: smart grid; forecasting; load; price; CNN; LR; ELR; RELM; ERELM

1. Introduction

For electricity generation and distribution, Traditional Grids (TGs) are used. The infrastructure of TG is getting obsolete, which results in energy loss and less efficient output. Due to the usage of outdated infrastructure, intensive power losses are being faced. This intensive power loss leads to load shedding, which is one of the major issues of today's world [1]. TGs use fossil fuels like coal, petrol, diesel, etc., for the combustion process of turbines. The extensive use of fossil fuels lead to natural resource depletion and increase in pollution. The literature has suggested to use Renewable Energy Sources (RES) and to modify the existing TGs by incorporating the latest technologies and updated infrastructure to overcome these issues. The new and modified form of TG is the Smart Grid (SG) [2]. The Information and Communication Technology (ICT) is integrated with TG to make SG. It provides bi-directional communication between consumers and utility. It monitors, protects and optimizes the generation, distribution and consumption of electric energy. It incorporates the latest technologies in TG: technical, control and communication technologies, to enable efficient energy transmission. With an ever increasing dilemma of energy shortage and cost inflation, people are attracted towards the SG. It provides the consumers with a reliable, economical, sustainable, secure and efficient energy as it uses intelligent methods. In SG, Demand Side Management (DSM) is used, which encourages the consumers to efficiently optimize the energy usage. DSM allows efficient load utilization by shifting maximum load from on-peak hours to off-peak hours. Thus, it reduced the cost of electricity. The differences between TG and SG are summarized in Table 1 [3].

Table 1. Differences between TG and SG.

TG	SG
Analogue	Digital
One way communication	Two way communication
Centralized power generation	Distributed power generation
Small number of sensors	Large number of sensors
Manual monitoring	Automatic monitoring
Difficult to locate failures	Easy to locate failures

Data analytics is the phenomenon of dealing with big data obtained from different sources. Big data is the term used for the datasets having large volume, velocity, variety and veracity. It has the problem of extreme complexity which makes the processing of data difficult. Data analytics techniques are the necessity for the processing of big data. Data analytics can be used in a number of fields. For example, handling the financial details of customers by a bank, dealing with the flight details of different passengers by an airline company, dealing with the electricity load and price forecasting of consumers, etc. In SG, data analytics is used to minimize the electricity cost and to improve the service quality of energy utilities. It is also used to predict the future patterns of electricity consumption. Forecasting is done to schedule the load consumption from on-peak hours to off-peak hours for next day, week or month to reduce the electricity cost and enhance user comfort [4].

The terms forecasting and prediction are used interchangeably in this article. The case with load and consumption is similar. The sole purpose of this work is to increase the accuracy of load and price forecasting. Two techniques are proposed to solve the aforementioned objectives, i.e., ELR and ERELM. Furthermore, two types of datasets are used, i.e., uni-variate and multi-variate. UCI is the uni-variate dataset. Uni-variate dataset contains one variable, i.e., load in this paper. However, real-time data has a number of variables. Thus, multi-variate dataset is required to handle multiple variables to achieve a better understanding. In this paper, multi-variate dataset, i.e., UMass Electric Dataset is used to predict the load and price. Two types of scenarios are considered in this paper, i.e., residential load and smart meters load. The proposed techniques outperformed existing techniques in terms of forecasting load and price. Consequently, energy prediction assists in energy management on the residential and utility side. List of abbreviations that are used in this paper is given in Table 2. Whereas, Table 3 shows complete list of symbols.

Table 2. List of abbreviations.

Abbreviation	Full Form
AEMO	Australia Electricity Market Operators
AI	Artificially Intelligent
ANN	Artificial Neural Network
ARIMA	Auto Regressive Integrated Moving Average
ARMAX	Auto Regressive Moving Average with Exogenous variables
BP	Back Propagation
CART	Classification and Regression Technique
CNN	Convolutional Neural Network
DAE	Deep Auto Encoders
DE-SVM	Differential Evolution Support Vector Machine
DNN	Deep Neural Network
DRN	Deep Residual Network
DSM	Demand Side Management
DWT	Discrete Wavelet Transform
ELM	Extreme Learning Machine
EPEX	European Power Exchange
ELR	Enhanced Logistic Regression
ERELM	Enhanced Recurrent Extreme Learning Machine
FFNN	Feed Forward Neural Network
GCA	Gray Correlation Analysis
GWO	Grey Wolf Optimization
GRU	Gated Recurrent Unit
ISO NECA	Independent System Operator New England Control Area
KELM	Kernel Extreme Learning Machine
KPCA	Kernel Principal Component Analysis
LR	Logistic Regression
LSTM	Long Short Term Memory
MAE	Mean Absolute Error
MAPE	Mean Absolute Percentage Error
MISO	Midcontinent Independent System Operator
MLP	Multi Layer Perceptron
MLR	Multi Linear Regression
MSE	Mean Square Error
NLS-SVM	Nonlinear Least Square Support Vector Machine
NN	Neural Network
NYISO	New York Independent System Operator
OS-ELM7	Online Sequential Extreme Learning Machine
PJM	Pennsylvania–New Jersey–Maryland
PSO	Particle Swarm Optimization
RBM	Restricted Boltzmann Machine
RELM	Recurrent Extreme Learning Machne
ReLU	Rectified Linear Unit
RES	Renewable Energy Sources
RFE	Recursive Feature Elimination
RMSE	Root Mean Square Error
RNN	Recurrent Neural Network
SARIMA	Seasonal Auto Regressive Integrated Moving Average
SBELM	Sparse Bayesian Extreme Learning Machine
SDA	Stacked De-noising Autoencoders
SG	Smart Grid
SLFN	Single Layer Feedforward Network
SM	Smart Meters
TG	Traditional Grid
TVC-ABC	Time Varying Coefficients Artificial Bee Colony

Table 3. List of symbols.

Symbol	Description
x	Actual value
x′	Predicted value
t	Time slot
y	Output
h	Step size
m	Mean
A_v	Actual value
F_v	Forecasted value
T	Total time duration
N	Total number of samples
α	Fittest wolf 1
β	Fittest wolf 2
δ	Fittest wolf 3
ω	Remaining wolves

The rest of the paper is organized as: Section 2 deals with related work, Section 3 contains the detailed description of techniques used in this paper. Section 4 covers the proposed system models. Results and their discussion are given in Section 5, whereas Section 6 consists of evaluation of the proposed models using the performance metrics. Conclusion and future studies are discussed in Section 7.

1.1. Motivation

The authors in [5] used Multi Layer Perceptron (MLP) and Artificial Neural Network (ANN) to solve the load and price forecasting problem. We proposed an enhanced technique to increase the accuracy of load and price forecasting based on a modified loss function. In Reference [6], authors used ELM and RELM to predict electricity load. We proposed an enhanced technique to optimize weights and biases of network for efficient load forecasting. Furthermore, two scenarios are considered and two different types of datasets are used to predict the load and price efficiently.

1.2. Problem Statement

Data of SGs are increasing dramatically so an efficient technique is required to predict the load and price of electricity. Authors in [6] used Recurrent Extreme Learning Machine (RELM) to predict the electricity load. However, in RELM, weights and biases are randomly assigned which leads to drastic variations in prediction results. An enhanced technique is proposed to solve the aforesaid issue. In Reference [7], authors used Convolutional Neural Networks (CNNs) for predicting the energy demand. However, CNN involves tuning of a number of layers which makes it spatio-temporal complex.

In this paper, two enhanced techniques are proposed to increase the accuracy of load and price of electricity efficiently. Uni-variate and multi-variate datasets are used for both techniques. Furthermore, analysis of both residential and utility data is performed collectively.

1.3. Contributions

The following are the contributions of this paper:

- Feature engineering is performed using Recursive Feature Elimination (RFE), Classification And Regression Technique (CART) and Relief-F
- Two new classification techniques are proposed, i.e., Enhanced Logistic Regression (ELR) and Enhanced Recurrent Extreme Learning Machine (ERELM)
- In ELR, the loss function of Logistic Regression (LR) is modified to increase the prediction accuracy

- The Grey Wolf Optimizer (GWO) learning algorithm is used with Recurrent Extreme Learning Machine (RELM) to optimize weights and biases in order to improve the forecasting accuracy
- The proposed techniques predict the electricity load and price efficiently
- ELR is used to predict the load and price of a smart home, whereas ERELM is used for forecasting the load of smart meters
- Cross validation is performed using *K*-Fold and Monte Carlo methods for assigning the fixed optimal values to weights and biases. This further increases the efficiency of GWO
- The accuracy of the proposed techniques is evaluated using the performance metrics, i.e., Mean Absolute Error (MAE), Mean Square Error (MSE), Mean Absolute Percentage Error (MAPE) and Root Mean Square Error (RMSE)

2. Related Work

Many forecasting techniques have been used in the past for load and price forecasting. These techniques can be categorized in three main groups: data driven, classical and Artificial Intelligence (AI). Data driven techniques consider past data to predict the desired outcomes. Classical methods comprise of the statistical and mathematical methods like Autoregressive Integrated Moving Average (ARIMA), Seasonal Autoregressive Integrated Moving Average (SARIMA), Random Forest (RF), etc., whereas AI methods mimic the behaviour of biological neurons like Feed Forward Neural Network (FFNN), Convolutional Neural Network (CNN), Long Short Term Memory (LSTM), etc.

2.1. Electricity Load Forecasting

In Reference [8], behavioural analytics are performed using Bayesian network and Multi Layer Perceptron (MLP). A number of experiments were performed using the data obtainedfrom the smart meters. Both short-term and long-term forecasting was performed. In Reference [9], Multiple Linear Regression (MLR) is used for forecasting purpose. However, it has the limitation that it can not be used for long term prediction. The authors in [10] used residual network for forecasting load on the basis of weather data. The authors in [11] used Restricted Boltzmann Machine (RBM) to train the data and Rectified Linear Unit (ReLU) to predict the electricity load. In Reference [12], Discrete Wavelet Transform (DWT) and Inconsistency Rate (IR) methods are proposed to select the optimal features from the feature set which helps in dimensionality reduction. Sperm Whale Algorithm (SWA) helps to optimize the parameters of SVM. Authors in [13] proposed a model for Short Term Load Forecasting (STLF). Mutual Information (MI) is used for feature selection whereas, better forecasting results are achieved by modifying the Artificial Neural Network (ANN). In Reference [14], authors predicted 24 h ahead cooling load of buildings using deep learning. The results show that deep learning techniques enhanced the load prediction. Similarly in [15], authors used Recurrent Neural Network (RNN), which groups the consumers into pool of inputs. The proposed model is implemented using Tensorflow package and it achieved better results.

ELM is a generalized single hidden layer FFNN learning algorithm that is proposed by the authors in References [16,17]. It proved to be effective in both regression and classification methods. In References [18,19], authors used the NNs for achieving better load prediction. In ELM learning processes, input weights and biases are randomly assigned, whereas output weights are calculated using the Moore–Penrose generalized inverse technique. In Reference [20], authors used Sparse Bayesian ELM for multi-classification purposes. The authors in [21] used Particle Swarm Optimization (PSO) and Discrete Particle Swarm Optimization (DPSO) techniques for efficient load forecasting. Authors in [22] implemented GWO with NNs to optimize weights and biases. It is proved that optimization of weights and biases increases the efficiency of network. In References [23], ELM is trained using back propagation by using context neurons as input to hidden and input layers. Accuracy is improved by further adjusting weights using previous iteration errors, whereas biases and neurons selection affect prediction accuracy as already discussed in [6].

2.2. Electricity Price Forecasting

In Reference [24], different models are used for price forecasting. These models belong to the class of deep learning. Based on the simulation results, it is proved that deep learning models perform better than the statistical models. In this paper, Gated Recurrent Unit (GRU) is used which is a variant of RNN. GRU outperformed LSTM and many other statistical models in terms of accuracy. In Reference [25], price forecasting is done using a variant of Auto Regressive Moving Average Model (ARMAX), i.e., Hilbertian ARMAX which uses the exogenous variables. The functional parameters used are modeled as the linear combinations of the sigmoid functions. These parameters are then optimized using a Quasi Newton (QN) algorithm. In Reference [26], two AI networks: CNN and LSTM are used for price forecasting in PJM electricity market. In Reference [27], Deep Neural Network (DNN) is used to extract complex patterns from the price dataset of Belgium. In Reference [28], Gray Correlation Analysis (GCA) is used along with Kernel Principal Component Analysis (KPCA) to deal with the dimensionality reduction issue. For prediction, Support Vector Machine (SVM) is used in combination with Differential Evolution (DE), where DE is used to tune the parameters of SVM. In Reference [29], a variant of autoencoder is used. This method comprises of encoder and decoder. First, the data is encoded to deal with space complexity. Once the output is obtained, it is decoded into original form. The authors in [30] implemented an enhanced form of Artificial Bee Colony (ABC) known as Time Varying Coefficients Artifical Bee Colony (TVC-ABC) for parameter tuning of Nonlinear Least Square Support Vector Machine (NLS-SVM). Inputs are first fed to ARIMA and then the output of ARIMA is given to NLS-SVM. This ARIMA + TVC-ABC NLS-SVM is a Multi Input Multi Output (MIMO) forecast engine. Limitations of gradient decent methods led researchers to evolve ELM based upon local minima, learning rate, stopping condition and iterations of learning [31]. ELM performance is different from traditional learning algorithms because it gives comparatively less forecasting error as well as proposing better generalization performance in [32].

Different versions of ELM have also been proposed by researchers such as Kernel Based Extreme Learning Machine (KELM). Robust classification is done in this paper. It is inspired by Mercer condition [33]. Related work is summarized in Table 4.

Table 4. Summary of related work.

Technique	Features/DataSet	Region	Contributions	Limitations
Bayesian, MLP [8]	Load/UKDale	UK	Forecasting done using behavioural analytic	Requires intensive training
MLR, BaggedT, NN [9]	Load and weather/Beijing	Beijing, China	Comparison between techniques done to overcome the limitations	Not suitable for long term forecasting
DRN [10]	Load and weather/ISO-NE	New England	Load forecasting done using weather data	Over fitting
RBM, ReLU [11]	Load/Korea Electric Company	Korea	Two stage forecasting performed	Long-term forecasting not supported
DWT-IR, SVM, SWA [12]	Load/NYISO, AEMO	Australia, US	Dimensionality reduction and paramater optimization done	Time complexity
Modified MI, ANN [13]	Load/PJM	US	Two stage forecasting is done	Time complexity
DAE [14]	Load/Hong Kong	Hong Kong	Cooling load prediction done	Time and space consuming
Pooling Deep RNN [15]	Load/IRISH	Ireland	Pooling of consumers done for aggregated prediction	Difficult to train
ELM [16,17]	Load/USF	US	Long, medium and short term forecasting done	Over fitting
SLFN [18,19]	Load/Marine Resources Division	Australia	Optimization of weights	Overfitting by using moore-penrose inverse
Sparse Bayesian ELM [20]	Electricity Load/Harvard medical college	USA	Optimization of weights and biases using BP	Require intensive training
PSO, DPSO [21]	Load/US	USA	Compact ANNs are produced	Large computational time
GWO with NN [22]		USA	Weights and biases optimization	Time complexity
RELM [6,23]	Electricity Load/Bench mark UCI machine	Portugal	Use of context neurons	Computationally expensive
LSTM, DNN, GRU [24]	Load and price/EPEX	Belgium	Comparison between different models	Over fitting
Hilbertian, ARMAX [25]	Price/EPEX	Spain, Germany	Optimization of functional parameters for price forecasting	Non linearity
CNN, LSTM [26]	Price/PJM	US	Two NNs are used for price forecasting	Computationally expensive

Table 4. *Cont.*

Technique	Features/DataSet	Region	Contributions	Limitations
DNN [27]	Price/EPEX	Belgium	Complex patterns are extracted for prediction	Space complexity
GCA, KPCA, DE-SVM [28]	Price/ISO NE-CA	New England	Dimensionality reduction is removed using hybrid of KPCA and GCA	Over fitting
SDA [29]	Price/MISO	Arkansas, Texas and Indiana	Variant of autoencoder used	Computationally expensive
ARIMA, TVC-ABC, NLS-SVM [30]	Price/PJM, NYISO, AEMO	Australia, US	Parameter tuning of SVM done using TV-ABC	Computationally expensive
ANN with meta heuristics optimization methods [31]	Load and price/Commercial load of building in China, Taiwan regional electricity load	Taiwan	Various paramater calculations done for accuracy	Accuracy of models depend on nature of dataset
OS-ELM with kernel [32]	Load and price/Sylva bench mark	US	Comparison of different algorithms done	Restrict to the computation of streamed data
ELM in multi class scenario [33]	Load and price/University of California Irvine	Canada	Robust classification	Computational cost overhead
Enhanced Logistic Regression	Load and Price/UMass Electric Dataset	USA	ELR beats conventional techniques	Large computational time
RELM enhanced using GWO	Load and Price/UCI Dataset	USA	Optimized weights and biases leads to better prediction	Large computational time is required for weight optimization

In literature, short term load and price forecasting using the conventional techniques is performed on individual basis mostly, whereas we used short term load and price forecasting simultaneously using enhanced techniques which surpasses the conventional techniques in terms of accuracy. The first proposed technique, i.e., ELR outperformed LR in terms of prediction accuracy, whereas the second proposed technique ERELM outperformed ELM and RELM using GWO and performs much better due to weights and biases optimization.

3. Existing and New Techniques

In this section, the existing and the proposed techniques are discussed.

3.1. Classification and Regression Technique (CART)

CART is a type of decision tree algorithm which consists of both classification and regression procedures and is used to predict the continuous and discrete variables, respectively. CART uses historical data to build decision trees. These newly built trees are then used for classification of data. It is a binary recursive process. Binary process has only two output values, i.e., 0 and 1. The algorithm will search for all possible values and variables before performing the split operation [34]. The CART method has three main parts:

- Construction of maximum tree,
- Choice of right tree size,
- Classification of data using the constructed tree.

The construction of a maximum tree refers to splitting of the tree till the last set of observations. This is the most time-consuming phase in CART. Constructing the maximum trees is a complex method which can have more than hundred levels. Therefore, the trees must be optimized before being used for classification of the data. The classification problems are the ones which involve discrimination between entities, e.g., discrimination among students to decide which student will be awarded with the degree this year. On the other hand, regression uses historical data patterns to predict the future values, e.g., load and price prediction of homes. The steps of CART are stated below:

- Problem definition,
- Variable selection,
- Specifying the accuracy criteria,
- Selecting split size,
- Determine the threshold to stop splitting,
- Selection of the best tree.

3.2. Recursive Feature Elimination (RFE)

RFE is a feature extraction process. It selects a set of most important features which are least redundant. As the name is self defining, it is an iterative process which keeps running in a loop unless all the best features are selected. The selected features are then ranked in the order they are being removed from the feature set. The computation time depends upon the number of features which need to be eliminated [35]. The pseudocode of RFE is given in Algorithm 1.

Algorithm 1: Pseudocode of RFE

Input Initialization
Tuning the model using the training set
Calculating the performance
Calculating variable importance
for *(Each subset size S(i), i = 1...S)* **do**
    Selecting the most important variables from S(i)
    Preprocessing the data
    Tuning the model using predictions
    Calculating the performance
    Recalculation of the rankings
end
Establish the performance profile using S(i)
Determine the number of important variables
Use the model corresponding to optimized S(i)
End

3.3. Relief-F

Relief-F is an extensively used method for feature selection. This method randomly selects an instance R and then find its nearest hits and miss instances. The nearest hits are the *k*-nearest neighbors of the selected random instance R. Afterwards, the average of all the weights of the nearest hits and miss is calculated to select the next instance. The pseudocode of Relief-F is discussed in Algorithm 2 [36].

Algorithm 2: Pseudocode of Relief-F

Input Initialization
Assign weights to all attributes (A): W[A]=0
for *(i = 1 to m)* **do**
    Randomly select an instance Ri
    Find k-nearest hits, Hj
end
for *(All other classes C != class (Ri))* **do**
    Find k nearest misses, Mj(C)
end
for *(A = 1 to a)* **do**
    Update weight of all attributes using Equation (1)

$$W[A] = W[A] - \sum_{j=1}^{k} - \frac{diff(A,Ri,Hj)}{m,k} + \sum_{C!=class(Ri)} \frac{P(C)}{1-P(class(Rj))} \sum_{j-1}^{k} \frac{diff(A,Ri,Mj(C))}{m,k} \quad (1)$$

end
Perform feature selection
End

3.4. Convolutional Neural Network (CNN)

CNN is a type of NN. It is built from neurons and work like the biological neurons. Each neuron is fed with some input, and then it performs a dot product and finally gives the output. It consists of more than one convolutional layer; followed by a multilayer NN. The basic type of CNN is a 2D network and is mostly used for images. The layers in CNN are: pooling layer, dense layer, dropout layer and convolutional layer. For forecasting data, 1D CNN can also be used. It also has an activation function like Sigmoid, ReLU, Tanh, etc. When new inputs are given to CNN, it does not know the exact feature mapping. Therefore, it creates a convolutional layer and then convolves this layer to find the correct feature mapping. The pooling layer in CNN has the ability of shrinking the large inputs. The most widely used activation function for CNN is ReLU. Its working is simple; whenever a negative number occurs, it is replaced by 0. Hidden layers are also present in CNN. The error minimization is performed using these layers.

3.5. Logistic Regression (LR)

LR is a type of statistical model used for regression. It is used to analyze a given dataset and then perform predictions using the independent variables. The outcome of LR is in the binary form. The main aim of LR is to describe a pattern between independent and dependent variables. There are two main parameters of LR: loss function and sigmoid function. The features should be in the binary form to use the LR method. Hence, normalization of data is required before implementing the LR model on the available data. The sigmoid function used in LR is given in following equation [37]:

$$sigmoid(t) = \frac{1}{1+e^{-t}}. \tag{2}$$

Logistic loss function is given by the following equation, which is taken from [37]:

$$lossfunction(t) = \frac{1}{m}(-y^t log(h) - (1-y)^t log(1-h)). \tag{3}$$

3.6. Enhanced Logistic Regression (ELR)

ELR is proposed in this paper. It is an enhanced form of LR technique. In ELR, a new loss function is used. Loss function is a group of objective functions that need to be minimized. It is a measure of how good a prediction model performs in predicting the outcome. Minimizing the value of the loss function increases the prediction accuracy. In this paper, the loss function is being minimized to enhance the prediction accuracy. The equation for the loss function of ELR is given below:

$$newlossfunction(t) = \frac{0.1}{m}(-y^t log(h) - (1-y)^t log(1-h)). \tag{4}$$

ELR is used to predict electricity load and price efficiently for a smart home and load of smart meters. Two different datasets, i.e., UMass Electric Dataset and UCI Dataset are used to test the proposed technique.

3.7. Grey Wolf Optimizer (GWO)

In this section, GWO technique is discussed in detail. In the proposed model, the metaheuristic technique GWO is used. It follows the social leadership and hunting mechanism of grey wolves as shown in Figure 1. The population is based on groups, i.e., alpha (α), beta (β), gamma (γ) and omega (ω). α , β and γ are considered as the fittest wolves who guide other wolves (ω) in search space. Grey wolves update their location according to the positions of the three fittest wolves, i.e., α, β and γ [22].

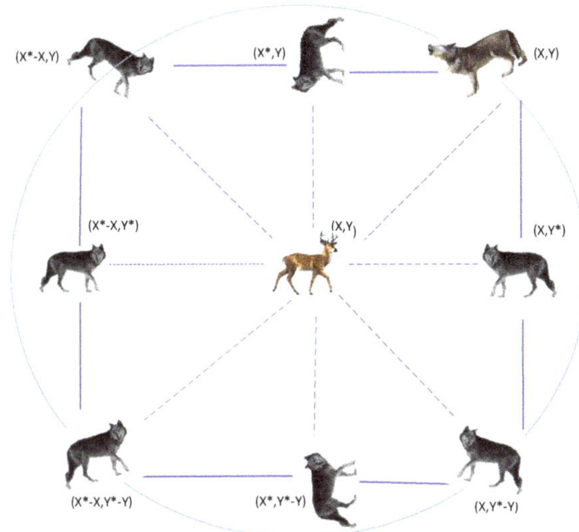

Figure 1. Grey wolf social hierarchy.

The general steps that are followed in GWO are:

- Parameters of grey wolves are initialized such as maximum number of iterations, the population size, upper and lower bounds of search space,
- Calculate fitness value to initialize the position of each wolf,
- Select three best wolves, i.e., α, β and γ,
- Calculate the positions of the remaining wolves (ω),
- Repeat from step 2 if current solution is not satisfied,
- The fittest solution is taken as α.

The pseudocode of GWO is given in Algorithm 3.

Algorithm 3: Pseudocode of GWO

Input Initialize population of Grey wolves $X_i(1, 2, 3, ...n)$, a, A and C
Calculate fitness value of each search agent X_α, X_β, X_γ respectively
Output Predicted desired value
for *(t < max number of iterations)* **do**
  for *(each search agent)* **do**
    Update position of each agent as per formulated problem
  end
  Update a, A and C
  Calculate fitness value of each agent
  Update $X_\alpha, X_\beta and X_\gamma$ $t = t + 1$
end
return X_α

3.8. Recurrent Extreme Learning Machine (RELM)

RELM is a single hidden layer neural network (SHLRN). It is a feedback intra network that uses output or hidden layers as given in Equation (5) [6]:

$$y = \sum_{j=1}^{m} \beta_j g\left(\sum_{i=1}^{n} w_{i,j} x_i + \sum_{n=i+1}^{n+r} W_{i,j} \delta(t-1+n) + b_j \right), \tag{5}$$

where δ represents delay, t shows current iteration and r indicates total number of context neurons. Context neurons are connected backward from output to input. These neurons perform similar to input neurons and hold delayed values of output neurons. The learning method to update weights and biases of ELM and RELM is similar to that shown in Figure 2.

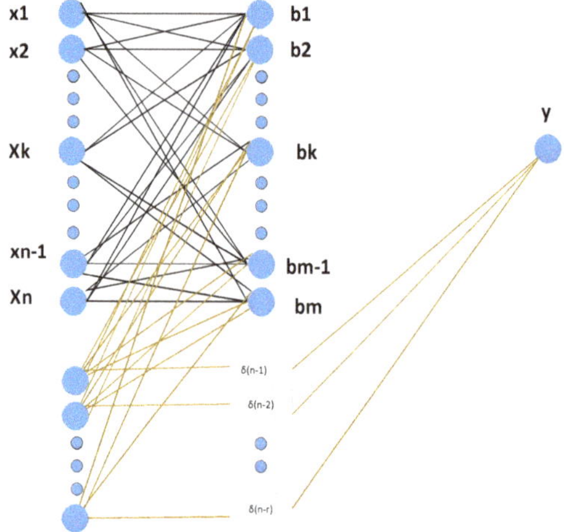

Figure 2. Functioning of RELM.

Weights and biases are decided randomly. Optimal results against weight and biases are utilized in RELM on a random basis. Training dataset is used to calculate the unknown weights of hidden layer. The unknown weights of hidden layer are calculated using a Moore–Penrose generalized inverse technique.

3.9. Enhanced Recurrent Extreme Learning Machine (ERELM)

ERELM is an enhanced form of RELM, whereas RELM is an enhanced form of ELM. ERELM is a single layer FFNN. In RELM, weights and biases are decided randomly, whereas the output weights are determined analytically. The output weights are determined using a simplified generalized inverse operation. The issue with ELM is that the classification boundary is not well defined and usually misclassifies some samples. To overcome this shortcoming, a new technique is proposed.

In the proposed technique, i.e., ERELM, weights and biases are decided after optimization using GWO algorithm. GWO finds the optimized solution which minimizes RMSE. Cross validation in ERELM is done using Monte Carlo and *K*-Fold methods. The Monte Carlo technique is used to model the probabilistic nature of the random variables. It performs risk analysis using the probability distribution. The common probability distributions used with Monte Carlo are: normal, uniform, triangular, discrete, etc. In *K*-Fold cross validation process, the entire dataset is divided into batches of *K* samples. The value of *K* could be any positive integer. Most commonly used *K*-Fold method is 10-Fold cross validation method, in which the value of *K* is 10. Each batch formed after splitting of data in the validation process is termed as fold. The pseudocode of ERELM is discussed in Algorithm 4.

Algorithm 4: Pseudocode of ERELM

Input Original dataset of N sample, objective function
Output Predicted desired value
begin
  Assign the input weights w_i and biases b_i as received from GWO
  Calculate the hidden layer output matrix **H**, where $H = h_{ij}$ $(i = 1, ..., N), j = (1, ..., K)$ and $h_{ij} = g(w_j.x_i + b_j)$
  Calculate the output weight matrix as $\beta = H^+T$, where H^+ shows Moore-Penrose generalized inverse of **H** matrix
  Updated weights are given as context neurons to input and hidden layers
end

4. Proposed System Models

Two system models are proposed in this section. The description of these models are given below.

4.1. Proposed System Model 1

The proposed system model consists of residential load and price data of a SH. The SH under consideration consists of six rooms and eight heavy appliances. The proposed model consists of four basic steps, i.e., normalization of data, feature selection using CART and RFE, feature extraction using Relief-F and finally forecasting of load and price using CNN, LR and ELR. ELR is a proposed technique which outperformed CNN and LR in terms of prediction accuracy. In this model, short term forecasting is performed to make decisions for efficient load and price scheduling for the near future.

The first proposed model is shown in Figure 3.

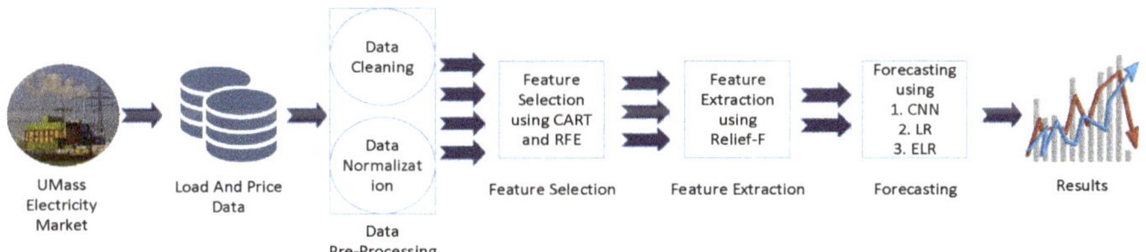

Figure 3. Proposed system model 1.

4.2. Proposed System Model 2

The second proposed system model is shown in Figure 4.

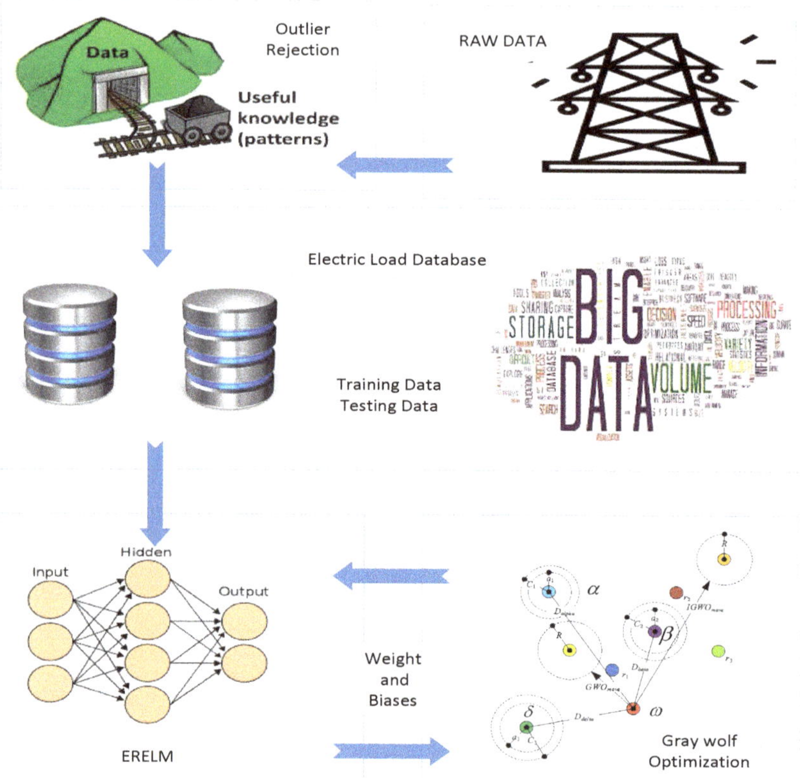

Figure 4. Proposed system model 2.

In the second system model, a load of 10 smart meters is taken. Subsequently, comparison is performed with multivariate residential data. The first step in this model is the preprocessing of data; after the data is preprocessed, the best parameters are selected using RELM. The optimization of RELM is performed using GWO. GWO optimizes biases and weights to improve the accuracy. Thereafter, the proposed technique ERELM reduces forecasting error. Cross validation is performed using Monte Carlo and *K*-Fold methods.

The simulation results and the assessment of both proposed models is done on the basis of four different performance metrics: MAPE, MAE, RMSE and MSE. The results show that the proposed techniques beat the existing techniques in terms of prediction accuracy.

5. Simulation Results and Discussion

This section covers the simulation results of the proposed models. The results are given in this section along with their discussion. The simulations are performed in Spyder (Python 3.6 package) provided by Anaconda (a data science platform manufactured by Anaconda, Inc. located in Austin, Texas, USA) on HP 450G ProBook, having 1 TB Hard Drive and 8 GB RAM.

5.1. Simulation Results and Discussion of Proposed System Model 1

The simulation results and discussion of proposed system model 1 is given below.

5.1.1. Data Description

The first dataset is taken from UMass Electric Dataset [38]. It is a multivariate dataset used for forecasting purpose. Half-yearly and yearly data is taken for the year 2016 to address scalability. The dataset contains the half-hourly load and price values of a single home. The dataset is divided into a 70:30 ratio, i.e., seventy percent data is used for training, whereas the remaining thirty percent is used for testing. Preprocessing of the dataset is done to remove the Not a Number (NaN) and blank values. UMass dataset is used for both proposed system models. Though, it performs much better when used with ELR.

Table 5 shows the features of UMass Electric Dataset excluding the target features. The targeted features are "Load" in case of load prediction and "Price" in case of price prediction. The values are given in standard units, i.e., kW for load and cents/kWh for price. The dataset is normalized in the range [0–1].

Table 5. Features in UMass Electric Dataset.

Original Features
Air Conditioner (AC), Furnace, Cellar lights, Washer, First floor lights, Utility room + Basement, Garage outlets, Master bed + Kids bed, Dryer, Panels, Home office, Dining room, Microwave, Fridge

5.1.2. CART

Table 6 shows the results of CART used to predict load and price using UMass Electric Dataset. CART gives respective values for different features.

Table 6. Results of CART for UMass Electric Dataset.

Features	Load Values	Price Values
AC	0.6653	0.6633
Furnace	0.0103	0.0101
Cellar lights	0.0018	0.0011
Washer	0.0029	0.0029
First floor	0.0032	0.0026
Utility + Basement	0.0615	0.0670
Garage	0.0036	0.0070
M. bed + K. bed	0.0080	0.0059
Dryer	0.1890	0.1927
Panels	0.0033	0.0030
Home office	0.0826	0.0083
Dining room	0.0079	0.0084
Microwave	0.0296	0.0262

5.1.3. RFE

After using the CART technique, RFE is implemented for feature selection. RFE keeps on iterating unless model is left with only the most prominent features. The choice of features depend upon requirements. The results of RFE are given in Table 7.

RFE assigns two values to the features, i.e., True and False. In the proposed model, number of selected features through RFE is 8, when used for UMass Electric Dataset. The RFE selected features are: AC, cellar lights, washer, garage, master bed + kids bed, panels, dining room and microwave. For a UCI Dataset, RFE did not give any output as it is a uni-variate dataset.

Table 7. RFE features for a UMass Electric Dataset.

Type	Number of Features
Original	16
Selected	8
Rejected	8

5.1.4. Relief-F

Relief-F is used for feature extraction. The threshold for Relief-F is 10. Table 8 shows the Relief-F features for UMass Electric Dataset. It did not give any output when used with UCI Dataset because it is uni-variate.

Table 8. Relief-F features for UMass Electric Dataset.

Parameters	Values
Threshold	10
Selected features	5
Nearest Neighbors	3

5.1.5. Load Forecasting

Figure 5a,b show the load prediction comparison of three different techniques for one day using two different hourly datasets. Similarly, Figure 6a,b show the load prediction comparison for one week using two different hourly datasets. From Figure 7a,b, monthly load prediction comparison can be observed. In this case, to avoid the cluttering of the graphs, data is taken after every four hours. These figures show that the proposed technique ELR outperformed LR and CNN for both datasets. It can be envisioned that the load prediction with ELR is close to the actual data. The prediction results obtained using a UMass Electric Dataset are better than the UCI Dataset.

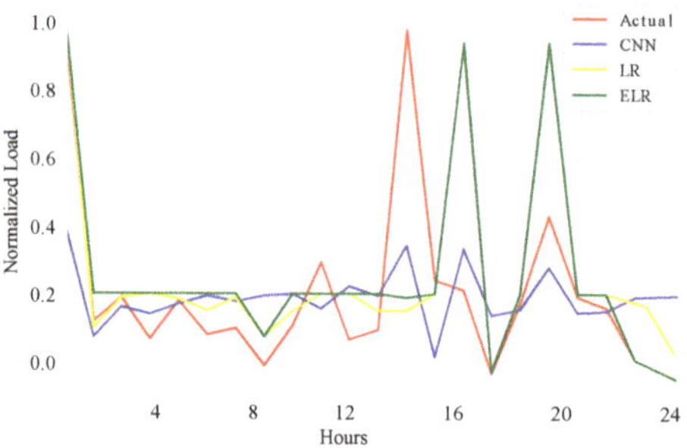

(**a**) Load prediction using UMass

Figure 5. *Cont.*

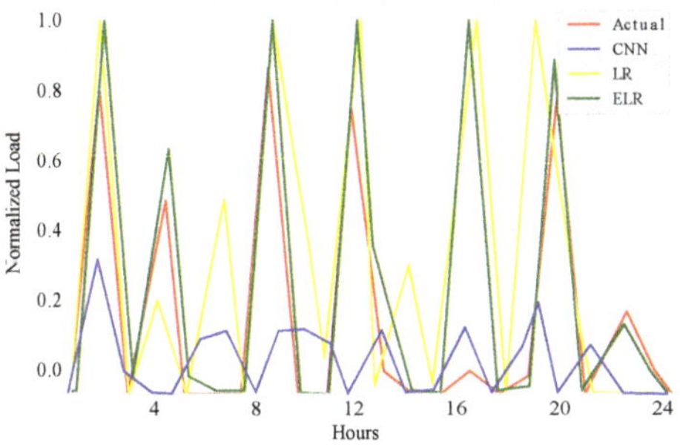

(**b**) Load prediction using UCI

Figure 5. One day load prediction.

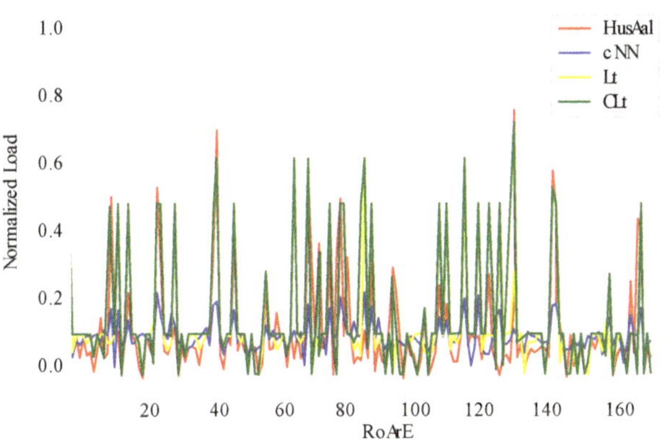

(**a**) Load prediction using UMass

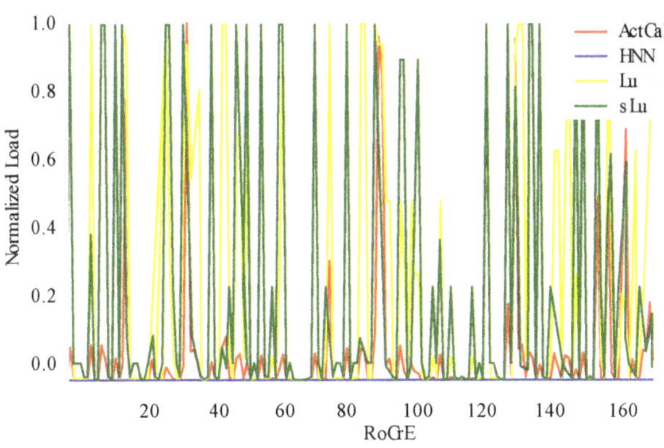

(**b**) Load prediction using UCI

Figure 6. One week load prediction.

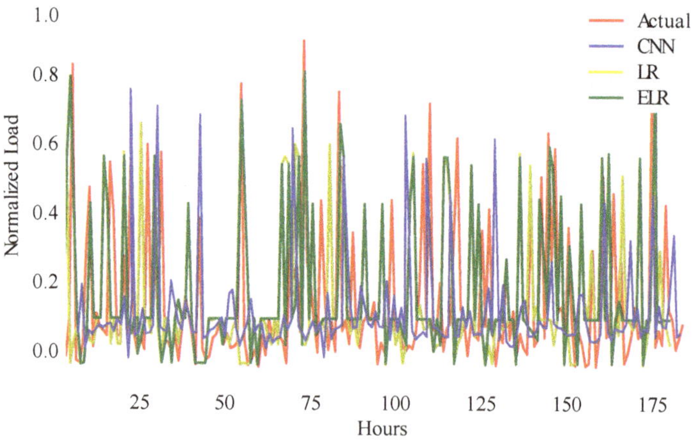

(**a**) Load prediction using UMass

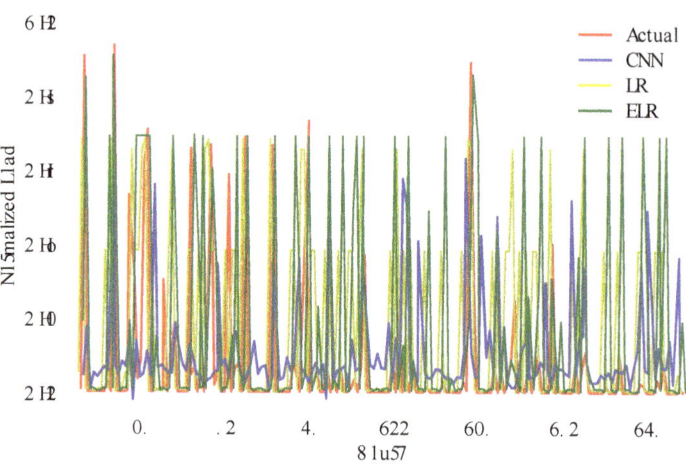

(**b**) Load prediction using UCI

Figure 7. One month load prediction.

5.1.6. Price Forecasting

Figure 8 shows the price prediction comparison of three different techniques for one day using UMass Electric Dataset. Similarly, Figure 9 shows the price prediction comparison for one week using UMass Electric Dataset. From Figure 10, monthly price prediction comparison can be observed. In this case, data is taken every four hours. These figures show that the proposed technique ELR outperformed LR and CNN for UMass Electric Dataset in terms of price prediction. It can be envisioned that the price prediction with ELR is close to the actual data.

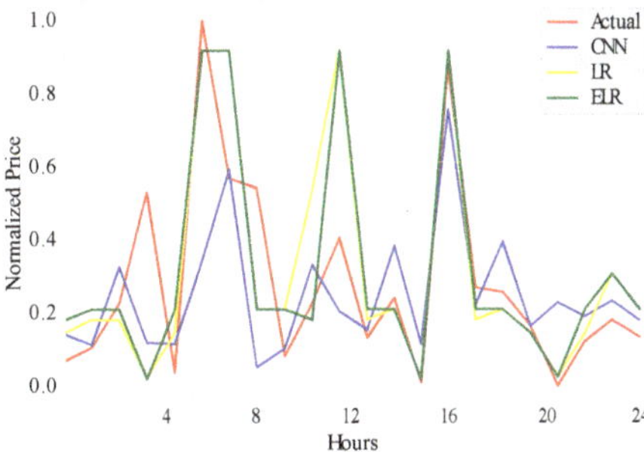

Figure 8. One day price prediction using UMass.

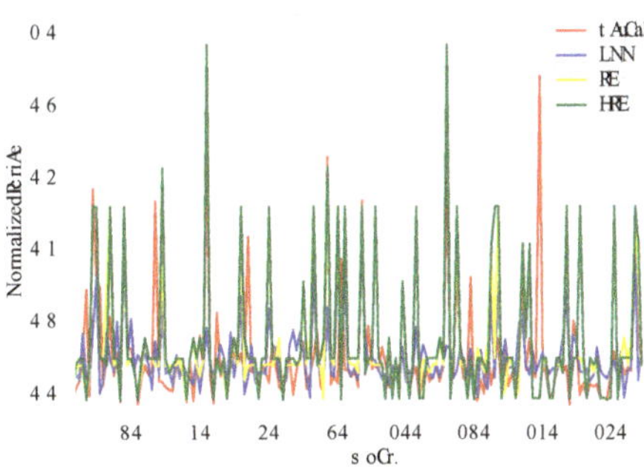

Figure 9. One week price prediction using UMass.

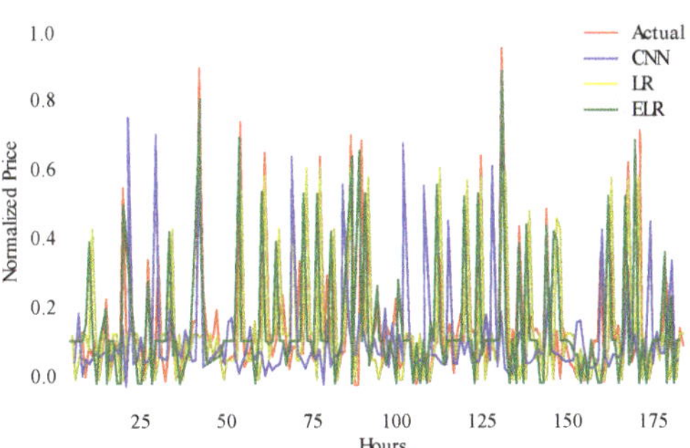

Figure 10. One month price prediction using UMass.

5.2. *Simulation Results and Discussion of Proposed System Model 2*

The simulation results and discussion of proposed system model 2 are given in this section.

5.2.1. Data Description

The second dataset is taken from the UCI machine learning repository. It is a uni-variate dataset developed by Artur Trindade [39]. Consumption of 370 substations is taken under consideration to analyze the load of smart meters. Daily data of meter ID: 166, 168, 169, 171, 182, 225, 237, 249, 250 and 257 substation is shown in Figure 11. The periodicity of load consumption can be observed. Pattern of intervals give trend of load consumption that later helps in prediction of future electricity load.

UCI Dataset is used for both proposed system models. In order to analyze scalability, half-yearly and yearly datasets are used. It performs well for smart meters because the only targeted feature is load. The values of load are given in kilo-Watts. This dataset is also normalized in the range [0–1].

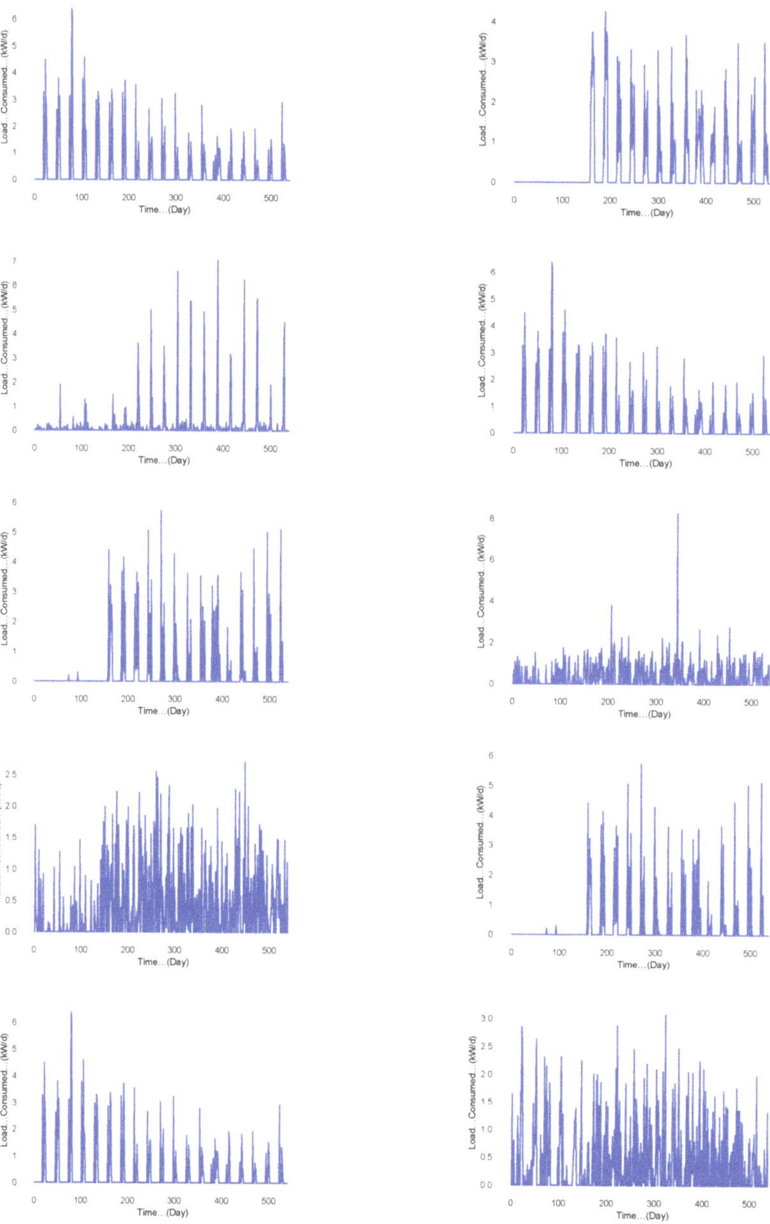

Figure 11. Daily load consumption of 10 different substations.

5.2.2. Results Discussion

Multiple approximation function is used in order to find optimal forecasting accuracy. These functions include hard limit, sine, tanh and sigmoid function. Number of neurons and context neurons are assumed as 2 and 5. The *MT166* dataset is selected to finalize functions that are producing optimal results. The dataset is normalized and scaled before use to remove spikes and noise in data. ELM, RELM and ERELM are tested on all functions one by one as given in Table 9. Sigmoid approximation function performed better than other functions. The simulations for the second proposed model are carried out on both datasets using the sigmoid approximation function.

Table 9. Obtained RMSE using ELM, RELM and ERELM.

Transfer Function	Forecasting Approach	Training	Testing
Hard Limit	ELM	0.0532	0.0535
	RELM	0.0412	0.0423
	ERELM	0.0332	0.0345
Sin	ELM	0.0513	0.0525
	RELM	0.0352	0.362
	ERELM	0.0321	0.0523
Tanh	ELM	0.0634	0.0673
	RELM	0.0453	0.0463
	ERELM	0.0341	0.0321
Sigmoid	ELM	0.0423	0.473
	RELM	0.0341	0.0381
	ERELM	0.0214	0.0235

In Table 10, simulation results of both datasets are given using six months of data. Cross validation is done using Monte Carlo and *K*-Fold. Simulations show that the proposed technique outperformed the conventional techniques in perspective of forecasting and gives minimum RMSE. Monte Carlo gives better results as compared to *K*-Fold. Similarly, Table 11 addresses the scalability of the proposed system and proves that the prediction accuracy increases with the increase in size of dataset. The Figure 12a,b show regression line produced by predicted and actual load using ELM. Similarly, Figure 13a,b show greater RMSE as compared to a proposed technique in regression plot using RELM. Figure 14a,b show plots produced by ERELM, where the regression line shows actual and predicted electricity load with minimum RMSE.

Table 10. Obtained RMSE for half-yearly testing data using ELM, RELM and ERELM by Monte Carlo and *K*-Fold cross validation.

Datasets	ERELM		RELM		ELM		RNN		LR	
	Monte Carlo	*K*-Fold	Monte Carlo	*K*-Fold	Monte Carlo	*K*-Fold	Monte Carlo	*K*-Fold	Monte Carlo	*K*-Fold
MT166	0.0235	0.0265	0.0734	0.0788	0.0824	0.0883	0.08234	0.0852	0.0853	0.0873
MT168	0.0134	0.0162	0.0421	0.0462	0.0524	0.0423	0.0854	0.0862	0.0756	0.0763
MT169	0.0153	0.02352	0.0531	0.0353	0.0382	0.0463	0.0853	0.0873	0.0735	0.0762
MT171	0.0354	0.0423	0.0524	0.0552	0.0634	0.0643	0.0854	0.0852	0.072	0.0712
MT182	0.0242	0.0353	0.0252	0.0352	0.0835	0.0952	0.0753	0.0776	0.0934	0.0952
MT235	0.0153	0.0142	0.0344	0.0397	0.0634	0.0643	0.0865	0.934	0.981	0.0991
MT237	0.0243	0.0297	0.0534	0.0535	0.0752	0.0795	0.0756	0.0762	0.0795	0.08255
MT249	0.0143	0.0163	0.0524	0.0532	0.0624	0.0693	0.0862	0.0891	0.0753	0.0778
MT250	0.0342	0.0452	0.0535	0.0562	0.0853	0.0873	0.0764	0.0784	0.0874	0.0894
MT257	0.0242	0.0215	0.0413	0.0413	0.0642	0.0683	0.0753	0.0794	0.0893	0.0934
UMass Electric	0.0398	0.0315	0.0534	0.0563	0.0681	0.0685	0.0712	0.0891	0.0888	0.0913

Table 11. Obtained RMSE for yearly testing data using ELM, RELM and ERELM by Monte Carlo and K-Fold cross validation.

Datasets	ERELM		RELM		ELM		RNN		LR	
	Monte Carlo	K-Fold	Monte Carlo	K-Fold	Monte Carlo	K-Fold	Monte Carlo	K-Fold	Monte Carlo	K-Fold
MT166	0.0224	0.0242	0.0651	0.0665	0.0756	0.0801	0.08732	0.0792	0.0862	0.0851
MT168	0.0124	0.0151	0.0634	0.0732	0.0521	0.0410	0.0831	0.0731	0.0701	0.0678
MT169	0.0144	0.0224	0.0501	0.0553	0.0424	0.0431	0.0812	0.0912	0.0741	0.0872
MT171	0.0142	0.0401	0.0421	0.0538	0.0512	0.0682	0.0781	0.0792	0.0712	0.0882
MT182	0.0182	0.0200	0.0242	0.0250	0.0743	0.0791	0.0824	0.0701	0.0892	0.0822
MT235	0.0142	0.0152	0.0301	0.0362	0.0582	0.0602	0.0852	0.0892	0.0889	0.0986
MT237	0.0224	0.0267	0.0513	0.0521	0.0623	0.0632	0.0701	0.0789	0.0862	0.0813
MT249	0.0132	0.0157	0.0421	0.0613	0.0523	0.0623	0.0782	0.0802	0.0671	0.0742
MT250	0.0242	0.0273	0.0412	0.0501	0.0602	0.0692	0.0682	0.0772	0.0785	0.0864
MT257	0.0324	0.0472	0.0401	0.0513	0.0602	0.0744	0.0623	0.0702	0.0876	0.0882
UMass Electric	0.0332	0.0412	0.0542	0.0552	0.0582	0.0603	0.0701	0.0771	0.0821	0.0891

It is clearly visible that predicted values are very close to the actual electricity load. Table 12 gives computational time comparison for execution of training and testing data of ELM, RELM and ERELM. ERELM has great computational time as compared to ELM and RELM due to its metaheuristic behaviour. Thus, there is a tradeoff between accuracy and computational time.

Table 12. Computational time comparison of ERELM, RELM and ELM execution.

Forecasting Technique	Training Time (s)	Testing Time (s)
ERELM	0.653	0.0346
RELM	0.0043	0.0012
ELM	0.00124	0.001076

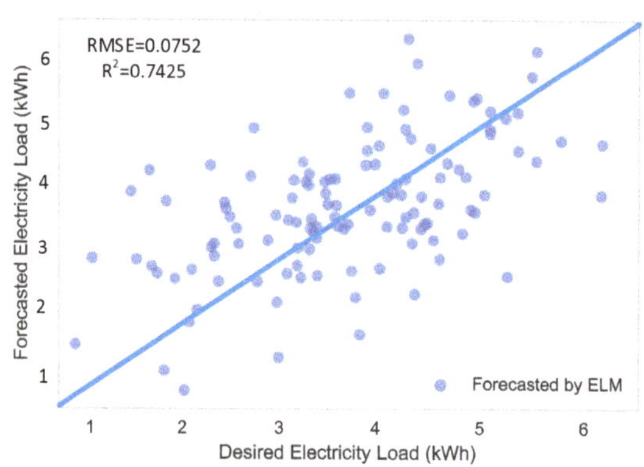

(a) ELM regression line plot for UMass

Figure 12. *Cont.*

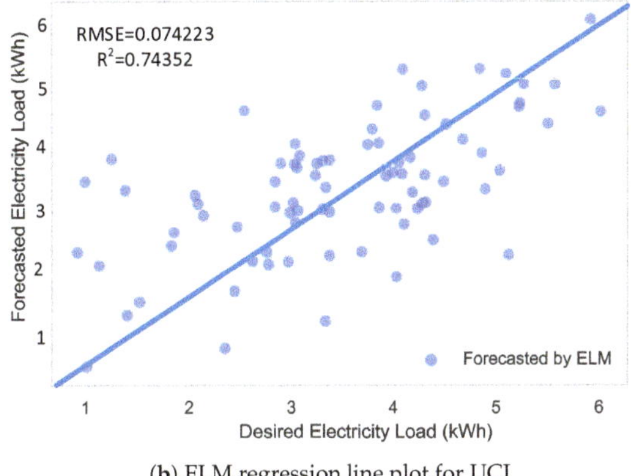

(**b**) ELM regression line plot for UCI

Figure 12. Regression line plots using ELM.

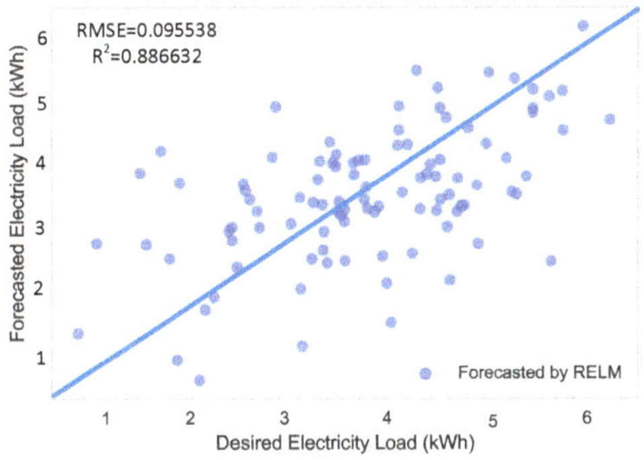

(**a**) RELM regression line plot for UMass

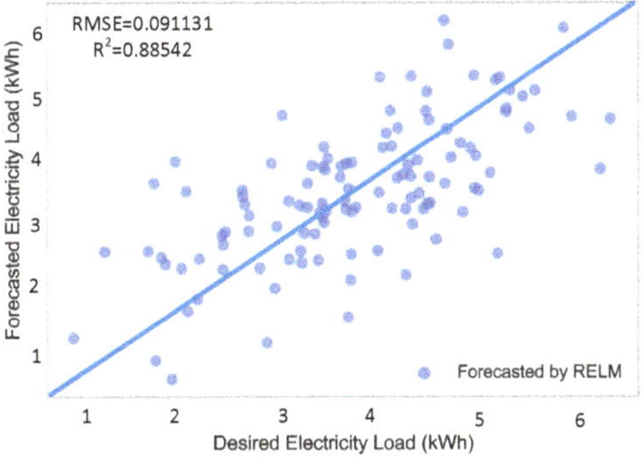

(**b**) RELM regression line plot for UCI

Figure 13. Regression line plots using RELM.

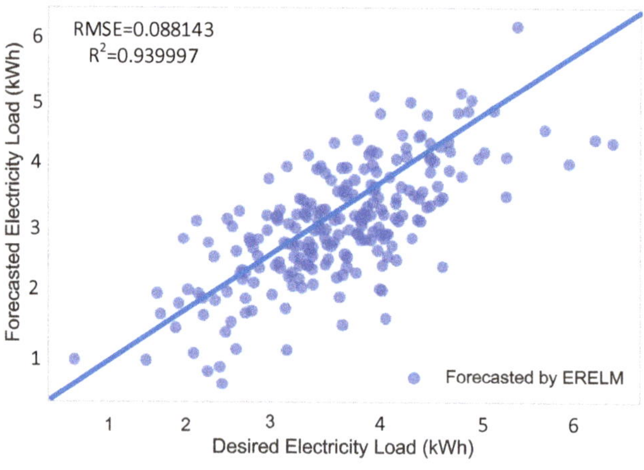

(**a**) ERELM regression line plot for UMass

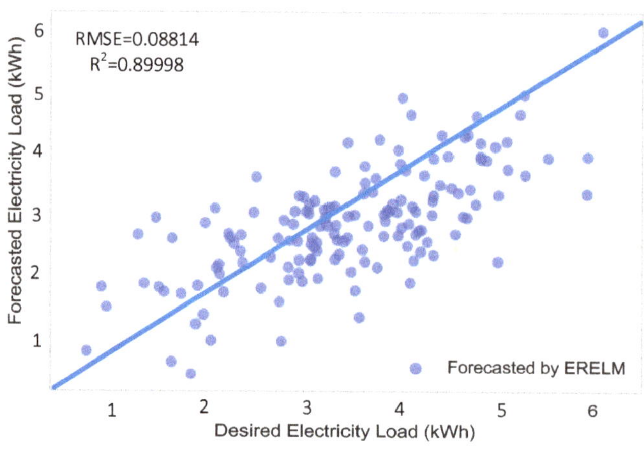

(**b**) ERELM regression line plot for UCI

Figure 14. Regression line plots using ERELM.

6. Performance Metrics

The performance of the proposed system models is evaluated on basis of four performance metrics. These performance metrics are: MAE, MSE, RMSE and MAPE. Out of these four, MAPE is given in terms of percentage whereas, the other three are given as absolute values:

$$MAPE = \frac{1}{T} \sum_{tm=1}^{TM} |\frac{A_v}{F_v}| * 100, \tag{6}$$

$$RMSE = \sqrt{\frac{1}{T}\sum_{tm=1}^{TM}(A_v - F_v)^2}, \tag{7}$$

$$MSE = \frac{1}{T}\sum_{tm=1}^{TM}(A_v - F_v)^2, \tag{8}$$

$$MAE = \frac{\sum_{n=1}^{N}|(F_v - A_v)|}{N}. \tag{9}$$

The accuracy of the model is calculated using the following equation:

$$Accuracy = 100 - RMSE. \tag{10}$$

Tables 13–15 show the load performance metrics comparison for half-yearly and yearly data to address the scalability issue. The dataset being used is UMass Electric Dataset. Similarly, Tables 16–18 show the price performance metrics comparison for half-yearly and yearly data to address the scalability issue using the UMass Electric Dataset.

Table 13. Load performance metrics comparison for one day using the UMass Electric Dataset.

Metrics	Half-Yearly Data			Yearly Data		
	CNN	LR	ELR	CNN	LR	ELR
MSE (abs. val)	8.84	5.84	4.77	10.2	8.97	6.8
MAE (abs. val)	9.24	5.25	4.34	8.75	6.25	4.24
RMSE (abs. val)	10.62	7.64	6.18	10.4	6.6	5.2
MAPE (%)	25.44	22.45	18.48	36.3	33.3	30.5
Accuracy (%)	89.38	92.35	93.82	89.6	93.4	94.8

Table 14. Load performance metrics comparison for one week using the UMass Electric Dataset.

Metrics	Half-Yearly Data			Yearly Data		
	CNN	LR	ELR	CNN	LR	ELR
MSE (abs. val)	20.2	17.97	12.97	18.9	16.5	11.3
MAE (abs. val)	8.82	6.87	5.28	8.75	6.25	5.19
RMSE (abs. val)	16.25	13.12	10.18	15.76	12.98	9.64
MAPE (%)	25.65	22.19	17.41	33.2	25.9	22.8
Accuracy (%)	83.75	86.88	89.81	84.24	87.02	91.36

Table 15. Load performance metrics comparison for one month using the UMass Electric Dataset.

Metrics	Half-Yearly Data			Yearly Data		
	CNN	LR	ELR	CNN	LR	ELR
MSE (abs. val)	25.82	21.82	17.37	24.02	20.56	14.25
MAE (abs. val)	10.55	8.23	6.79	10.35	8.15	5.33
RMSE (abs. val)	17.85	14.77	11.79	12.55	9.98	6.52
MAPE (%)	29.13	27.13	23.39	25.45	21.2	20.6
Accuracy (%)	82.15	85.72	88.21	87.45	90.02	93.48

Table 16. Price performance metrics comparison for one day using the UMass Electric Dataset.

Metrics	Half-Yearly Data			Yearly Data		
	CNN	LR	ELR	CNN	LR	ELR
MSE (abs. val)	15.08	14.39	10.09	12.58	10.85	8.6
MAE (abs. val)	8.25	7.65	6.01	7.85	7.05	5.88
RMSE (abs. val)	16.63	14.99	12.98	15.05	11.52	9.85
MAPE (%)	19.02	18.59	17.11	20.05	18.85	15.55
Accuracy (%)	83.37	85.01	87.02	84.95	88.48	90.15

Table 17. Price performance metrics comparison for one week using the UMass Electric Dataset.

Metrics	Half-Yearly Data			Yearly Data		
	CNN	LR	ELR	CNN	LR	ELR
MSE (abs. val)	14.05	12.80	11.22	15.02	13.45	11.25
MAE (abs. val)	7.55	6.04	5.03	8.02	7.05	5.25
RMSE (abs. val)	13.05	11.30	9.47	12.55	10.45	8.64
MAPE (%)	14.25	13.71	13.03	16.45	15.75	15.25
Accuracy (%)	86.95	88.70	90.53	87.45	89.55	91.36

Table 18. Price performance metrics comparison for one month using the UMass Electric Dataset.

Metrics	Half-Yearly Data			Yearly Data		
	CNN	LR	ELR	CNN	LR	ELR
MSE (abs. val)	19.45	18.91	16.47	20.05	18.54	13.35
MAE (abs. val)	8.95	7.70	6.44	9.35	8.15	6.42
RMSE (abs. val)	14.78	13.75	11.48	12.44	11.02	9.45
MAPE (%)	21.44	20.54	18.89	23.36	22.55	19.45
Accuracy (%)	85.22	86.25	88.52	87.56	88.98	90.55

Tables 19–21 show the load performance metrics comparison for half-yearly and yearly data to address the scalability issue. The dataset being used is UCI Dataset.

Table 19. Load performance metrics comparison for one day using the UCI Dataset.

Metrics	Half-Yearly Data			Yearly Data		
	CNN	LR	ELR	CNN	LR	ELR
MSE (abs. val)	15.05	12.47	10.56	18.20	14.3	10.4
MAE (abs. val)	12.45	10.34	8.56	21.47	18.37	16.61
RMSE (abs. val)	18.52	15.54	13.45	16.22	13.78	10.16
MAPE (%)	28.24	25.34	20.45	27.05	23.25	20.05
Accuracy (%)	81.48	84.46	86.55	83.78	86.22	89.84

Table 20. Load performance metrics comparison for one week using the UCI Dataset.

Metrics	Half-Yearly Data			Yearly Data		
	CNN	LR	ELR	CNN	LR	ELR
MSE (abs. val)	25.20	19.66	15.77	13.25	8.34	7.2
MAE (abs. val)	11.35	10.01	8.24	13.98	12.47	11.39
RMSE (abs. val)	22.45	19.25	16.45	20.25	17.80	13.11
MAPE (%)	28.56	25.45	19.63	30.50	25.45	18.52
Accuracy (%)	77.55	80.75	83.55	79.75	82.20	86.89

Table 21. Load performance metrics comparison for one month using the UCI Dataset.

Metrics	Half-Yearly Data			Yearly Data		
	CNN	LR	ELR	CNN	LR	ELR
MSE (abs. val)	28.35	25.55	21.68	15.50	13.8	5.47
MAE (abs. val)	15.35	10.34	8.95	23.46	19.3	17.7
RMSE (abs. val)	23.97	20.87	17.69	20.02	17.17	14.49
MAPE (%)	31.23	29.43	25.67	27.45	25.35	20.02
Accuracy (%)	76.03	79.13	82.31	79.98	82.23	85.51

Tables 22 and 23 represent accuracy of proposed technique ERELM using RMSE, MSE and MAE, using half-yearly and yearly data. Results represent that ERELM outperformed in all performance metrics.

Table 22. Accuracy of ERELM using RMSE, MSE and MAE for half-yearly data.

Datasets	RMSE	MSE	MAE
MT166	0.0235	0.00055	0.0243
MT168	0.0134	0.00017	0.0135
MT169	0.0153	0.00023	0.0174
MT171	0.0354	0.00125	0.0352
MT182	0.0242	0.00058	0.0252
MT235	0.0153	0.00023	0.0253
MT237	0.0243	0.00059	0.0254
MT249	0.0143	0.00020	0.0153
MT250	0.0342	0.001169	0.0342
MT257	0.0242	0.00058	0.0253
UMass Electric	0.0256	0.00071	0.0623
Arithmetic Mean	0.0227	0.00055	0.024218
Standard Deviation	0.00600	0.000350	0.006515

Table 23. Accuracy of ERELM using RMSE, MSE and MAE for yearly data.

Datasets	RMSE	MSE	MAE
MT166	0.0224	0.00041	0.0215
MT168	0.0124	0.00016	0.0142
MT169	0.0144	0.00015	0.0162
MT171	0.0142	0.00124	0.0221
MT182	0.0200	0.00042	0.0224
MT235	0.0142	0.00047	0.0201
MT237	0.0224	0.00015	0.0241
MT249	0.0132	0.00012	0.0142
MT250	0.0242	0.00102	0.0163
MT257	0.0324	0.00045	0.0177
UMass Electric	0.0332	0.00061	0.0546
Arithmetic Mean	0.02027	0.00047	0.02212
Standard Deviation	0.00712	0.00034	0.01077

7. Conclusions and Future Work

In this paper, electricity load and price forecasting are performed using two techniques. UMass Electric Dataset is used to predict day ahead, week ahead and month ahead load and price of a SH. Six months of hourly data are considered for day ahead and week ahead prediction, whereas four hours of data are considered for month ahead prediction. It is a multi-variate dataset. The data is first normalized and split into a training set and testing set. Feature engineering is then performed using three different techniques: RFE, CART and Relief-F. For efficient load and price prediction, a new technique, i.e., ELR is proposed. ELR outperformed CNN and LR in terms of prediction accuracy. ELR is used for UCI Dataset as well. It is a uni-variate dataset having data of smart meters of different substations. The results show that the first proposed model works well with UMass Electric Dataset. The techniques used are then accessed on the basis of four different performance metrics, i.e., MAPE, MAE, MSE and RMSE. The simulation results show that ELR outperformed LR and CNN for both datasets.

For accurate short term load forecasting, a new technique, i.e., ERELM is proposed. Short term forecasting is performed to ensure efficient load scheduling and price reduction. Parameter optimization of RELM is done using GWO. GWO optimizes biases and weights to improve the accuracy. Prediction accuracy is further increased using Monte Carlo and *K*-Fold. ERELM is used with both datasets. The results show that ERELM works well for UCI Datasets. It is observed that ERELM outperformed ELM and RELM for both datasets. The phenomenon of scalability is also addressed using both proposed techniques. Results prove that the prediction accuracy increases with the increase in size of dataset.

In future, the proposed methods will be used to perform mid-term and long-term forecasting. Weights and biases of ERELM will be further optimized using better methods. Furthermore, efficient work is required to reduce the computational time of ELR and ERELM.

Author Contributions: All authors contributed equally.

Acknowledgments: The authors extend their appreciation to the Deanship of Scientific Research at King Saud University for funding this work through research group NO (RG-1438-034).

Conflicts of Interest: The authors declare no conflicts of interest.

References

1. Ipakchi, A.; Albuyeh, F. Grid of the future. *IEEE Power Energy Mag.* **2009**, *7*, 52–62. [CrossRef]
2. Yoldaş, Y.; Önen, A.; Muyeen, S.M.; Vasilakos, A.V.; Alan, İ. Enhancing smart grid with microgrids: Challenges and opportunities. *Renew. Sustain. Energy Rev.* **2017**, *72*, 205–214. [CrossRef]
3. Shaukat, N.; Ali, S.M.; Mehmood, C.A.; Khan, B.; Jawad, M.; Farid, U.; Ullah, Z.; Anwar, S.M.; Majid, M. A survey on consumers empowerment, communication technologies, and renewable generation penetration within Smart Grid. *Renew. Sustain. Energy Rev.* **2018**, *81*, 1453–1475. [CrossRef]
4. Zhou, K.; Fu, C.; Yang, S. Big data driven smart energy management: From big data to big insights. *Renew. Sustain. Energy Rev.* **2016**, *56*, 215–225. [CrossRef]
5. Nazar, M.S.; Fard, A.E.; Heidari, A.; Shafie-khah, M.; Catalão, J.P.S. Hybrid model using three-stage algorithm for simultaneous load and price forecasting. *Electr. Power Syst. Res.* **2018**, *165*, 214–228. [CrossRef]
6. Ertugrul, Ö.F. Forecasting electricity load by a novel recurrent extreme learning machines approach. *Int. J. Electr. Power Energy Syst.* **2016**, *78*, 429–435. [CrossRef]
7. Muralitharan, K.; Sakthivel, R.; Vishnuvarthan, R. Neural network based optimization approach for energy demand prediction in smart grid. *Neurocomputing* **2018**, *273*, 199–208. [CrossRef]
8. Shailendra, S.; Yassine, A. Big Data Mining of Energy Time Series for Behavioral Analytics and Energy Consumption Forecasting. *Energies* **2018**, *11*, 452. [CrossRef]
9. Ahmad, T.; Chen, H. Short and medium-term forecasting of cooling and heating load demand in building environment with data-mining based approaches. *Energy Build.* **2018**, *166*, 460–476. [CrossRef]
10. Kunjin, C.; Kunlong, C.; Qin, W.; Ziyu, H.; Jun, H.; He, J. Short-term Load Forecasting with Deep Residual Networks. *IEEE Trans. Smart Grid* **2018**, *99*. [CrossRef]
11. Seunghyoung, R.; Noh, J.; Kim, H. Deep neural network based demand side short term load forecasting. *Energies* **2016**, *10*, 3.
12. Liu, J.P.; Li, C.L. The short-term power load forecasting based on sperm whale algorithm and wavelet least square support vector machine with DWT-IR for feature selection. *Sustainability* **2017**, *9*, 1188. [CrossRef]
13. Ahmad, A.; Javaid, N.; Guizani, M.; Alrajeh, N.; Khan, Z.A. An accurate and fast converging short-term load forecasting model for industrial applications in a smart grid. *IEEE Trans. Ind. Inform.* **2017**, *13*, 2587–2596. [CrossRef]
14. Fan, C.; Xiao, F.; Zhao, Y. A short-term building cooling load prediction method using deep learning algorithms. *Appl. Energy* **2017**, *195*, 222–233. [CrossRef]

15. Shi, H.; Xu, M.; Li, R. Deep learning for household load forecasting—A novel pooling deep RNN. *IEEE Trans. Smart Grid* **2018**, *9*, 5271–5280. [CrossRef]
16. Huang, G.B.; Zhu, Q.Y.; Siew, C.K. Extreme learning machine: theory and applications. *Neurocomputing* **2006**, *70*, 489–501. [CrossRef]
17. Huang, G.B.; Zhou, H.; Ding, X.; Zhang, R. Extreme learning machine for regression and multiclass classification. *IEEE Trans. Syst. Man Cybern. Part B* **2012**, *42*, 513–529. [CrossRef] [PubMed]
18. Fallah, S.N.; Deo, R.C.; Shojafar, M.; Conti, M.; Shamshirband, S. Computational Intelligence Approaches for Energy Load Forecasting in Smart Energy Management Grids: State of the Art, Future Challenges, and Research Directions. *Energies* **2018**, *11*, 596. [CrossRef]
19. Zeng, Y.R.; Zeng, Y.; Choi, B.; Wang, L. Multifactor-influenced energy consumption forecasting using enhanced back-propagation neural network. *Energy* **2018**, *127*, 381–396. [CrossRef]
20. Luo, J.; Vong, C.M.; Wong, P.K. Sparse Bayesian extreme learning machine for multi-classification. *IEEE Trans. Neural Netw. Learn. Syst.* **2014**, *25*, 836–843. [PubMed]
21. Yu, J.; Wang, S.; Xi, L. Evolving artificial neural networks using an improved PSO and DPSO. *Neurocomputing* **2008**, *71*, 1054–1060. [CrossRef]
22. Mirjalili, S.; Mirjalili, S.M.; Lewis, A. Grey wolf optimizer. *Adv. Eng. Softw.* **2014**, *69*, 46–61. [CrossRef]
23. Saremi, S.; Mirjalili, S.Z.; Mirjalili, S.M. Evolutionary population dynamics and grey wolf optimizer. *Neural Comput. Appl.* **2015**, *26*, 1257–1263. [CrossRef]
24. Lago, J.; De Ridder, F.; De Schutter, B. Forecasting spot electricity prices: deep learning approaches and empirical comparison of traditional algorithms. *Appl. Energy* **2018**, *221*, 386–405. [CrossRef]
25. González, J.P.; San Roque, A.M.; Perez, E.A. Forecasting functional time series with a new Hilbertian ARMAX model: Application to electricity price forecasting. *IEEE Trans. Power Syst.* **2018**, *33*, 545–556. [CrossRef]
26. Kuo, P.H.; Huang, C.J. An Electricity Price Forecasting Model by Hybrid Structured Deep Neural Networks. *Sustainability* **2018**, *10*, 1280. [CrossRef]
27. Wang, K.; Xu, C.; Zhang, Y.; Guo, S.; Zomaya, A. Robust big data analytics for electricity price forecasting in the smart grid. *IEEE Trans. Big Data* **2017**, *5*, 34–45. [CrossRef]
28. Lago, J.; De Ridder, F.; Vrancx, P.; De Schutter, B. Forecasting day-ahead electricity prices in Europe: the importance of considering market integration. *Appl. Energy* **2018**, *211*, 890–903. [CrossRef]
29. Long, W.; Zhang, Z.; Chen, J. Short-Term Electricity Price Forecasting with Stacked Denoising Autoencoders. *IEEE Trans. Power Syst.* **2017**, *32*, 2673–2681.
30. Ghasemi, A.; Shayeghi, H.; Moradzadeh, M.; Nooshyar, M. A novel hybrid algorithm for electricity price and load forecasting in smart grids with demand-side management. *Appl. Energy* **2016**, *177*, 40–59. [CrossRef]
31. Huang, G.B.; Chen, L.; Siew, C.K. Universal approximation using incremental constructive feedforward networks with random hidden nodes. *IEEE Trans. Neural Netw.* **2006**, *17*, 879–892. [CrossRef] [PubMed]
32. Bartlett, P.L. For valid generalization the size of the weights is more important than the size of the network. In *Advances in Neural Information Processing Systems*; MIT Press: Cambridge, MA, USA, 1997; pp. 134–140.
33. Scardapane, S.; Comminiello, D.; Scarpiniti, M.; Uncini, A. Online sequential extreme learning machine with kernels. *IEEE Trans. Neural Netw. Learn. Syst.* **2015**, *26*, 2214–2220. [CrossRef] [PubMed]
34. Loh, W.Y. *Classification and Regression Trees. Wiley Interdisciplinary Reviews: Data Mining and Knowledge Discovery*; John Wiley and Sons Inc.: Hoboken, NJ, USA, 2011; Volume 1, pp. 14–23.
35. Recursive Feature Elimination. Available online: https://topepo.github.io/caret/recursive-feature-elimination.html (accessed on 10 November 2018).
36. Durgabai, R.P.L. Feature selection using ReliefF algorithm. *Int. J. Adv. Res. Comput. Commun. Eng.* **2014**, *3*, 10, 8215–8218.
37. Logistic Regression. Available online: https://ml-cheatsheet.readthedocs.io/en/latest/logistic-regression.html (accessed on 10 November 2018).
38. UMass Electric Dataset. Available online: http://traces.cs.umass.edu/index.php/Smart/Smart (accessed on 10 November 2018).
39. Lichman, M. *UCI Machine Learning Repository*; University of California: Irvine, CA, USA, 2013.

Article

Insulator Detection Method in Inspection Image Based on Improved Faster R-CNN

Zhenbing Zhao *, Zhen Zhen, Lei Zhang, Yincheng Qi, Yinghui Kong and Ke Zhang

School of Electrical and Electronic Engineering, North China Electric Power University, Baoding 071003, China; zz_hbdl@sina.com (Z.Z.); ai_zhanglei1212@126.com (L.Z.); qiych@126.com (Y.Q.); kongyhbd2015@ncepu.edu.cn (Y.K.); zhangke41616@126.com (K.Z.)

* Correspondence: zhaozhenbing@ncepu.edu.cn; Tel.: +86-159-3377-3709

Received: 20 February 2019; Accepted: 25 March 2019; Published: 28 March 2019

Abstract: The detection of insulators in power transmission and transformation inspection images is the basis for insulator state detection and fault diagnosis in thereafter. Aiming at the detection of insulators with different aspect ratios and scales and ones with mutual occlusion, a method of insulator inspection image based on the improved faster region-convolutional neural network (R-CNN) is put forward in this paper. By constructing a power transmission and transformation insulation equipment detection dataset and fine-tuning the faster R-CNN model, the anchor generation method and non-maximum suppression (NMS) in the region proposal network (RPN) of the faster R-CNN model were improved, thus realizing a better detection of insulators. The experimental results show that the average precision (AP) value of the faster R-CNN model was increased to 0.818 with the improved anchor generation method under the VGG-16 Net. In addition, the detection effect of different aspect ratios and different scales of insulators in the inspection images was improved significantly, and the occlusion of insulators could be effectively distinguished and detected using the improved NMS.

Keywords: insulator; Faster R-CNN; object detection; RPN; deep learning

---

1. Introduction

As one of the most important infrastructures in power systems, insulators play an important role in the safe operation of transmission lines and substations [1]. However, insulators mostly work outdoors, which makes them prone to becoming dirty, cracked, and damaged, and thus threatens the safety and stability of power grids. For the insulator state detection and fault diagnosis thereafter, the accurate detection of insulators in power transmission and transformation inspection images provides a foundation and is of great importance.

In recent years, the continuous development of helicopters, unmanned aerial vehicles (UAVs), and other high-altitude operation platforms has brought new opportunities to power transmission and transformation inspection work [2]. However, there are the following limitations in the process of manually judging patrol images: Firstly, it relies on inspection personnel with rich professional experience to avoid misjudgments or omissions; secondly, the flood of generated images or video data makes the maintenance speed too slow and the cost too high when only using manual judgment [3]. Therefore, the application of computer vision technologies for the detection of insulators in patrol inspection images has great significance for the running state of intelligent detection of insulators, and it can greatly save manpower and materials whilst also improving monitoring efficiency [4].

At present, the existing insulator image detection methods are mainly divided into two categories: One based on pixels or artificial features, the other based on the deep learning model. In the detection methods of insulators based on pixels or artificial features, the idea of threshold segmentation was adopted in [5], which basically relies on the saturation of the image to extract insulator objects;

however, the relatively poor ability to distinguish objects with similar saturation makes it unpractical. In the insulator detection methods based on the deep learning model, Tao G [6] learned insulator characteristics through the convolution neural network in complex aerial images, and then those characteristics were applied to identify a variety of insulators. However, the detection effect of this method is not satisfactory. Deep convolutional neural networks [7] were adopted in [8] to realize insulator status detection. The experimental results showed that a result obtained by the pre-trained model for classification was more accurate than the shallow features by hand-crafted models, which verifies the effectiveness of deep learning to extract object features. The fast region-convolutional neural network (R-CNN) [9] model was adopted in [10] to realize insulator detection in the background of complex aerial photography. This method can only detect the insulators in visible images with a relatively low accuracy. The faster R-CNN [11] was introduced in [12] to detect the spacer bar, pressure sharing ring, and shockproof hammer in the images in power systems; but it is only a direct application of faster R-CNN. The faster R-CNN was introduced in [13] to detect insulators in the images with complex backgrounds, but no improvements were made on the basis of the external characteristics of the insulators and the accuracy was low.

The method based on the deep learning model is rarely applied in insulator detection [14]. When setting the parameters, such as the shape of the object and the object scales, it only satisfies the detection requirements of general objects, and so is not suitable for the insulator object when applying the faster R-CNN directly to insulator image detection as it causes false detection, missed detection, and other problems. Therefore, further study is needed into the faster R-CNN model which also considers the intrinsic characteristics of insulators. In view of this, a method for insulator detection in inspection images based on an improved faster R-CNN is presented in this paper.

2. Description of Problem

It is extremely difficult to detect insulators accurately in transmission lines with deep learning. The difficulties mainly lie in the following two aspects: First, in the power system, the insulators have different aspect ratios and dimensions due to the special functions of the insulators and the different distances away of the inspection machines; second, the insulators have serious occlusion problems due to the complexity of the outdoor environment. These problems greatly increase the difficulty of insulator detection based on the deep learning detection model.

2.1. Insulator Detection with Different Aspect Ratios and Different Scales

In the classic faster R-CNN models, the setting of an anchor in the region proposal network (RPN) only meets the general object detection requirements; when it comes to the detection of an insulator in inspection images, the detection results are not satisfactory. Some cases of poor insulator detection results in power transmission and transformation inspection images after fine-tuning of faster R-CNN are demonstrated in Figure 1. There are two horizontal insulators whose color is close to the background in Figure 1a. Only a small part of one insulator can be detected, and the size of the output box does not fit the insulator well. In Figure 1b, there are two tilting insulators. The shape of the left insulator umbrella disc is not obvious and it is close to a column. The application of the classical faster R-CNN model fails to detect it. In Figure 1c, the insulators are of various scales. For the insulator in the lower-right corner, because it is small, it is missed in the detection frequently. In Figure 1d, there are two nearly vertical insulators, but the complex background and the mutual-occlusion cause a failure in the detection of both insulators, and furthermore, a false detection is generated. Therefore, it is far from enough to fine-tune the faster R-CNN model by using the constructed professional database. The anchor generation method in the RPN needs to be improved according to the characteristics of insulators from a deeper level.

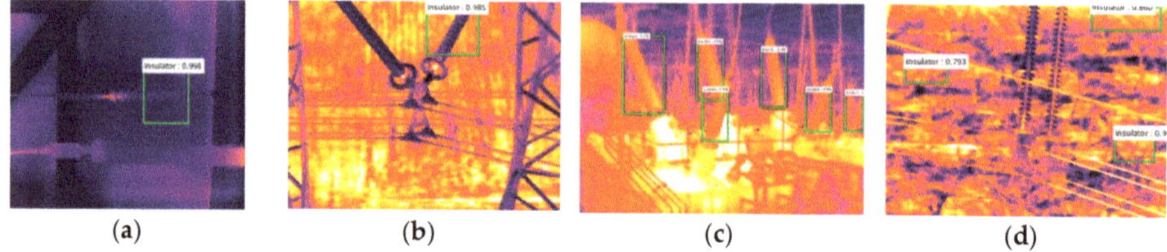

Figure 1. Insulator detection images obtained by fine-tuning the faster R-CNN: (**a**) Detection of insulators with color close to background; (**b**) detection of tilting insulators; (**c**) detection of insulators with different scales; (**d**) detection of insulators with occlusion problems.

It can be observed that the insulators have a specific outer shape which is generally slender, and their aspect ratio is numerically slightly larger than general objects. As shown in Figure 2, the aspect ratio of the insulators varies from 2.5:1 to 8.5:1. This inherent characteristic of insulators makes it necessary to adjust the aspect ratio of the anchor in the parameter setting part of the RPN.

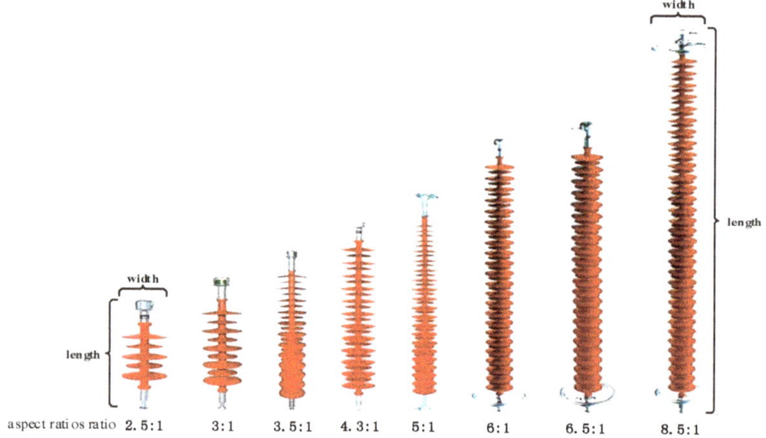

Figure 2. Schematic diagram of insulators with different aspect ratios.

In addition, most of the insulator images from the field also vary in size. The reference anchor with the size of 16 × 16 in the original RPN does not meet the requirements for accurate detection of smaller-scale insulators in power transmission and transformation inspection images. Furthermore, after a magnification of the benchmark anchor for three scales (8, 16, 32), the ability to detect small-scale insulators is reduced.

In view of the above two points, to achieve accurate detection of insulators with different aspect ratios and different scales in power transmission and transformation inspection images, the inherent characteristics should be considered and incorporated into the faster R-CNN, and the anchor generation method in the RPN should be modified according to the insulator characteristics.

2.2. Insulator Detection Under Mutual Occlusion

Non-maximum suppression (NMS) is a widely applied post-processing algorithm to perform redundant removal in object detections. The traditional NMS which performs greedy clustering based on a fixed threshold is manually designed. It greedily selects the detection results with high scores and deletes those adjacent results that exceed the threshold. Specifically, the bounding boxes are sorted according to the score, and the one with the highest score is retained; while the boxes whose overlapping area with the retained one is larger than a certain ratio are removed. However, as can be seen in Figure 3, this method has the following problem: Only one output box remains for the two insulators blocking each other in the figure after operation of the traditional NMS, that is to say, the two insulators are not accurately detected.

Figure 3. Result of removing redundant boxes by original non-maximum suppression (NMS).

The traditional NMS method mainly suppresses the bounding boxes through the intersection over union (IoU) threshold, but the IoU threshold needs to be manually set. If the IoU threshold is set too high, the bounding box suppression may be insufficient; on the contrary, too low a threshold may lead to multiple correct bounding boxes coming back together. In insulator images obtained in power transmission and patrol inspection, the insulator superposition and occlusion problem often occurs. Aiming specifically at this phenomenon, this paper proposes a more reasonable suppression method based on the traditional NMS algorithm to solve the problem of mutual occlusion of insulators in the image.

3. Insulator Detection Method Based on Improved faster R-CNN

By analyzing the insulator characteristics and the faster R-CNN model, we propose a detection model for the insulator features. The anchor generation method and NMS in the RPN of the faster R-CNN model are improved, respectively. The specific improvement method is as follows.

3.1. Construction of Transmission and Transformation Insulation Equipment Detection Database

Deep learning, as a multi-layer and hidden layer representation learning method, is constructed by simple nonlinear modules [15]. These nonlinear modules not only convert the input representation into a more abstract representation but also automatically learn very complex functions [16]. Deep learning relies on a large number of sample images for the training of the models. Therefore, in order to achieve higher insulator detection accuracy, a large number of insulator images are needed as a training set, which enables the neural network to learn features in line with the object characteristics. At present, the design and application of deep learning models is mostly aimed at public data sets of visible light, such as MNIST [17], ImageNet [18], and PASCAL VOC [19], but transformation devices in power system [20] are not included. In order to better understand the characteristics of the insulator, this paper first constructs a professional transmission and transformation electrical insulation equipment detection dataset that only contains transmission and transformation line insulation equipment. It provides a base for studying the method of insulator object detection.

The database of power transmission and transformation equipment constructed in this paper contains 2535 infrared transmission images, with 7343 insulators being labeled. The images acquired on-site are infrared images, which meet the needs of practical applications. Various situations were considered during the selection of the images. About 33 transmission lines, four substations, one converter station, and several indoor high-voltage laboratories of 110~500 kV of different grades across the country were chosen. As regards the materials, the insulators made of glass, porcelain, and composite were all considered; besides, insulators with different angles and different scales, such as top view, bottom view, and head view were selected as well. Figure 4 demonstrates the diversity of the constructed dataset. Altogether, 85% of the total images were used as a training set, and for the remaining images, a total of 380 images and 200 visible images were used as test sets. It is worth noting that we had previously conducted a large number of experiments, indicating that the model trained only with infrared images has no effect on the detection of visible light images, so only the infrared images were trained to avoid repeated work [21]. Using the constructed database to fine-tune the faster R-CNN under the VGG-16 Net [22], the insulator detection effect in the power transmission and patrol images was improved significantly. The average precision (AP) value of the faster R-CNN in the VGG-16 Net reached 0.640. However, the improvement in the method with only fine-tuning does not meet the requirements of insulator detection in the inspection images. How to further improve the detection accuracy on this basis became the issue we focused on.

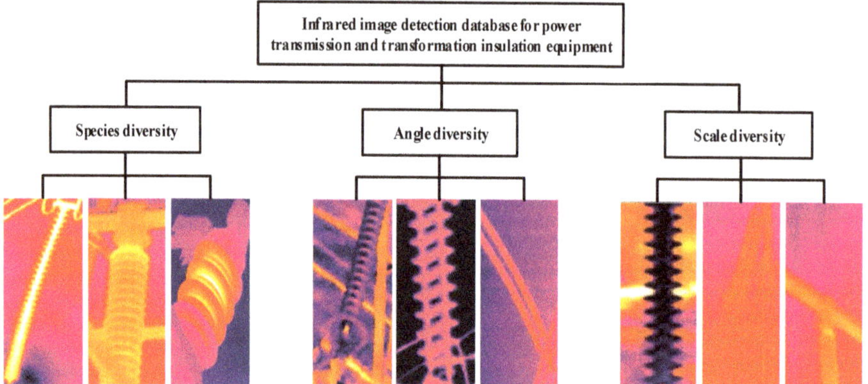

Figure 4. Diversity display of infrared image detection database for power transmission and transformation insulation equipment.

3.2. Framework of Method

In view of the two problems mentioned above, this paper proposes the faster R-CNN model, which can be considered as a combination of the RPN and the fast R-CNN. Firstly, the RPN and the fast R-CNN part were fine-tuned using the constructed power transmission and transformation equipment detection database. After that, considering the inherent characteristics of the insulator, the anchor generation method and the NMS in the RPN were improved in the fine-tuned RPN. Finally, the improved RPN and the fine-tuned fast R-CNN part were combined to form the improved faster R-CNN model. At this time, the power transmission and patrol images were applied as the test images for insulator detection. In such a way, a more accurate detection effect was achieved. The block diagram of the insulator image detection method based on the improved faster R-CNN model is demonstrated in Figure 5.

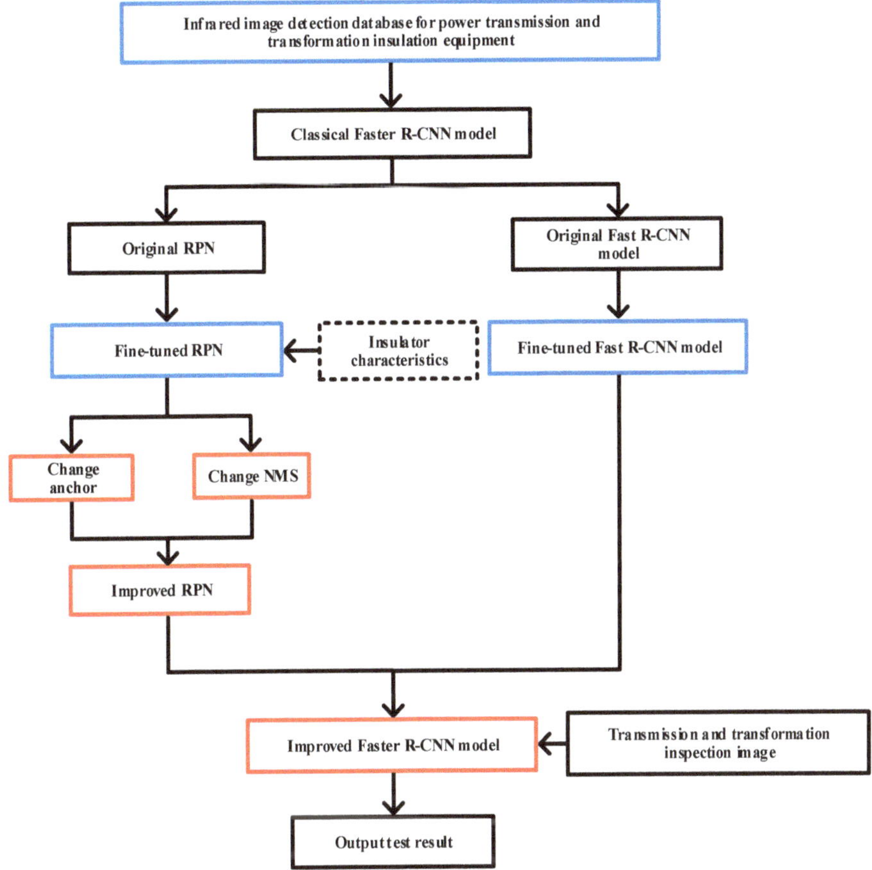

Figure 5. Framework of insulator images detection method based on improved the faster region-convolutional neural network (R-CNN).

3.3. Anchor Generation Method Improvement

The core of the faster R-CNN model is the use of the RPN to generate candidate object regions. Unlike traditional multi-scale sliding window approaches, the faster R-CNN applies the sliding window to the convolutional feature map generated by the convolutional neural network in the RPN. In order to effectively deal with object detection with different scales, the RPN uses three different types of anchors. The aspect ratios of each anchor are 1:1, 1:2, and 2:1, respectively. Meanwhile, three different scales are applied to each anchor for zooming. To be specific, a reference anchor is set first, and the size of this reference anchor in the faster R-CNN model is 16 × 16. Then, with the reference anchor area staying unchanged, its aspect ratio is set to 1:1, 1:2, and 2:1. The anchors of the three different aspect ratios are then zoomed in by three scales (8, 16, 32), and finally a total of nine anchors are obtained. Figure 6 shows the structure of the generation of region proposals using the RPN, and Figure 7 shows an example of nine different-scale anchors.

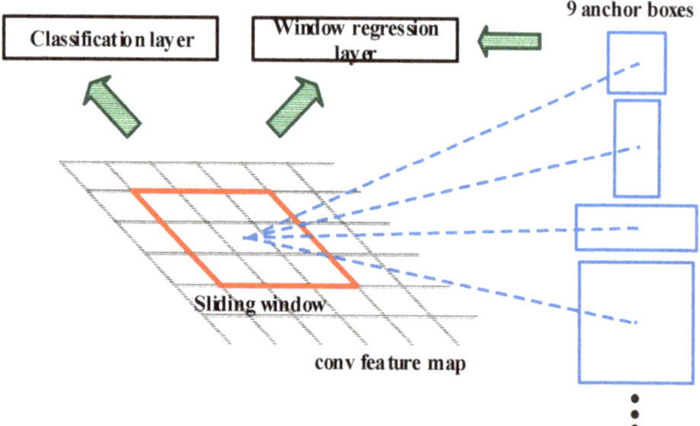

Figure 6. Process of generate candidate object area structure using RPN.

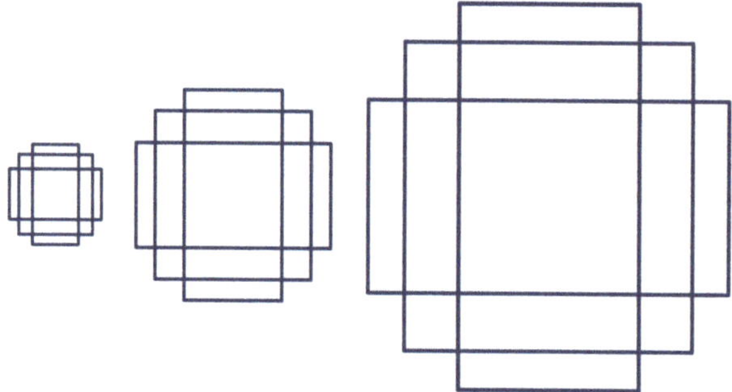

Figure 7. Anchor examples with nine different scales.

Considering the two problems found when detecting the insulators with different aspect ratios and different scales in power transmission and transformation inspection images, to make the detection method more suitable for insulators, the new RPN adopts five different aspect ratios to combine six scaling scales in the improvement of the RPN in the faster R-CNN model. Thus, a total of 30 anchors were obtained. The five aspect ratios are 1:1, 1:2, 1:3, 2:1, and 3:1, and the six scaling scales are 2, 4, 8, 16, 32, and 64, respectively. In order to fit the new scaling scale better, the new anchor size of the benchmark changes from 16 × 16 to 8 × 8.

3.4. NMS Improvements

Differing from traditional NMS methods, the improved NMS in this paper reorders the scores according to the function to suppress the obtained bounding box. For the regions generated by faster R-CNN and their corresponding score s_i, the bounding box M with the highest score is first sorted, and then the remaining bounding box b_i and the IoU value of the bounding box iou(M, b_i) are calculated (iou(M, b_i) ∈ [0,1]). The conventional NMS method reserves a bounding box in which iou(M, b_i) is smaller than the threshold according to the manually set IoU threshold N_t, and deletes the bounding box in which iou(M, b_i) exceeds the threshold, as shown in Equation (1). In the improvement of the NMS, instead of discarding the box with the IoU value greater than the threshold in the traditional NMS, the scores are reduced by a certain function and then reordered after the transformation, and the final detection result is obtained.

$$s_i = \begin{cases} s_i, \mathrm{iou}(M, b_i) < N_t \\ 0, \mathrm{iou}(M, b_i) \geq N_t \end{cases} \quad (1)$$

$$s_i = \begin{cases} s_i, \text{iou}(M, b_i) < N_t \\ s_i e^{-\frac{\text{iou}(M,b_i)^2}{2}}, \text{iou}(M, b_i) \geq N_t \end{cases} \tag{2}$$

Specifically, the bounding boxes score s_i is transformed according to Equation (2). It can be seen from Equation (2) that through the improved transformation, the higher the overlapping area of the to-be-processed box b_i and the highest score box M, the lower the score s_i of the b_i; and the smaller the iou(M, b_i) value, the slower the s_i decreases. Similarly, the larger the iou(M, b_i) value, the faster the s_i decreases. This stops the remaining insulators from being directly removed when there are multiple mutually occluded insulators in the input image, even if the IoU values of the two boxes are too large. At the same time, the IoU value score can be appropriately reduced so that the output result can be displayed as the final detection result.

4. Experimental Results and Analysis

In order to verify the effectiveness of our method, several test pictures in the infrared image detection database of the power transmission and transformation equipment were used to test and observe the experimental results. The analysis of the experimental results was carried out both qualitatively and quantitatively.

4.1. Detection Experiment of Insulators with Different Aspect Ratios

In order to verify the improved RPN, the detection experiments of different aspect ratio insulators were carried out. A comparison of the detection results of the insulators in power transmission and transformation inspection images by the traditional RPN and the improved RPN is demonstrated in Figure 8. Figure 8a,c,e are the experimental results obtained by traditional RPN while Figure 8b,d,f are the experimental results obtained by the improved RPN. In Figure 8a,c, the aspect ratio of the output boxes is 1:1 or 1:2, which is not a suitable size for insulators. In Figure 8b,d, the aspect ratio of the final output boxes is significantly larger, and the narrow insulator object almost completely fits. In Figure 8e, to some extent, the insulator object is blocked by the framing frame of the infrared camera so that the final output of the output boxes does not completely contain the insulator. In Figure 8f, the final test results effectively overcome the isolation of the infrared camera frame and the detection accuracy is significantly improved.

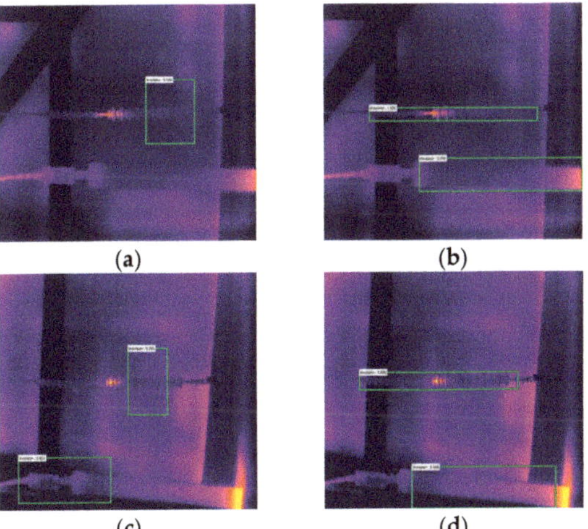

Figure 8. *Cont.*

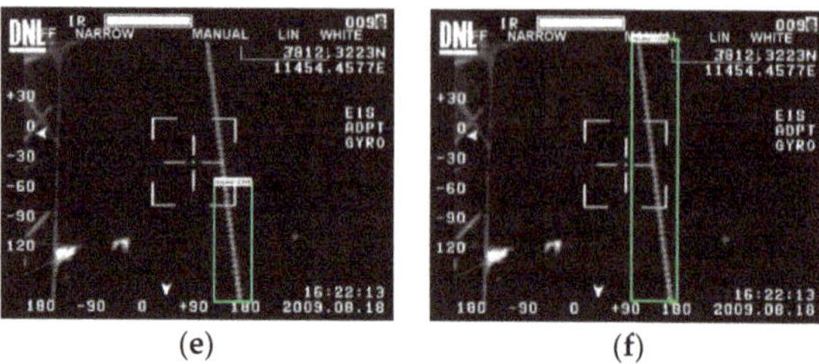

Figure 8. Detection results of insulators with different aspect ratios. (**a**,**c**,**e**) Results obtained by traditional RPN; (**b**,**d**,**f**) results obtained by improved RPN.

4.2. Detection Experiment of Insulators with Different Scales

In order to verify the detection effect for different scale insulators by using the improved RPN, experiments on insulators with different scales were carried out. A comparison of the detection results of the insulators using the traditional RPN and the improved RPN in power transmission and transformation inspection images was performed and is demonstrated in Figure 9; dealing specifically with insulators with different scales. Figure 9a,c,e,g are show the experimental results obtained by traditional RPN, while Figure 9b,d,f,h show the experimental results obtained by the improved RPN. As can be seen in Figure 9a,c,e,g, the faster R-CNN model can detect larger-scale insulators more accurately, but the missed detection rate for small-scale insulators is extremely high. In Figure 9b,d,f,h, after the application of the small-scale reference anchor in the RPN, the detection problem for small-scale insulators is greatly improved and small insulators in the image can also be detected accurately.

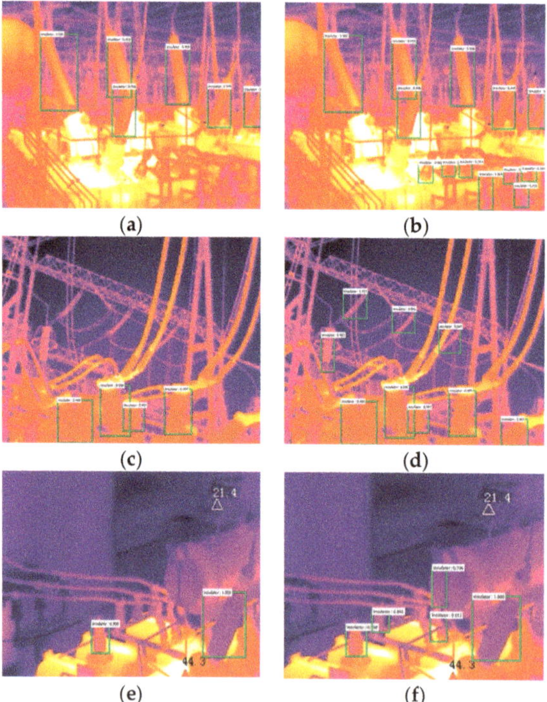

Figure 9. *Cont.*

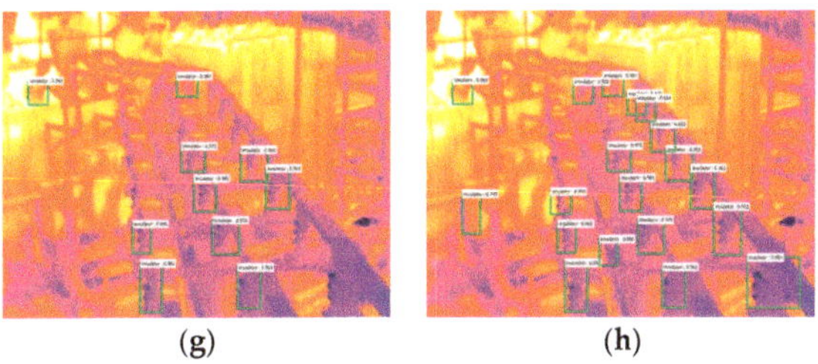

Figure 9. Detection results of insulators with different scales. (**a**,**c**,**e**,**g**) Results obtained by traditional RPN; (**b**,**d**,**f**,**h**) results obtained by improved RPN.

4.3. Detection Experiment of Insulators Under Mutual Occlusion

In order to verify the detection effect of the NMS on the mutual occluded insulators, a mutual occluded insulator test was carried out. Figure 10 is the score distribution of each bounding box output by Figure 3. Figure 10a represents for the distribution of the bounding box scores before the improvement of the NMS, while Figure 10b is the distribution of the bounding box scores after the improvement of the NMS. For the sake of easy observation, the scores of the top 100 test boxes out of 300 were selected to clearly show the trend of the score change. It can be seen from the score distribution that after the improvement of the score mapping relationship in the traditional NMS, the improved bounding box score reduces the redundancy of error detection under the top 10 boxes with the highest scores. At the same time, the two insulators in the middle of Figure 3 can be detected separately. The final detection results are shown in Figure 11.

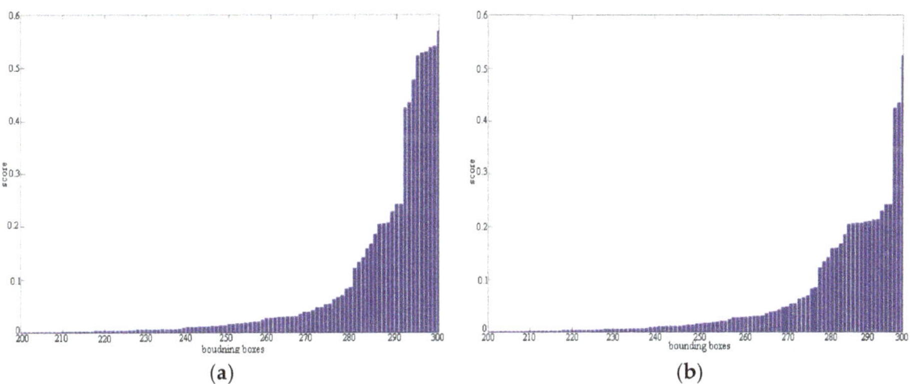

Figure 10. Score distribution of bounding boxes. (**a**) Before improvement; (**b**) after improvement.

Figure 11. Results of eliminating redundant boxes with improved NMS.

4.4. Comparison Experiment with Other R-CNN Object Detection Models

In the R-CNN series model, the region-based fully convolutional networks (R-FCN) [23] and single shot multibox detector (SSD) [24] are two new object detection models proposed after the faster R-CNN model. R-FCN uses a region-based full convolutional network for object detection, which can realize fully shared computation between the outputs of each convolutional layer. SSD is based on the idea of regression and is combined with the anchor generation mechanism in the faster R-CNN model. It makes use of the multi-scale region features at various locations of the full figure to make regressions while generating candidate object regions.

A comparison of the detection results among the RPN in the faster R-CNN model, the R-FCN, and the SSD methods are made in Figures 12 and 13. Figure 12a shows the experimental results obtained by the method proposed in this paper, while Figure 12b,c shows the experimental results obtained by the R-FCN and SSD methods, respectively. Figure 13a shows the recall-accuracy results obtained by the method proposed herein, while Figure 13b,c shows the recall-accuracy results obtained by the R-FCN and SSD methods, respectively. It can be seen from the figures that through the improvement of the RPN and NMS method of the faster R-CNN model, the detection accuracy of the insulator in power transmission and transformation inspection images is higher than that of the R-FCN and SSD methods. The method in this paper effectively solves the problem of insulator detection with different aspect ratios and different scales. At the same time, the missed detection rate is also significantly reduced.

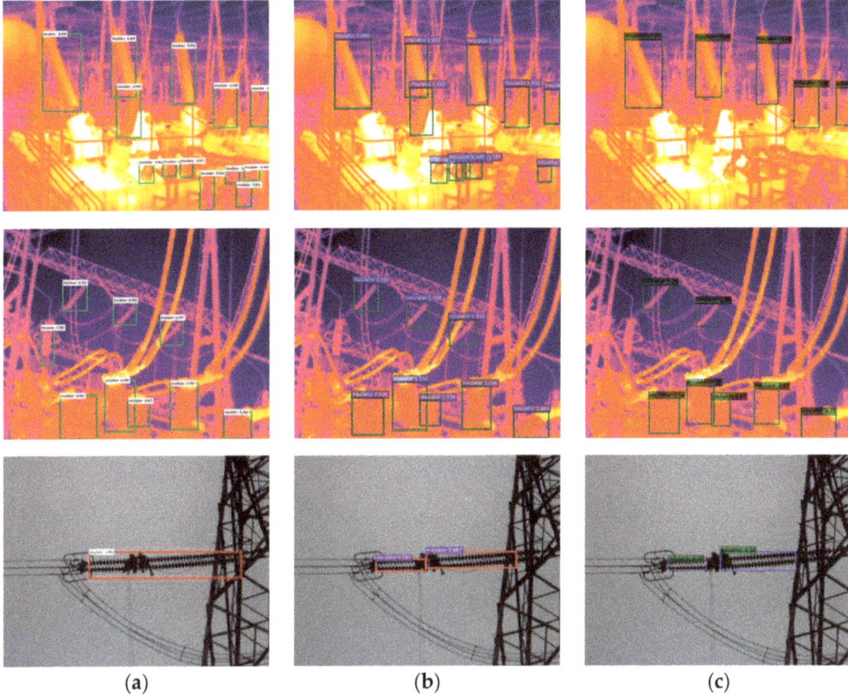

(a) (b) (c)

Figure 12. The detection results of our method, and the R-FCN and SSD methods. (**a**) Proposed method; (**b**) R-FCN; (**c**) SSD.

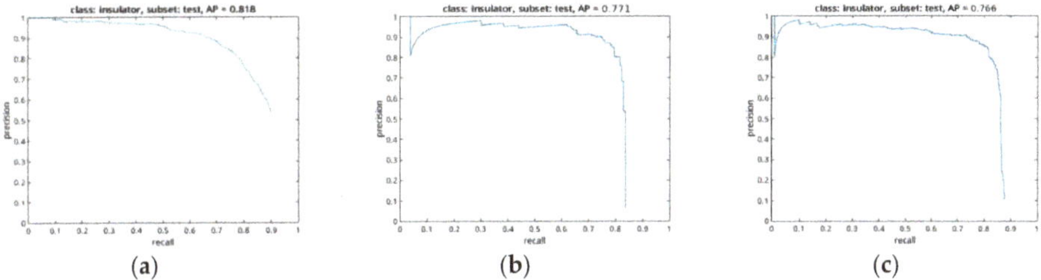

(a) (b) (c)

Figure 13. Result comparison of the proposed method, and the R-FCN and SSD methods. (**a**) Proposed method; (**b**) R-FCN; (**c**) SSD.

4.5. Quantitative Analysis on Experimental Results

In addition to the qualitative observation of the experimental results, this paper also makes a quantitative analysis of the performance difference before and after the improvement from a numerical perspective. The experimental results were quantitatively analyzed mainly by precision recall (PR) curve and AP value. The horizontal axis of the PR curve is the recall rate of the object detection, while the vertical axis is the accuracy of the object detection, which demonstrates the relationship between accuracy and recall rate. The AP value is the area surrounded by the PR curve and the horizontal and vertical axis. Figure 14 shows the PR and the AP values for the faster R-CNN model using the VGG-16 Net before and after modification by the proposed method. After using the transmission and transformation electrical insulation equipment detection database to fine-tune the faster R-CNN model, the AP value of the faster R-CNN model using the VGG-16 Net reached 0.640. On this basis, after the improvement of the anchor generation method and NMS in the RPN, the AP value of faster R-CNN using the VGG-16 Net is increased to 0.818, which is 27.81% higher than before the improvement. The PR curve is also smoother and closer to the upper-right corner, which proves the effectiveness of the method.

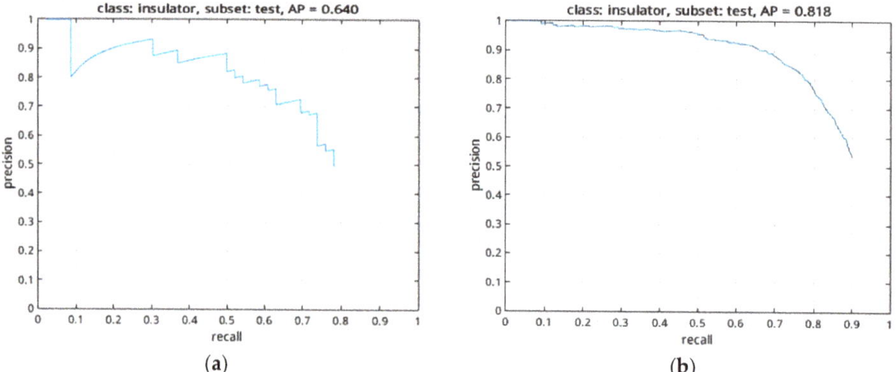

Figure 14. Precision recall (PR) curve and average precision (AP) value of the faster R-CNN model under VGG-16 Net. (**a**) Before improvement; (**b**) after improvement.

5. Conclusions

This paper proposes an insulator detection method based on the detection data set of insulation equipment trained by a fine-tuned faster R-CNN model. It can detect insulators with different aspect ratios and different scales as well as mutually occluded ones in the power transmission and inspection images effectively. In this paper, the anchor generation method and NMS method in the faster R-CNN model RPN are improved, respectively. The experimental results show that the improvement of the anchor generation method increases the AP value of the faster R-CNN model to 0.818 by using VGG-16 Net. The detection effect of different aspect ratios and different scales of insulators in inspection images is improved significantly. Meanwhile, the problem of missed insulators as a result of occlusion is also resolved with the proposed NMS improvement method. It can effectively realize the separate detection of the insulators in the image, which lays a solid foundation for further state detection and fault diagnosis of insulators.

It is noted that different types of insulators are not distinguished in our method. In order to provide a more reliable guarantee for state detection and fault diagnosis of insulators, further research on high-quality detection methods for different types of insulators will be done in the future.

Author Contributions: Z.Z. (Zhenbing Zhao), L.Z. and Y.Q. developed the system modeling and algorithm design; Z.Z. (Zhen Zhen) and L.Z. performed the experiments; Y.K. and K.Z. analyzed the experimental data; Z.Z. (Zhenbing Zhao), Z.Z. (Zhen Zhen), and L.Z. wrote the paper.

Funding: This research is supported in part by the National Natural Science Foundation of China (NSFC) under grant number 61871182, 61401154, 61773160, 61302163, by Beijing Natural Science Foundation under grant number 4192055, by the Natural Science Foundation of Hebei Province of China under grant number F2016502101, F2017502016, F2015502062, by the Fundamental Research Funds for the Central Universities under grant number 2018MS095, 2018MS094, and by the Open Project Program of the National Laboratory of Pattern Recognition (NLPR).

Conflicts of Interest: The authors declare no conflict of interest.

References

1. Wang, M.; Du, Y.; Zhang, Z.R. Study on power transmission lines inspection using unmanned aerial vehicle and image recognition of insulator defect. *J. Electron. Meas. Instrum.* **2015**, *26*, 1862–1869.
2. Huang, X.B.; Liu, X.H.; Zhang, W. Classification and identification method of aerial insulator based on red-blue difference and improved K-means algorithm. *High Volt. Technol.* **2018**, *44*, 1528–1534.
3. Huang, X.O.; Li, J.Q.; Zhang, Y.; Zhang, F. The recognition and detection technology of ice-covered insulators under complex environment. *High Volt. Eng.* **2017**, *43*, 891–899.
4. Tong, W.G.; Yuan, J.S.; Li, B.S. Application of Image Processing in Patrol Inspection of Overhead Transmission Line by Helicopter. *Power Syst. Technol.* **2010**, *34*, 204–208.
5. Huang, X.N.; Zhang, Z.L. A Method to Extract Insulator Image from Aerial Image of Helicopter Patrol. *Power Syst. Technol.* **2010**, *34*, 194–197.
6. Tao, G.; Fengxiang, C.; Wei, W.; Ping, S.; Lei, S.; Tianzhu, C. Electric insulator detection of UAV images based on depth learning. In Proceedings of the IEEE 2017 2nd International Conference on Power and Renewable Energy (ICPRE), Chengdu, China, 20–23 September 2017; pp. 37–41.
7. Krizhevsky, A.; Sutskever, I.; Hinton, G.E. Imagenet classification with deep convolutional neural networks. In Proceedings of the Advances in Neural Information Processing Systems, Lake Tahoe, NV, USA, 3–6 December 2012; pp. 1097–1105.
8. Zhao, Z.; Xu, G.; Qi, Y.; Liu, N.; Zhang, T. Multi-patch deep features for power line insulator status classification from aerial images. In Proceedings of the IEEE 2016 International Joint Conference on Neural Networks (IJCNN), Vancouver, BC, Canada, 24–29 July 2016; pp. 3187–3194.
9. Girshick, R. Fast R-CNN. In Proceedings of the IEEE International Conference on Computer Vision (ICCV), Araucano Park, Chile, 11–18 December 2015; pp. 1440–1448.
10. Chen, Q.; Yan, B.; Ye, R.; Zhou, X.J. Insulator detection and recognition of explosion fault based on convolutional neural networks. *J. Electron. Meas. Instrum.* **2017**, *31*, 942–953.
11. Ren, S.; He, K.; Girshick, R.; Sun, J. Faster R-CNN: towards real-time object detection with region proposal networks. In Proceedings of the International Conference on Neural Information Processing Systems (NIPS), Palais des congrès de Montréal, Montreal, QC Canada, 7–12 December 2015; pp. 91–99.
12. Wang, W.G.; Tian, B.; Liu, Y.; Liu, L.; Li, J.X. Study on the Electrical Devices Detection in UAV Images based on Region Based Convolutional Neural Networks. *J. Geo-Inf. Sci.* **2017**, *19*, 256–263.
13. Ma, L.; Xu, C.; Zuo, G.; Bo, B. Detection Method of Insulator Based on Faster R-CNN. In Proceedings of the IEEE 2017 7th Annual International Conference on CYBER Technology in Automation, Control, and Intelligent Systems (CYBER), Sheraton Princess Kaiulani, HI, USA, 31 July–4 August 2017; pp. 1410–1414.
14. Li, J.F.; Wang, Q.R.; Li, M. Image recognition of power equipment combined with deep learning and random forest. *High Volt. Eng.* **2017**, *43*, 3705–3711.
15. Li, S.M.; Lei, G.Q.; Fan, R. Depth Map Super-resolution Reconstruction Based on Convolutional Neural Network. *Acta Opt. Sin.* **2017**, *37*, 1210002.
16. Fu, W.B.; Sun, T.; Liang, L.; Yan, B.W.; Fan, F.X. A Review of the Principles and Applications of Deep Learning. *Comput. Sci.* **2018**, *45*, 11–15.
17. Deng, L. The MNIST Database of Handwritten Digit Images for Machine Learning Research. *IEEE Signal Process. Mag.* **2012**, *29*, 141–142. [CrossRef]
18. Deng, J.; Dong, W.; Socher, R.; Li, L.-J.; Li, K.; Li, F.-F. ImageNet: A large-scale hierarchical image database. In Proceedings of the Computer Vision and Pattern Recognition (CVPR), Xiami Beach, FL, USA, 20–25 June 2009; pp. 248–255.
19. Everingham, M.; Gool, L.V.; Williams, C.K.I.; Winn, J.; Zisserman, A. The Pascal Visual Object Classes (VOC) Challenge. *Int. J. Comput. Vis.* **2010**, *88*, 303–338. [CrossRef]
20. Fu, B.; Jiang, Y.; Wang, H.G.; Jiang, W.D.; Song, Q.F.; Wang, C.C.; Chu, Q.L.; Zhao, Y.P. Intelligent Diagnosis System for Patrol Check Images of Power Transmission Lines. *CAAI Trans. Intell. Syst.* **2016**, *11*, 70–77.
21. Xu, G.Z. *Research on Infrared Image Positioning Method of Insulator Based on Depth Feature Expression*; North China Electric Power University: Beijing, China, 2017.
22. Simonyan, K.; Zisserman, A. Very Deep Convolutional Networks for Large-Scale Image Recognition. In Proceedings of the International Conference on Learning Representations (ICLR), San Diego, CA, USA, 7–9 May 2015; pp. 1–14.

23. Dai, J.F.; Li, Y.; He, K.M.; Sun, J. R-FCN: Object Detection via Region-based Fully Convolutional Networks. In Proceedings of the Conference and Workshop on Neural Information Processing Systems (NIPS), Barcelona, Spain, 5–10 December 2016; pp. 379–387.
24. Liu, W.; Anguelov, D.; Erhan, D.; Szegedy, C.; Reed, S.; Fu, C.; Berg, A.C. SSD: Single Shot MultiBox Detector. In Proceedings of the European Conference on Computer Vision (ECCV), Amsterdam, The Netherlands, 8–16 October 2016; pp. 21–37.

Article

Data-Driven Framework to Predict the Rheological Properties of $CaCl_2$ Brine-Based Drill-in Fluid Using Artificial Neural Network

Ahmed Gowida [1], Salaheldin Elkatatny [1,*], Emad Ramadan [2] and Abdulazeez Abdulraheem [1]

1 Petroleum department, King Fahd University of Petroleum & Minerals, Dhahran 31261 Box 5049, Saudi Arabia; g201708730@kfupm.edu.sa (A.G.); aazeez@kfupm.edu.sa (A.A.)
2 Information & Computer Science Department, King Fahd University of Petroleum & Minerals, Dhahran 31261 Box 5049, Saudi Arabia; eramadan@kfupm.edu.sa
* Correspondence: elkatatny@kfupm.edu.sa; Tel.: +966-594663692

Received: 18 April 2019; Accepted: 13 May 2019; Published: 17 May 2019

Abstract: Calcium chloride brine-based drill-in fluid is commonly used within the reservoir section, as it is specially formulated to maximize drilling experience, and to protect the reservoir from being damaged. Monitoring the drilling fluid rheology including plastic viscosity, *PV*, apparent viscosity, *AV*, yield point, Y_p, flow behavior index, *n*, and flow consistency index, *k*, has great importance in evaluating hole cleaning and optimizing drilling hydraulics. Therefore, it is very crucial for the mud rheology to be checked periodically during drilling, in order to control its persistent change. Such properties are often measured in the field twice a day, and in practice, this takes a long time (2–3 h for taking measurements and cleaning the instruments). However, mud weight, *MW*, and Marsh funnel viscosity, *MF*, are periodically measured every 15–20 min. The objective of this study is to develop new models using artificial neural network, ANN, to predict the rheological properties of calcium chloride brine-based mud using *MW* and *MF* measurements then extract empirical correlations in a white-box mode to predict these properties based on *MW* and *MF*. Field measurements, 515 points, representing actual mud samples, were collected to build the proposed ANN models. The optimized parameters of these models resulted in highly accurate results indicated by a high correlation coefficient, *R*, between the predicted and measured values, which exceeded 0.97, with an average absolute percentage error, *AAPE*, that did not exceed 6.1%. Accordingly, the developed models are very useful for monitoring the mud rheology to optimize the drilling operation and avoid many problems such as hole cleaning issues, pipe sticking and loss of circulation.

Keywords: mud rheology; drill-in fluid; artificial neural network; Marsh funnel; plastic viscosity; yield point

1. Introduction

Drilling Fluids are considered a key element in the drilling operation. Conventional drilling fluids are water-based, oil-based or synthetic-based fluid systems, which are used in the drilling process to give the best performance under certain temperatures and pressures experienced downhole [1]. Drilling the section from the sea-bed/land to the top of the reservoir is different, regarding the economic value of the final project, compared to the reservoir section. As in the top sections, the concerns are to seal the permeable formations, and help sustain the wellbore stability.

Special measures will be taken into consideration while drilling the reservoir section to avoid damaging the reservoir and plugging the reservoir pores. For that target, special drilling fluids are used, called reservoir drill-in fluids (RDFs), which are specially formulated to maximize drilling experience and protect the reservoir from being damaged until the completion process is proceeded [2].

There are many types of RDFs with different chemical compositions, but the concern of this study is about the clear brine-based mud which is often used within completions, as the presence of solids is a major contributor to formation damage [3]. However, when used as drilling fluid, the solids-free nature of brine operationally improves the rate of penetration (*ROP*), stabilization of sensitive formations, density, and abrasion or friction [4]. Clear brine fluids properties are easier to maintain than conventional solids-laden fluid systems, so that when properly run, clear systems require very little maintenance, because many functional issues are inherently solved by the dissolved salts. Clear brine fluids also allow for drill site cost reductions because of the ability to reuse the fluid [5].

Brine fluids can be prepared with one salt or a combination of salts. All salts provide unique properties to the base fluid. Saturated brines fluids provide excellent inhibitive properties and lubricity, as compared to conventional aqueous fluids. With optimal heat transference characteristics, they can greatly improve bit life, and increase the rate of penetration in hard rock drilling. Among the different salts used for clear-brine systems, calcium chloride has been selected, as it is considered one of the most economic brine systems, with its broad range of densities (from 9.0 to 11.6 *ppg*), availability, low cost, and its ability to reduce the water activity of the fluid [6].

1.1. Drilling Fluid Rheology

Drilling fluid rheology plays a key role in optimizing drilling performance [7]. These properties significantly affect the efficiency of the hole cleaning and the drilling rate [8], which are critical factors controlling the performance of drilling operation [9]. These rheological properties include mud density to provide the control on the formation pressure, while PV, Y_P, AV, n and k are used for controlling hole cleaning and optimizing the drilling performance [10]. Plastic viscosity of the drilling fluid is crucial for optimizing the drilling operation [11]. It is an indication of the solid content in the drilling fluid which may negatively affect the drilling performance when it exceeds critical limits, and can cause many problems like pipe sticking and decreasing the rate of penetration [12]. Yield point can be simply stated as the attractive forces among colloidal particles in drilling fluid [11]. The optimization of yield point is central to controlling the efficiency of hole cleaning [10,13]. Moreover, apparent viscosity is considered a key factor in the optimization of mud hydraulics while drilling [8]. In addition, the parameters k and n can be used for evaluating hole cleaning during the drilling operation [14].

The rheological properties can be measured in the laboratory using mud balance and viscometer. The mud balance is used to measure the mud weight while the rheometer is used to measure (PV, Y_P, AV, n and k). However, this process takes a relatively long time (2–3 h for taking measurements and cleaning the instruments) which makes it difficult to be performed periodically and practically in the field. Therefore, it is taken as a common procedure that only density and Marsh funnel viscosity are measured periodically every 15–20 min, using mud balance and Marsh funnel devices. On the other hand, a complete mud test (including all the drilling fluid properties), using the mud balance and viscometer, is performed twice a day. Marsh funnel viscosity provides an indication of the changes in the rheology of the drilling fluid. This funnel was first introduced by Marsh [15]. This tool is cheap and takes a short time, so it can be utilized to give field measurements frequently and estimate some parameters like yield stress [16]. Based on the literature, there are two models developed to predict the drilling fluid viscosity from mud density and Marsh funnel measurements. These two measurements were used as inputs to calculate the effective viscosity of the drilling fluid as stated by Pitt [17] in Equation (1). Then a modification on the previous model was introduced by Almahdawi et. al. [18], who figured out that changing the value of the constant to 28 in Equation (2) instead of 25 presented by Pitt [17], is more effective give more accurate results, as compared to Equation (1).

$$AV = D(T - 25) \tag{1}$$

$$AV = D(T - 28) \tag{2}$$

where AV is the apparent viscosity in cP, D is the fluid density in g/cm^3, T is the Marsh funnel time in sec.

Several mathematical models have been mentioned in the literature for estimating the fluid rheological properties using Marsh funnel devices. Some of them suggested using the temporal variations in the fluid height in the funnel to determine different rheological parameters such as PV, Y_P and AV [19–22]. They introduced a methodology to determine the shear rate and the shear stress on the walls of the Marsh funnel from the measured discharged fluid volume of the Marsh funnel at different points. Then several rheological parameters have been related to the obtained shear rate and shear stress. Abdulrahman et al. [23] investigated different water-based drilling fluids using the Marsh funnel and showed that PV and AV can be estimated using consistency plots and the methodology described in [19]. However, these models showed considerable discrepancies between the results obtained from the Marsh funnel and the standard viscometers. Other studies tried to model the fluid volume flow in the Marsh funnel with higher order polynomial functions, rather than the simplified functions used in the previously mentioned studies [24,25]. This attempt was to simulate the fluid temporal height in the Marsh funnel more properly, and to get closer results of rheological parameters to those obtained from the standard viscometers.

The objective of this work is to develop new models using artificial neural networks, ANN, to predict the rheological properties of the $CaCl_2$ brine-based drilling fluid depending on frequent measurements of MW and MF. The real-time measurements of these parameters are very helpful for identifying the efficiency of the hole cleaning, optimizing the drilling fluid hydraulics, equivalent circulating density calculations and swab and surge pressure determination.

1.2. Artificial Neural Network (ANN)

Artificial intelligence, (AI), can be simply defined as the computer science branch for creating intelligent machines [26] to exhibit human brains to make predictions and help take the right decisions for the future scenarios [27]. Recently, different AI methods such as fuzzy logic, FL, support vector machine, SVM, genetic algorithm and artificial neural network, ANN, have been applied in petroleum engineering, and specifically in the field of drilling fluid engineering. Some of these applications include fluid flow patterns prediction in wellbore annulus [28], stuck pipe prediction [29], drilling hydraulics optimization [30], frictional pressure loss estimation [31], hole cleaning and prediction of cutting concentration [32], estimation of the static Poisson's ratio from log data [33].

ANN is one of the most common AI techniques which has the ability to deal with different engineering problems with high complexity that exceed the computational capability of classical mathematics and procedures [34]. It is based on analogy with biological neural networks to simulate the performance of the human biological neural system [35]. The elementary units for ANN are neurons [36]. The structure of the ANN consists of three main types of layers. The first one is for the input parameters. The second one is called hidden layers, which include the neurons assigned with the transfer functions between the inputs and the outputs. The third type is for the outputs. These layers, with the suitable training algorithm, describe the nature of the problem [37]. The performance of the network is controlled by key parameters including the number of neurons, weights and biases [38]. To optimize the weight and biases, the network is trained using different algorithms to achieve the lowest possible error. Among these algorithms is Levenberg-Marquardt (LM), which is an iterative, curve fitting algorithm. This algorithm proved its outstanding performance in solving non-linear least-squares problems [39].

There are many ANN applications of ANN in the field of drilling fluid in the last few years. Some of these researches are the prediction of filtration volume and mud cake permeability of water-based mud (WBM) [40], drill cutting settling velocity prediction [41], prediction of differential pipe sticking [42], lost circulation prediction [43], hole cleaning efficiency of foam fluid [44], rheological properties of invert emulsion mud [45], invert emulsion mud rheology [46] and spud mud rheology prediction [47], generating geomechanical well logs [48], prediction of oil PVT properties [49].

Based on the literature, more than 50 percent of the applications in the drilling fluid area used ANN for the predictions and got high accurate results. Accordingly, ANN has been selected for building the proposed models in this study [26].

2. Methodology

2.1. Data Description

A typical sample of the data for $CaCL_2$ brine-based drill-in fluid (515 field data for actual mud samples) is listed in Table 1, including $\left(MW,\ MF,\ PV,\ \text{and}\ Y_p\right)$. The drilling fluid samples are collected after the mud was cleaned from the cuttings by using the shale shaker MW and MF are measured in the field using a mud balance and Marsh funnel, respectively. The rheometer is used for measuring the rheology of the mud, namely PV, and Y_p at atmospheric pressure and 120°F. The collected data have a wide range as follows: MW ranges from 43 to 119 Ib/ft^3, MF ranges from 26 to 135 $s/quart$, PV ranges from 10 to 54 cP, and Y_P ranges from 8 to 41 $Ib/100\ ft^2$. Figure 1 shows that MW has R of 0.36 and 0.76 with Y_P and PV respectively while MF has R of 0.86 with Y_P and 0.36 with PV.

Table 1. A typical sample of the $CaCl_2$ brine-based drilling fluid collected data.

MW, Ib/ft^3	MF, $s/quart$	PV, cP	Y_P, $Ib/100\ ft^2$
78	62	8	21
80	45	8	22
88	62	12	18
73	50	11	20
65	48	11	21
75	44	12	20

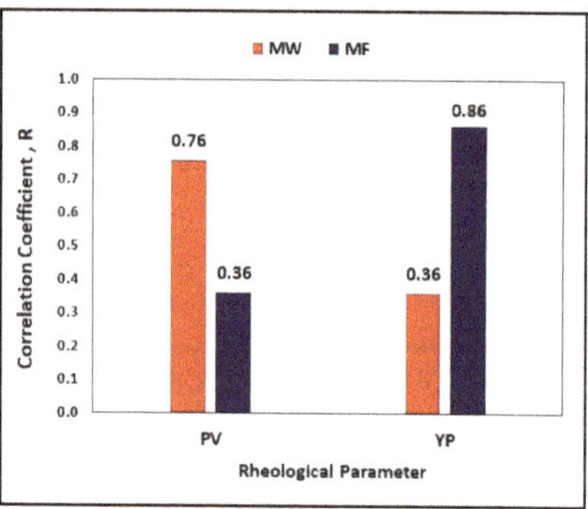

Figure 1. The relative importance of MW and MF with the rheological properties (Y_P and PV) of $CaCL_2$ brine-based mud in terms of the correlation coefficient, R.

For better prediction using AI models, data should be analyzed and filtered [50]. Therefore, the selected data have been cleaned from any noise and false values for higher representation quality. The filtration process included eliminating all the values that cannot be representative, like negative values. Finally removing the outliers that show significant deviation from the other values of a variable, the outliers were removed using a box and whisker plot, in which top whisker represents the upper limit of the data, and the bottom whisker represents the lower limit of the data, then any value beyond these limits is considered an outlier and removed [51]. These limits are determined by dividing the data into four equal divisions (quartiles) along with using the minimum, maximum, mean and median parameters [52] obtained from the statistical analysis of the data listed in Table 2.

Table 2. Statistical Analysis of the $CaCl_2$ brine-based mud collected data.

Parameter	*MW, Ib/ft*^{3}	*MF, s/quart*	*PV, cP*	Y_P*, Ib/*100 *ft*^{2}
Min.	43	26	10	8
Max.	119	135	54	41
Mean	85.45	56.33	22.59	25.88
Mode	76	109	44	33
Range	72	50	19	24
Skewness	0.50	1.43	1.29	0.83

2.2. Development of ANN Models

The collected data were used to calculate R_{600} and R_{300} (rheometer readings at 600 and 300 rpm, respectively) using Equations (3) and (4). These two parameters are very crucial for identifying fluid properties and flow regimes. Then, the apparent viscosity, AV, flow behavior index, n, and flow consistency index, k, are calculated using Equations (5)–(7) respectively.

$$R_{600} = P_V + R_{300} \tag{3}$$

$$R_{300} = P_V + Y_p \tag{4}$$

$$AV = \frac{R_{600}}{2} \tag{5}$$

$$n = 3.32 \times \log\left(\frac{R_{600}}{R_{300}}\right) \tag{6}$$

$$k = \frac{R_{600}}{1022^n} \tag{7}$$

For all the upcoming developed models, different scenarios have been performed to optimize the ANN variables to reach the highest accuracy with the lowest possible error for prediction using different combinations of the available options of the ANN variables. The optimized parameters obtained from the tuning process of these parameters are summarized in Table 3. The chosen architecture for the developed models includes three layers:

- Input layer: It contains input features which are MW and MF.
- (One) Hidden layer: It contains the optimized number of neurons which was found to be 20 neurons.
- Output layer: It contains the output parameters, which are (PV, Y_P, AV, n and k individually).

The network was trained using the Levenberg-Marquardt (LM) algorithm to get the optimized weights and biases. The neurons are arranged to be trained using a learning rate of 0.12. Activation function of the tan-sigmoidal type (tansig) was assigned between the input and hidden layers while the pure-linear function was assigned between the hidden and output layers. Figure 2 shows a typical schematic of the architecture of the developed ANN model.

Table 3. Summary of the optimized parameters for the developed ANN models.

Neural Network Parameter	Types and Range
Training Algorithm	Levenberg Marquardt
Number of neurons	20
Number of hidden layer(s)	1
Learning rate	0.12
The hidden layer transfer function	Tan-sigmoidal
The outer layer transfer function	Pure-linear
Training ratio	70%
Testing ratio	30%

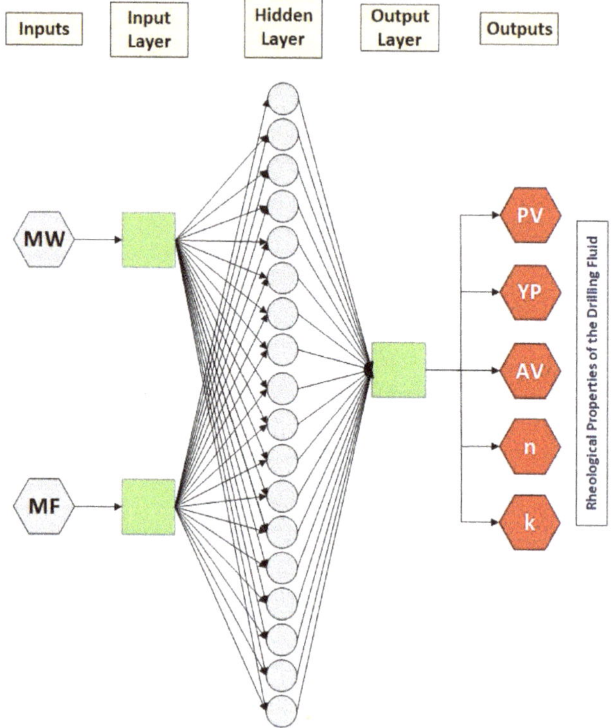

Figure 2. Typical schematic for the architecture of the developed ANN models.

3. Results and Discussion

3.1. Yield Point (Y_P) Model

An ANN-Based model was developed using *MW* and *MF* as inputs to predict the Y_P values. The obtained data were divided into ratios 70:30 for training and testing the model, respectively. Figure 3 shows the high match between the measured and the predicted Y_P values from the developed ANN model in terms of *R* of 0.97 and *AAPE* of 3.9%. Thereafter, a new correlation has been developed using the ANN model to predict Y_P based on *MW* and *MF*. First, the inputs should be normalized using Equations (8) and (9) to substitute the values MW_n and MF_n in Equation (10); where MW_n refers to the first normalized input, and MF_n refers to the second normalized input.

$$MW_n = 0.036(MF - 64) + 1 \tag{8}$$

$$MF_n = 0.133(MF - 26) + 1 \tag{9}$$

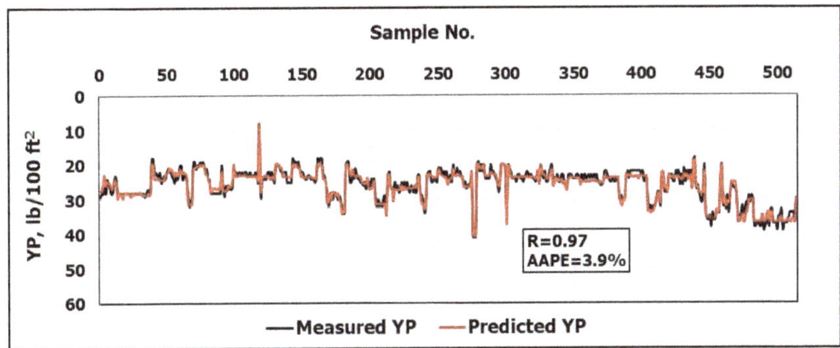

Figure 3. Measured Y_P vs. Predicted Y_P from the ANN model.

Then, the normalized value Y_{Pn} is calculated using Equation (9) with its optimized coefficients listed in Table 4.

$$Y_{P_n} = \left[\sum_{i=1}^{N} w_{2,i}\left(\frac{2}{1+\exp\left(-2\left(MW_n \times w_{1_{i,1}} + MF_n \times w_{1_{i,2}} + b_{1,i}\right)\right)} - 1\right)\right] + b_2 \quad (10)$$

where i is the index of the neuron in the hidden layer, N is the optimized number of neurons for only one hidden layer, which is found to be 20, w_1 is the weight vector linking the input and the hidden layer, w_2 is the weight vector linking the hidden and output layer, b_1 is the biases vector for the input layer, and b_2 for the output layer.

Finally, the required Y_P value can be obtained by denormalizing Y_{Pn} using Equation (11).

$$Y_p = 16.5(Yp_n + 1) + 8 \quad (11)$$

Table 4. The optimized coefficients for estimating the normalized Y_{Pn} in Equation (10).

Neuron Index	Input Layer Weights		Hidden Layer Weights	Input Layer Biases	Output Layer Bias
i	$w_{1_{i,1}}$	$w_{1_{i,2}}$	$w_{2,i}$	$b_{1,i}$	b_2
1	−4.251	5.585	−0.950	5.530	−0.508
2	−0.709	−6.857	−0.155	4.936	-
3	2.631	5.539	0.168	−4.752	-
4	−0.743	6.411	1.005	3.899	-
5	−4.986	5.903	−0.932	3.661	-
6	−5.203	−0.250	−1.022	3.135	-
7	4.859	4.645	−0.410	−2.792	-
8	−1.185	−6.192	0.721	2.879	-
9	4.188	−3.646	−1.697	−3.015	-
10	3.238	−5.080	0.297	−0.585	-
11	0.708	−7.849	−0.380	0.213	-
12	−4.893	−9.220	−0.428	−1.489	-
13	2.227	−6.971	−1.173	2.051	-
14	3.101	5.504	−1.046	1.840	-
15	−6.059	−1.558	−0.030	−3.063	-
16	−5.020	3.702	−0.902	−3.873	-
17	2.892	5.503	−0.260	4.287	-
18	−0.736	−5.668	0.190	−5.818	-
19	4.290	−4.592	0.639	5.571	-
20	−4.290	4.570	−0.686	−6.252	-

3.2. Apparent Viscosity (AV) Model

Similarly, AV was predicted using ANN, based on MW and MF. The model was trained using 70% of the available data, while 30% of the data were used for testing the model. Figure 4 shows the high R between the predicted and the measured AV values, which is 0.99 with $AAPE$ of 3.2%. Afterward, a new correlation for predicting AV was extracted from the developed ANN model. To use this correlation, the inputs should be normalized at first using Equations (12) and (13) to substitute the values MW_n and MF_n in Equation (14); where MW_n refers to the first normalized input, and MF_n refers to the second normalized input.

$$MW_n = 0.036(MF - 64) + 1 \tag{12}$$

$$MF_n = 0.053(MF - 35) + 1 \tag{13}$$

Then, the normalized value AV_n is calculated using Equation (14) with its optimized coefficients listed in Table 5.

$$AV_n = \left[\sum_{i=1}^{N} w_{2,i}\left(\frac{2}{1+\exp\left(-2\left(MW_n \times w_{1_{i,1}} + MF_n \times w_{1_{i,2}} + b_{1,i}\right)\right)} - 1\right)\right] + b_2 \tag{14}$$

Finally, AV can be predicted by denormalizing AV_n using Equation (15).

$$AV = 27(AV_n + 1) + 19 \tag{15}$$

Table 5. The optimized coefficients for estimating the normalized AV_n in Equation (14).

Neuron Index	Input Layer Weights		Hidden Layer Weights	Input Layer Biases	Output Layer Bias
i	$w_{1_{i,1}}$	$w_{1_{i,2}}$	$w_{2,i}$	$b_{1,i}$	b_2
1	2.960	7.136	−0.950	−5.667	1.535
2	−5.586	−7.703	−0.155	2.751	-
3	4.037	5.355	0.168	−1.886	-
4	−3.362	3.743	1.005	2.292	-
5	6.295	−5.066	−0.932	−8.110	-
6	0.406	7.091	−1.022	6.492	-
7	−10.26	−8.654	−0.410	−0.995	-
8	−0.572	−9.022	0.721	−0.909	-
9	−7.565	4.329	−1.697	4.884	-
10	−4.256	3.855	0.297	2.094	-
11	6.458	1.765	−0.380	1.786	-
12	4.537	−4.152	−0.428	1.324	-
13	−5.410	3.103	−1.173	−2.553	-
14	−4.859	−2.202	−1.046	−1.924	-
15	−7.190	1.704	−0.030	−4.783	-
16	2.196	−5.993	−0.902	2.408	-
17	−0.576	6.113	−0.260	−4.569	-
18	−2.889	−4.782	0.190	−4.645	-
19	−3.799	−7.588	0.639	−2.337	-
20	−3.877	4.861	−0.686	−6.301	-

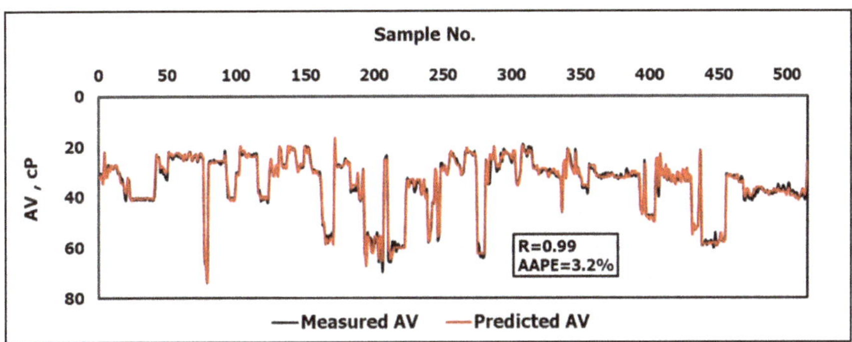

Figure 4. Measured AV vs. Predicted AV from the ANN model.

3.3. Plastic Viscosity (PV) Model

For PV, another ANN model was developed based on MW and MF. For building the model, the ratio of the training to testing points is 70:30. The model gave high accurate results indicated by a high R of 0.98 between the predicted and the measured PV values and maximum $AAPE$ of 6.1% as shown in Figure 5. A new correlation has been extracted from the model to predict PV without the need to run the ANN model.

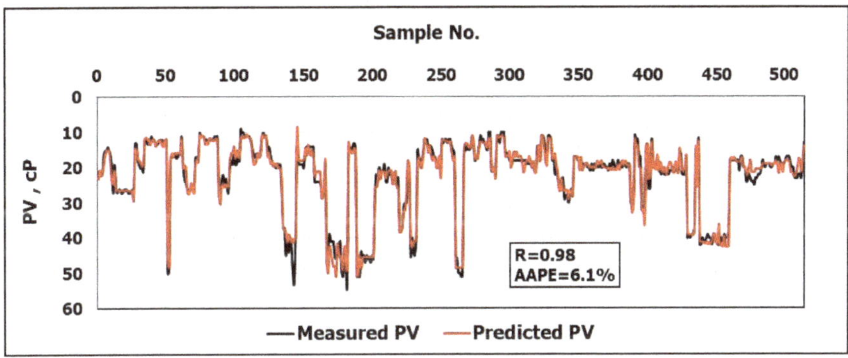

Figure 5. Measured PV vs. Predicted PV from the ANN model.

First, the inputs should be normalized using Equations (16) and (17) to substitute the values MW_n and MF_n in Equation (18); where MW_n refers to the first normalized input, and MF_n refers to the second normalized input.

$$MW_n = 0.037(MF - 64) + 1 \quad (16)$$

$$MF_n = 0.105(MF - 35) + 1 \quad (17)$$

Then, the normalized value PV_n is calculated using Equation (18) with its optimized coefficients listed in Table 6.

$$PV_n = \left[\sum_{i=1}^{N} w_{2,i}\left(\frac{2}{1 + \exp\left(-2\left(MW_n \times w_{1_{i,1}} + MF_n \times w_{1_{i,2}} + b_{1,i}\right)\right)} - 1\right)\right] + b_2 \quad (18)$$

Finally, PV can be predicted by denormalizing PV_n using Equation (19).

$$PV = 22(PV_n + 1) + 10 \quad (19)$$

Table 6. The optimized coefficients for estimating the normalized PV_n in Equation (18).

Neuron Index	Input Layer Weights		Hidden Layer Weights	Input Layer Biases	Output Layer Bias
i	$w1_{i,1}$	$w1_{i,2}$	$w_{2,i}$	$b_{1,i}$	b_2
1	−3.740	4.001	1.983	9.616	−1.831
2	−3.304	−5.296	−1.243	−5.482	-
3	−11.57	−3.788	3.380	7.537	-
4	−6.403	−2.376	−4.833	4.301	-
5	1.308	7.156	−0.307	2.069	-
6	0.457	−12.03	1.182	1.413	-
7	−3.684	10.384	−0.141	3.236	-
8	1.511	−3.887	1.414	−3.601	-
9	−7.490	−1.848	1.788	6.460	-
10	−5.945	−6.028	−0.985	5.412	-
11	2.211	−1.365	−1.087	−1.030	-
12	6.136	−3.409	1.093	3.100	-
13	−0.450	−2.759	0.560	1.993	-
14	15.104	15.336	0.612	1.763	-
15	8.423	2.774	0.501	7.032	-
16	−6.361	−2.459	−0.544	−1.495	-
17	−5.252	5.003	1.075	−1.282	-
18	−4.470	−4.547	1.533	−5.938	-
19	−3.457	6.210	1.761	−2.930	-
20	3.769	−5.543	1.014	5.828	-

3.4. Prediction Power Law Model Parameters (n and k)

Following the same procedure, another two models have been developed using ANN to predict n and k based on MW and MF. For the prediction of n, the R between the measured and the predicted values was 0.98 with $AAPE$ of 2.4% as shown in Figure 6. While for the prediction of k, the R was 0.99 with $AAPE$ of 3.6%, as indicated in Figure 7. Then new correlations for estimating n and k were extracted from the developed ANN models.

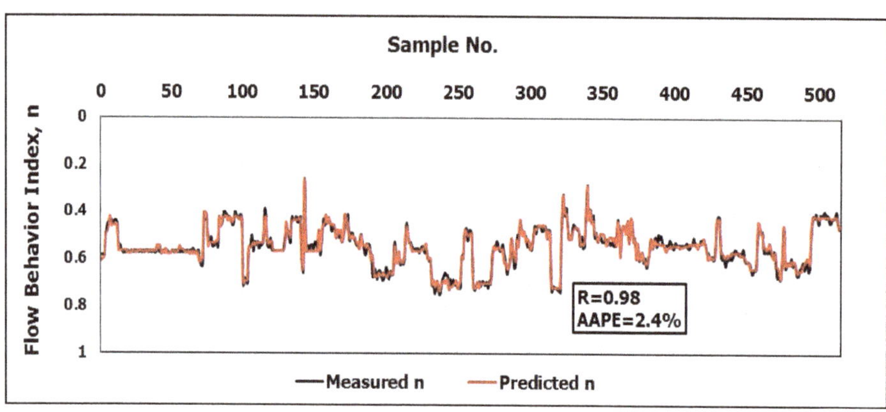

Figure 6. Measured n vs. Predicted n from the ANN model.

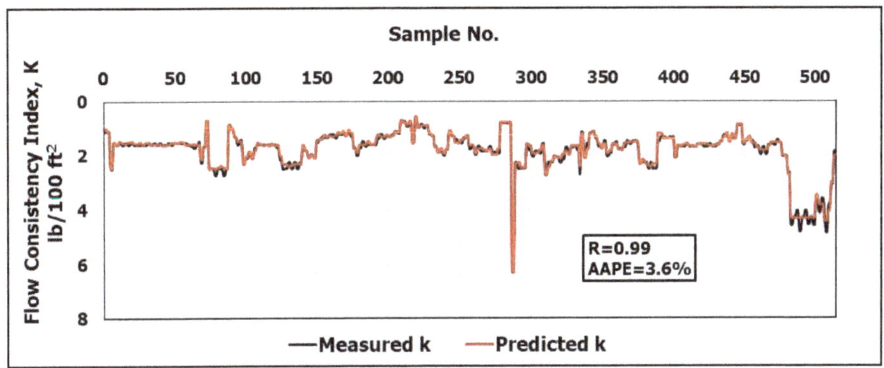

Figure 7. Measured k vs. Predicted k from the ANN model.

In the beginning, the inputs should be normalized using Equations (20) and (21) for the correlation of n and Equations (22) and (23) for the correlation of k in order to substitute the values MW_n and MF_n in Equations (24) and (25); where MW_n refers to the first normalized input and MF_n refers to the second normalized input.

For the Correlation of Parameter n:

$$MW_n = 0.026(MF - 43) + 1 \tag{20}$$

$$MF_n = 0.018(MF - 26) + 1 \tag{21}$$

For the Correlation of Parameter k:

$$MW_n = 0.036(MF - 64) + 1 \tag{22}$$

$$MF_n = 0.022(MF - 30) + 1 \tag{23}$$

Subsequently, the normalized values n_n and k_n can be estimated using Equations (24) and (25), respectively, with their optimized coefficients listed in Tables 7 and 8, respectively.

$$n_n = \left[\sum_{i=1}^{N} w_{2,i}\left(\frac{2}{1+\exp\left(-2\left(MW_n \times w_{1_{i,1}} + MF_n \times w_{1_{i,2}} + b_{1,i}\right)\right)} - 1\right)\right] + b_2 \tag{24}$$

$$k_n = \left[\sum_{i=1}^{N} w_{2,i}\left(\frac{2}{1+\exp\left(-2\left(MW_n \times w_{1_{i,1}} + MF_n \times w_{1_{i,2}} + b_{1,i}\right)\right)} - 1\right)\right] + b_2 \tag{25}$$

Eventually, the predicted values of n and k can be estimated using Equations (26) and (27), respectively.

$$n = 0.244(n_n + 1) + 0.263 \tag{26}$$

$$k = 2.78(k_n + 1) + 0.731 \tag{27}$$

Table 7. The optimized coefficients for estimating the normalized n_n in Equation (24).

Neuron Index	Input Layer Weights		Hidden Layer Weights	Input Layer Biases	Output Layer Bias
i	$w1_{i,1}$	$w1_{i,2}$	$w_{2,i}$	$b_{1,i}$	b_2
1	11.343	−1.862	−0.798	−9.530	−0.867
2	2.093	−7.313	−1.517	−7.827	-
3	−0.197	7.842	−0.937	5.778	-
4	6.518	−3.319	2.163	−5.132	-
5	4.967	1.743	−1.977	−3.641	-
6	−5.643	−4.788	0.789	2.501	-
7	7.452	−2.491	1.194	−1.205	-
8	−8.873	1.168	1.495	0.426	-
9	−3.821	0.042	−8.516	2.781	-
10	−4.896	1.759	6.460	3.734	-
11	−6.355	−8.089	0.160	−0.850	-
12	−12.05	3.917	−1.539	−1.785	-
13	9.180	−0.935	−2.199	2.377	-
14	2.500	−6.815	−2.203	3.462	-
15	−3.105	4.973	−0.752	−3.526	-
16	−4.068	4.687	−1.196	−3.731	-
17	8.037	9.291	0.592	9.212	-
18	6.726	5.979	0.963	3.974	-
19	4.042	−5.058	1.376	5.431	-
20	4.585	−3.606	−0.777	6.701	-

Table 8. The optimized coefficients for estimating the normalized k_n in Equation (25).

Neuron Index	Input Layer Weights		Hidden Layer Weights	Input Layer Biases	Output Layer Bias
i	$w1_{i,1}$	$w1_{i,2}$	$w_{2,i}$	$b_{1,i}$	b_2
1	−6.753	−1.103	1.041	−9.530	−0.106
2	8.434	3.590	−2.501	−7.827	-
3	−5.541	2.870	0.111	5.778	-
4	−2.502	−4.938	−0.160	−5.132	-
5	1.257	−4.551	0.671	−3.641	-
6	−6.886	0.157	0.523	2.501	-
7	2.427	−3.904	1.129	−1.205	-
8	3.711	−4.231	−0.680	0.426	-
9	−4.383	−2.456	−4.228	2.781	-
10	3.781	2.628	0.781	3.734	-
11	4.197	−0.920	−0.658	−0.850	-
12	−5.986	7.171	−0.378	−1.785	-
13	5.429	4.213	−2.285	2.377	-
14	3.700	−7.289	−1.825	3.462	-
15	4.037	3.723	2.991	−3.526	-
16	5.432	2.211	−1.677	−3.731	-
17	7.672	−3.691	5.346	9.212	-
18	−1.757	6.114	2.971	3.974	-
19	2.719	2.906	2.767	5.431	-
20	−7.991	−2.637	1.733	6.701	-

3.5. Validation of the Apparent Viscosity (AV) Model vs the Models in the Literature

As mentioned in the introduction, Pitt [17] introduced a numerical model to calculate the apparent viscosity using Equation (1). After using the collected data, the results obtained using Equation (1) showed a coefficient of determination R^2 of 0.5 and *AAPE* of 32.2%. Also, Almahdawi et al. [18] concluded that Equation (2) using the constant 28 is more appropriate than 25 in Equation (1), and the results obtained by applying Equation (2) to estimate *AV* using *MF* readings give R^2 of 0.5 and *AAPE* of 23.6%, as shown in Figure 8. However, the developed correlation for the ANN model gives highly accurate results, as shown in Figure 9 with R^2 of 0.98, and *AAPE* does not exceed 3.2%.

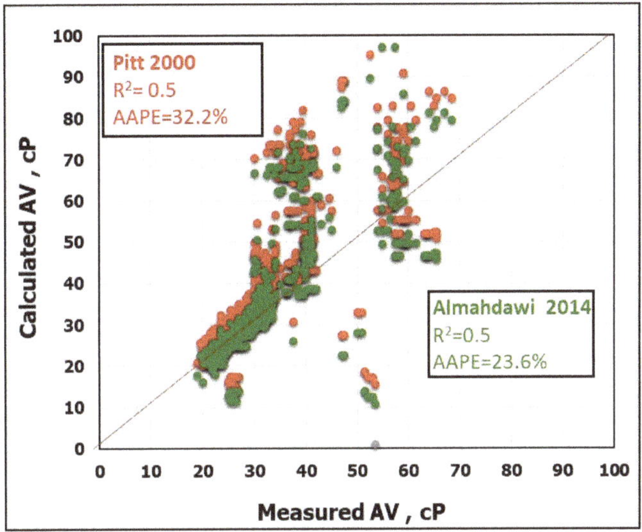

Figure 8. Prediction of *AV* using Pitt's [17] and Almahdawi's et al. [18] correlations.

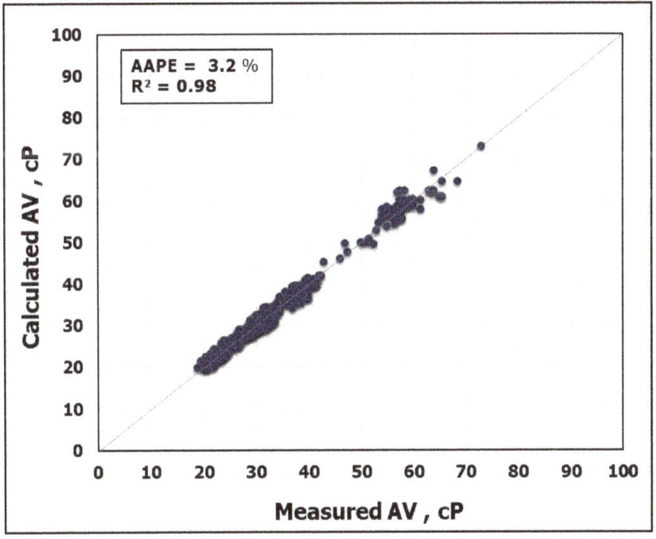

Figure 9. Prediction of *AV* using the developed ANN model (*AAPE* of 3.2%).

4. The Value of Predicting the Drilling Fluid Rheology in Real-Time

For drilling optimization, it is very important to have periodic monitoring of the parameters affecting the drilling process. Mud system design and hole cleaning processes are affected by the pressure losses within the system which rely on the properties of the drilling fluid used, and the efficiency of the cuttings removal from the hole. Pressure losses can be obtained once the parameters of the Bingham model Y_P and *AV* and power low model (n and k) are obtained. Annular pressure losses can be calculated by Equation (28) based on the real-time values of (Y_P and AV), which can be obtained from the developed ANN model.

Also, equivalent circulating density, ECD, can be calculated from Equation (29) [53] using the obtained pressure loss value, so that surge and swab pressures can be determined to help predict critical drilling problems such as pipe sticking and well control issues [54].

$$\Delta p = \left(\frac{PV \times v}{1000(d_2 - d_1)^2} + \frac{Y_p}{200(d_2 - d_1)} \right) L \tag{28}$$

$$ECD = MW + \frac{\Delta p}{0.052 \times h} \tag{29}$$

where Δp is the annular pressure loss (in *psi*), PV is the predicted plastic viscosity (*in cP*), Y_P is the predicted yield point (*in* $Ib/100ft^2$), v is the average annular velocity (in *ft/s*), d_1 is the inside diameter of the hole or casing, (in inches), d_2 is the outside diameter of the drill pipe, (inches), L is the drill pipe, or drill collar length (in *ft*), MW is the mud density (in *ppg*), h is the hole depth (in *ft*), and ECD is the equivalent circulation density (in *ppg*).

Accordingly, the ability of the prediction of the rheological properties in real-time can help avoid many problems during drilling with early detection of these problems by identifying the anomaly in normal behavior trends. This will optimize the drilling operation and save money by minimizing the drilling time.

5. Conclusions

In this work, new models have been developed using ANN to predict the rheological properties of $CaCL_2$ brine-based drill-in fluid in a real-time (15–20 min) including (PV, Y_P, AV, n and k) using 515 field data measurements of MW and MF in ratios 70:30 for training and validating the ANN models respectively. Accordingly, the following conclusions can be drawn:

(1) The new ANN models can predict the rheological parameters $\left(PV,\ Y_p,\ AV,\ n,\ and\ k\right)$ in real time based on MW and MF with high accuracy (R was greater than 0.97 and $AAPE$ was less than 6.1%).
(2) The optimization process for the ANN models showed that the optimized parameters yielding the highest accuracy and the lowest error were 20 neurons for only one hidden layer, the Levenberg-Marquardt algorithm of learning rate 0.12. The activation function linking the input and hidden layers was the tan-sigmoidal function, while a linear function was used for linking the hidden and output layers.
(3) The extracted correlations from the developed ANN models provide the ability to estimate the rheological properties of $CaCL_2$ brine-based mud directly without the need to run the models.
(4) These models are very helpful in the calculations of rig hydraulics, surge and swab pressures, and ECD.
(5) The developed correlations can help in predicting several drilling problems by providing the ability for real-time monitoring of the hole cleaning performance, and detecting any abnormal changes in the normal trends to avoid interrupting problems like sticking. As a result, this will save on the drilling cost, and it optimizes the drilling operation.

Author Contributions: Conceptualization, S.E. and A.G.; methodology, E.R., A.A.; validation, S.E., A.G.; formal analysis, A.G.; investigation, A.A.; resources, S.E.; data curation, A.G.; writing—original draft preparation, S.E. and A.G.; writing—review and editing, A.G., A.A.; visualization, A.A., E.R.; supervision, S.E., A.A.

Funding: This research received no external funding.

Acknowledgments: The authors wish to acknowledge King Fahd University of Petroleum and Minerals (KFUPM) for utilizing the various facilities in carrying out this research. Many thanks are due to the anonymous referees for their detailed and helpful comments.

Conflicts of Interest: The author declares no conflict of interest.

References

1. Caenn, R.; Darley, H.C.H.; Gray, G.R. *Composition and Properties of Drilling and Completion Fluids*, 6th ed.; Gulf Professional Publishing: Waltham, MA, USA, 2011.
2. Knox, D.A.; Bradshaw, R.J.; Svoboda, C.F.; Hodge, R.M.; Wolf, N.O.; Hudson, C.E. Reservoir Drilling Fluids-Designing for Challenging Wells in a Remote Location. In Proceedings of the SPE Annual Technical Conference and Exhibition, Dallas, TX, USA, 9–12 October 2005.
3. Conners, J.H. Use of Clear Brine Completion Fluids as Drill-In Fluids. In Proceedings of the SPE Annual Technical Conference and Exhibition, Las Vegas, NV, USA, 23–26 September 1979.
4. Doty, P.A. Clear Brine Drilling Fluids: A Study of Penetration Rates, Formation Damage, and Wellbore Stability in Full-Scale Drilling Tests. *SPE Drill. Eng.* **1986**, *1*, 17–30. [CrossRef]
5. Ballantine, W.T. Drill-site Cost Savings Through Waste Management. In Proceedings of the SPE 68th Annual Technical Conference and Exhibition of the Society of Petroleum Engineers, Houston, TX, USA, 3–6 October 1993.
6. Redburn, M.; Heath, G. Improved Fluid Characteristics with Clear Calcium Chloride Brine Drilling Fluid. In Proceedings of the Offshore Mediterranean Conference and Exhibition, Ravenna, Italy, 29–31 March 2017.
7. Power, D.; Zamora, M. Drilling fluid yield stress: Measurement techniques for improved understanding of critical drilling fluid parameters. In Proceedings of the AADE Technical Conference, Houston, TX, USA, 1–3 April 2003.
8. Guo, B.; Liu, G. Mud Hydraulics Fundamentals. In *Applied Drilling Circulation Systems*; Gulf Professional Publishing: Houston, TX, USA, 2011; pp. 19–59.
9. Mitchell, R.F.; Miska, S.Z. *Fundamentals of Drilling Engineering*; Society of Petroleum Engineers: Houston, TX, USA, 2011; pp. 87–90.
10. Feng, Y.; Gray, K.E. Review of fundamental studies on lost circulation and wellbore strengthening. *J. Pet. Sci. Eng.* **2017**, *152*, 511–522. [CrossRef]
11. Weikey, Y.; Sinha, S.L.; Dewangan, S.K. Role of additives and elevated temperature on rheology of water-based drilling fluid: A review paper. *Int. J. Fluid Mech. Res.* **2018**, *45*, 37–49. [CrossRef]
12. Paiaman, A.M.; Al-Askari, M.K.G.; Salmani, B.; Al-Anazi, B.D.; Masihi, M. Effect of drilling fluid properties on the rate of penetration. *NAFTA* **2009**, *60*, 129–134.
13. Saasen, A.; Løklingholm, G. The Effect of Drilling Fluid Rheological Properties on Hole Cleaning. In Proceedings of the IADC/SPE Drilling Conference, Dallas, TX, USA, 26–28 February 2002.
14. Robinson, l.; Morgan, M. Effect of hole cleaning on drilling rate performance. In Proceedings of the AADE Drilling Fluid Conference Held at the Radisson, Houston, TX, USA, 6–7 April 2004.
15. Marsh, H. Properties and treatment of rotary mud. *Trans. AIME* **1931**, *92*, 234–251. [CrossRef]
16. Balhoff, M.T.; Lake, L.W.; Bommer, P.M.; Lewis, R.E.; Weber, M.J.; Calderin, J.M. Rheological and yield stress measurements of non-Newtonian fluids using a Marsh funnel. *J. Pet. Sci. Eng.* **2011**, *77*, 393–402. [CrossRef]
17. Pitt, M.J. The Marsh Funnel and Drilling Fluid Viscosity: A New Equation for Field Use. *SPE Drill. Completion* **2000**, *15*, 3–6. [CrossRef]
18. Almahdawi, F.H.; Al-Yaseri, A.Z.; Jasim, N. Apparent viscosity direct from Marsh funnel test. *Iraqi J. Chem. Pet. Eng.* **2014**, *15*, 51–57.
19. Guria, C.; Kumar, R.; Mishra, P. Rheological analysis of drilling fluid using Marsh Funnel. *J. Pet. Sci. Eng.* **2013**, *105*, 62–69. [CrossRef]
20. Liu, K.; Qiu, Z.; Luo, Y.; Zhou, G. Measure rheology of drilling fluids with Marsh funnel viscometer. *Drill. Fluid Completion I Fluid* **2014**, *31*, 60–62. (In Chinese)
21. Schoesser, B.; Thewes, M. Marsh Funnel testing for rheology analysis of bentonite slurries for Slurry Shields. In Proceedings of the ITA WTC 2015 Congress and 41st General Assembl, Dubrovnik, Croatia, 22–28 May 2015.
22. Schoesser, B.; Thewes, M. Do we tap the full potential from Marsh funnel tests? Rheology testing for bentonite suspensions. *Tunn. J.* **2016**. [CrossRef]
23. Abdulrahman, H.A.; Jouda, A.S.; Mohammed, M.M.; Mohammed, M.M.; Elfadil, M.O. Calculation Rheological Properties of Water Base Mud Using Marsh Funnel. Ph.D. Thesis, Sudan University of Science and Technology, Khartoum State, Sudan, 2015.

24. Sedaghat, A.; Omar, M.A.A.; Damrah, S.; Gaith, M. Mathematical modelling of the flow rate in a marsh funnel. *J. Energy Technol. Res.* **2016**, *1*, 1–12. [CrossRef]
25. Sedaghat, A. A novel and robust model for determining rheological properties of Newtonian and non-Newtonian fluids in a marsh funnel. *J. Pet. Sci. Eng.* **2017**, *156*, 896–916. [CrossRef]
26. Agwu, O.E.; Akpabio, J.U.; Alabi, S.B.; Dosunmu, A. Artificial intelligence techniques and their applications in drilling fluid engineering: A review. *J. Pet. Sci. Eng.* **2018**, *167*, 300–315. [CrossRef]
27. Ali, J.K. Neural networks: A new tool for the petroleum industry? In Proceedings of the European Petroleum Computer Conference, Aberdeen, UK, 15–17 March 1994.
28. Oladunni, O.O.; Trafalis, T.B. Single-phase fluid ow classification via learning models. *Int. J. Gen. Syst.* **2011**, *40*, 561–576. [CrossRef]
29. Rostami, H.; Khaksar Manshad, A. A New Support Vector Machine and Artificial Neural Networks for Prediction of Stuck Pipe in Drilling of Oil Fields. *J. Energy Resour. Technol.* **2014**, *136*, 024502. [CrossRef]
30. Wang, Y.; Salehi, S. Application of Real-Time Field Data to Optimize Drilling Hydraulics Using Neural Network Approach. *J. Energy Resour. Technol.* **2015**, *137*, 062903. [CrossRef]
31. Shahdi, A.; Arabloo, M. Application of SVM algorithm for frictional pressure loss calculation of three phase flow in inclined annuli. *J. Pet. Environ. Biotechnol.* **2016**, *5*, 1. [CrossRef]
32. Al-Azani, K.; Elkatatny, S.; Abdulraheem, A.; Mahmoud, M.; Ali, A. Prediction of Cutting Concentration in Horizontal and Deviated Wells Using Support Vector Machine. In Proceedings of the SPE Kingdom of Saudi Arabia Annual Technical Symposium and Exhibition, Dammam, Saudi Arabia, 23–26 April 2018.
33. Elkatatny, S. Application of Artificial Intelligence Techniques to Estimate the Static Poisson's Ratio Based on Wireline Log Data. *J. Energy Resour. Technol.* **2018**, *140*, 072905. [CrossRef]
34. Shahin, M.A.; Jaksa, M.B.; Maier, H.R. Recent advances and future challenges for artificial neural systems in geotechnical engineering applications. *Adv. Artif. Neural Syst.* **2009**, *5*. [CrossRef]
35. Omosebi, A.; Osisanya, S.; Chukwu, G.; Egbon, F. Annular pressure prediction during well control. In Proceedings of the Nigeria Annual International Conference and Exhibition, Lagos, Nigeria, 6–8 August 2012.
36. Nakamoto, P. *Neural Networks and Deep Learning: Deep Learning Explained to Your Granny a Visual Introduction for Beginners Who Want to Make Their Own Deep Learning Neural Network (Machine Learning)*; CreateSpace Independent Publishing Platform: Scotts Valley, CA, USA, 2017.
37. Lippmann, R. An introduction to computing with neural nets. *IEEE ASSP Mag.* **1987**, *4*, 4–22. [CrossRef]
38. Hinton, G.; Osindero, S.; Teh, Y. A fast learning algorithm for deep belief nets. *Neural Comput.* **2006**, *18*, 1527–1554. [CrossRef] [PubMed]
39. Lourakis, M.I. A brief description of the Levenberg-Marquardt algorithm implemented by levmar. *Found. Res. Technol.* **2005**, *4*, 1–6.
40. Jeirani, Z.; Mohebbi, A. Artificial neural networks approach for estimating the filtration properties of drilling fluids. *J. Jpn. Pet. Inst.* **2005**, *49*, 65–70. [CrossRef]
41. Agwu, O.E.; Akpabio, J.U.; Alabi, S.B.; Dosunmu, A. Settling velocity of drill cuttings in drilling fluids: A review of experimental, numerical simulations and artificial intelligence studies. *Powder Technol.* **2018**, *339*, 728–746. [CrossRef]
42. Siruvuri, C.; Nagarakanti, S.; Samuel, R. Stuck pipe prediction and avoidance: A convolutional neural network approach. In Proceedings of the IADC/SPE Drilling Conference, Miami, FL, USA, 21–23 February 2006.
43. Moazzeni, A.R.; Nabaei, M.; Jegarluei, S.G. Prediction of lost circulation using virtual intelligence in one of Iranian oil fields. In Proceedings of the Annual SPE International Conference and Exhibition, Tinapa-Calabar, Nigeria, 31 July–7 August 2010.
44. Rooki, R. Estimation of pressure loss of Herschel Bulkley drilling fl uids during horizontal annulus using artificial neural network. *J. Dispers. Sci. Technol.* **2015**, *36*, 161–169. [CrossRef]
45. Elkatatny, S.; Tariq, Z.; Mahmoud, M. Real-Time Prediction of Drilling Fluid Rheological Properties Using Artificial Neural Networks Visible Mathematical Model (White Box). *J. Pet. Sci. Eng.* **2016**, *146*, 1202–1210. [CrossRef]
46. Elkatatny, S. Real-time prediction of rheological parameters of KCL water-based drilling fluid using artificial neural networks. *Arab. J. Sci. Eng.* **2017**, *42*, 1655–1665. [CrossRef]

47. Abdelgawad, K.; Elkatatny, S.; Moussa, T.; Mahmoud, M.; Patil, S. Real-Time Determination of Rheological Properties of Spud Drilling Fluids Using a Hybrid Artificial Intelligence Technique. *J. Energy Resour. Technol.* **2019**, *141*, 032908. [CrossRef]
48. Parapuram, G.; Mokhtari, M.; Ben Hmida, J. An Artificially Intelligent Technique to Generate Synthetic Geomechanical Well Logs for the Bakken Formation. *Energies* **2018**, *11*, 680. [CrossRef]
49. Elkatatny, S.; Moussa, T.; Abdulraheem, A.; Mahmoud, M. A Self-Adaptive Artificial Intelligence Technique to Predict Oil Pressure Volume Temperature Properties. *Energies* **2018**, *11*, 3490. [CrossRef]
50. Hemphill, T.; Bern, P.; Rojas, J.; Ravi, K. Field validation of drillpipe rotation effects on equivalent circulating density. In Proceedings of the SPE annual technical conference and exhibition, Anaheim, CA, USA, 11–14 November 2007.
51. Dawson, R. How Significant Is A Boxplot Outlier? *J. Stat. Educ.* **2011**, *19*. [CrossRef]
52. Chandrasegar, T.; Vignesh, M.; Balaji, R. Data Analysis Using Box and Whisker Plot for Lung Cancer. In Proceedings of the Innovations in Power and Advanced Computing Technologies (i-PACT) Conference, Vellore, India, 21–22 April 2017.
53. Guo, B.; Liu, G. *Applied Drilling Circulation Systems: Hydraulics, Calculations, and Models*; Gulf Professional Publishing: Houston, TX, USA, 2011.
54. Burkhardt, J.A. Wellbore pressure surges produced by pipe movement. *J. Pet. Technol.* **1961**, *13*, 595–605. [CrossRef]

Article

Estimation of Static Young's Modulus for Sandstone Formation Using Artificial Neural Networks

Ahmed Abdulhamid Mahmoud [1], Salaheldin Elkatatny [1,*], Abdulwahab Ali [2] and Tamer Moussa [1]

1 College of Petroleum Engineering and Geosciences, King Fahd University of Petroleum & Minerals, Dhahran 31261, Saudi Arabia; eng.ahmedmahmoud06@gmail.com (A.A.M.); engineertamer.2007@gmail.com (T.M.)

2 Center of Integrative Petroleum Research, King Fahd University of Petroleum & Minerals, Dhahran 31261, Saudi Arabia; awali@kfupm.edu.sa

* Correspondence: elkatatny@kfupm.edu.sa; Tel.: +966-594-663-692

Received: 4 May 2019; Accepted: 30 May 2019; Published: 3 June 2019

Abstract: In this study, we used artificial neural networks (ANN) to estimate static Young's modulus (E_{static}) for sandstone formation from conventional well logs. ANN design parameters were optimized using the self-adaptive differential evolution optimization algorithm. The ANN model was trained to predict E_{static} from conventional well logs of the bulk density, compressional time, and shear time. The ANN model was trained on 409 data points from one well. The extracted weights and biases of the optimized ANN model was used to develop an empirical relationship for E_{static} estimation based on well logs. This empirical correlation was tested on 183 unseen data points from the same training well and validated using data from three different wells. The optimized ANN model estimated E_{static} for the training dataset with a very low average absolute percentage error (AAPE) of 0.98%, a very high correlation coefficient (R) of 0.999 and a coefficient of determination (R^2) of 0.9978. The developed ANN-based correlation estimated E_{static} for the testing dataset with a very high accuracy as indicated by the low AAPE of 1.46% and a very high R and R^2 of 0.998 and 0.9951, respectively. In addition, the visual comparison of the core-tested and predicted E_{static} of the validation dataset confirmed the high accuracy of the developed ANN-based empirical correlation. The ANN-based correlation overperformed four of the previously developed E_{static} correlations in estimating E_{static} for the validation data, E_{static} for the validation data was predicted with an AAPE of 3.8% by using the ANN-based correlation compared to AAPE's of more than 36.0% for the previously developed correlations.

Keywords: static young's modulus; artificial neural networks; self-adaptive differential evolution algorithm; sandstone reservoirs

---

1. Introduction

Young's modulus is a measure of the sample stiffness against being subjected to a uniaxial load [1]. Static Young's modulus (E_{static}) is an essential parameter required to develop the geomechanical earth model [2] which is required for fracture mapping and designing [3]. A complete description of the in-situ stresses which requires assessment of different petrophysical and mechanical parameters is also needed during the drilling operations to ensure wellbore stability [4]. Several previous studies confirmed the impact of the E_{static} on both fracture design and wellbore stability [1,5].

Lithology is one of the main factors affecting the E_{static}. According to Howard and Fast [6] and Fjaer et al. [1], E_{static} for shale ranges from 0.1–1.0 MPsi; for sandstone it is between 2 and 10 MPsi; and for limestone it is between 8 and 12 MPsi [6]. These ranges confirm the very large variation in

E_{static} in different formations, as well as the wide range for same lithology type. These facts indicate the necessity to estimate E_{static} along the different sections of the drilled well.

Currently, two methods are available for assessing the rocks elastic parameters, these are, namely, (1) laboratory measurements, and (2) through applying empirical correlations. The elastic properties of a rock sample could be measured in the laboratory using either dynamic or static method. The dynamic method involves estimating the modulus from measurements of density, compressional and shear waves velocities while the static method directly measures the deformation in the rock caused by subjecting a sample to uniaxial or triaxial load [7]. In oil and gas fields, the shear and compressional wave velocities measured by the wireline logging [8]. The determined acoustic velocities are then used to calculate the dynamic Young's modulus ($E_{dynamic}$), Equation (1):

$$E_{dynamic} = \frac{\rho V_S^2\left(3V_P^2 - 4V_S^2\right)}{V_P^2 - V_S^2} \tag{1}$$

where ρ represents the bulk formation density in g/cm^3, V_S and V_P denote the shear and compressional wave's velocities in km/s, and $E_{dynamic}$ is the dynamic Young's modulus in GPa.

For the same rock, usually the laboratory-measured $E_{dynamic}$ is significantly greater than E_{static} [9–11]. $E_{dynamic}$ could be 1.5–3 times greater than E_{static} [12,13], and in some cases $E_{dynamic}$ could be up to ten times larger than E_{static} [14–16]. The difference is attributed to the strain amplitude between the two testing techniques, and it decreases with the increase in the strength of the rock [17].

The reservoir in situ stress-strain conditions are truly represented by the static elastic parameters [18], determination of these parameters requires retrieval of real core samples along the reservoir section which is a costly and time-consuming process [13,19]. To minimize the high cost of retrieving the core samples and performing laboratory tests; usually few cores samples are collected from the targeted (reservoir) interval, the laboratory evaluated properties of these core samples are used to develop empirical correlations based on the well log data, to evaluate the required core-derived properties. Dynamic elastic modulus could then be calibrated using these log-based correlations to predict the static modulus throughout the reservoir depths [3]. The applicability of log-derived correlations will be restricted to the formations used to develop these correlations, thus, because of the complexity of the heterogeneous formations, the log-derived correlations will not be able to capture the trend of the static parameters changes. To overcome this limitation different empirical correlations were developed to estimate E_{static} from the $E_{dynamic}$, every correlation is restricted for a specific formation type.

Eissa and Kazi [20] developed a generalized empirical equation to predict the E_{static} as a function of both Edynamic and formation density. The authors developed this correlation (Equation (2)) based on the regression analysis of 76 tests, with data collected from different sources, and they found that considering the formation density improved the predictability of the E_{static} considerably:

$$\log_{10} E_{static} = 0.02 + 0.77 \log_{10}\left(\gamma E_{dynamic}\right) \tag{2}$$

where E_{static} and $E_{dynamic}$ are in Gpa, and γ is the formation density in g/cm^3.

Canady [21] developed another generalized empirical correlation (Equation (3)) which could also be used effectively to estimate the E_{static} for any rock type. This correlation enabled prediction of the E_{static} where only $E_{dynamic}$ is known, the results of the E_{static} predicted with (Equation (3)) was compared to previously available correlations and it found to be well correlated to these models:

$$E_{static} = \frac{\ln\left(E_{dynamic} + 1\right) \times \left(E_{dynamic} - 2\right)}{4.5} \tag{3}$$

where E_{static} and $E_{dynamic}$ are in GPa.

Najibi et al [22] developed another simple correlation (Equation (4)) to evaluate the E_{static} for Sarvak and Asmari limestone based on only the compressional velocity (Vp). This model is very useful when the shear velocity (Vs) is not available:

$$E_{static} = 0.169 \times V_P{}^{3.24} \tag{4}$$

where E_{static} is in Gpa, and Vp is in km/s.

Recently, Fei et al. [23] developed an empirical correlation to predict E_{static} from $E_{dynamic}$ especially for sandstone formation. The developed equation (Equation (5)) is based on the triaxial tests conducted on 22 sandstone core samples:

$$E_{static} = \left(0.564\ E_{dynamic}\right) - 3.4941 \tag{5}$$

E_{static} and $E_{dynamic}$ are in GPa.

Mahmoud et al. [24] developed empirical correlations for E_{static} estimation for different rock types. The developed correlations do not require the knowledge of $E_{dynamic}$ and they directly evaluated E_{static} based on the bulk density, shear, and compressional time data.

It is clear from the literature that obtaining the E_{static} required retrieve cores from specific depth of the well which is costly and time consuming which required to perform the laboratory analysis. In addition, the analysis will be performed for specific well which cannot easily generalized through the entire field while using the developed empirical correlations had their own limitations such as core type, data range and accuracy. The main objective of this study is to develop an ANN model to predict E_{static} from the well logs using a real field data (592 core and log data points) which were collected from the whole sandstone field. Furthermore, a new empirical correlation will be developed for estimating E_{static} for sandstone reservoirs; the correlation is developed based on the extracted weights and biases of the optimized ANN model.

2. Uses of Artificial Intelligence in Estimating Rock Mechanical Parameters

The use of artificial intelligence (AI) techniques in many scientific fields, including the petroleum industry, started in the early 1990s. Since then many publications have treated various areas of petroleum engineering, including the prediction of the bubble point pressure, evaluation of drilling mud, interpretation of the well log data, reservoir characterization, recovery factor estimation, optimization of rate of penetration, and many more.

Recent publications (2016–2018) reported several studies that used AI in estimating rock mechanical parameters. These studies used various AI techniques to: predict failure parameter for carbonates [25]; compare ANN, ANFIS, and SVM in predicting static Poisson's ratio [26]; develop empirical correlation for static Young's modulus [27]; develop an ANN-based correlation to predict sonic transit time [28]; estimation of the unconfined compressive strength (UCS) based on the ANN [29]; and use ANN in estimating Young's modulus, Poisson's ratio, and UCS from log data [30].

3. Methodology

An artificial neural network (ANN) is an artificial intelligence technique developed to enable estimation, classification, identification, decision making by a machine program in various conditions or situations. Different ANN structures are currently available; the simplest ANN structure is called the multi-layered perceptron (MLP) which is used in this study. The MLP consists of a single input layer, one or several hidden layers (mid-layers) and one output layer.

The performance of the ANN depends on many design parameters, such as the training/testing dataset ratio, number of the hidden (training) layers, the number of neurons in each training layer, and the training and transferring functions. The optimization of different combinations of these design parameters requires a long computational time.

Differential evolution (DE) is an accurate, reliable, fast, and robust optimization technique, which has been used to solve effectively different numerical optimization problems. The main limitation for the DE is the need to set the values of the DE control parameters which is problem-dependent, thus, parameter tuning is time-consuming. Omran et al. [31] developed the self-adaptive differential evolution (SaDE) algorithm, which does not require parameter tuning.

In this study, the SaDE optimization algorithm will be used to speed up the optimization process to select the different design parameters of the ANN model to predict E_{static}. A new empirical correlation for estimating E_{static} for sandstone reservoirs will be developed based on the extracted weights and biases of the optimized ANN model.

3.1. Data Preparation

In this study, the ANN model was trained using well log data of bulk density, compressional time, and shear time as inputs to predict the core-derived E_{static} as an output. The input well log data has been selected based on their correlation coefficient with E_{static}, the importance of the input parameters considered in this study in estimating E_{static} is reported by several previous studies. Eissa and Kazi [20] confirmed the ability of improving E_{static} prediction by incorporating the formation density, and the necessity of $E_{dynamic}$, which is dependent on the compressional and shear transit times as reported by several previous studies [21–24].

Data collected in this study are from four wells: 598 data points from Well-A; 34 data points from Well-B; six data points from Well-C; and 11 data points from Well-D. The majority of the data belongs to Well-A, therefore, it will be used to build and test the ANN model which will then be used to develop an ANN-based correlation. The rest of the unseen data which was collected from Well -B, Well-C, and Well-D will be used to validate the developed ANN-based correlation. All the study data are collected from sandstone formations in the Middle East.

Data preparation and preprocessing are the most important steps to ensure a highly accurate prediction of the objective property using any of the AI techniques [32]. As stated earlier the input variables are log-derived, which will be used to predict a core-derived output. Thus, the first step in this study is to perform a depth matching between the core-derived E_{static} and the log data, gamma ray log was considered to perform the data matching. Then, statistical analysis was performed on the input and output parameters to remove data outliers. For the purpose of outlier removal, all parameter values without a range of ±3.0 standard deviation are considered an outlier and not considered to develop the ANN model. Six data points (outliers) from Well-A were removed in this process.

3.2. Training the ANN Model

The 592 data points of Well-A, log data, and their corresponding core-derived E_{static} were considered as valid data to build the ANN model. Sixty-nine percent (409 data points) of Well-A data, were randomly selected to train the ANN model.

Table 1 summarizes the statistical analysis of the training dataset. The analysis shows that the bulk density (ρ_b) for the input dataset ranges from 2.312–2.968 g/cm^3; the compressional time (ΔT_C) ranges between 44.3 and 77.8 µsec/ft; the shear time (ΔT_S) is between 73.2 and 136.1 µsec/ft; and E_{static} ranges from 7.5–92.8 GPa.

The relative importance of the input parameters is shown in Figure 1. The bulk density and compressional time are strong functions on E_{static} with correlation coefficients of 0.724 and −0.815, respectively, while the E_{static} dependence on the shear time is moderate with a correlation coefficient of 0.439.

Design parameters of the ANN model were optimized using the SaDE algorithm, the best combination of design parameters is the one that enables prediction of the E_{static} with the lowest average absolute percentage error (AAPE), as well as highest correlation coefficient (R) and coefficient of determination (R^2). During the optimization process, we evaluated the performance of different training functions such as Levenberg–Marquardt backpropagation (trainlm), gradient descent with momentum backpropagation (traingdm), Gradient descent with adaptive learning rate backpropagation (traingda),

Bayesian regularization backpropagation (trainbr), and conjugate gradient (traincgf); different transfer functions such as tan-sigmoid (tansig), log-sigmoid (logsig), and pure line (purelin); number of hidden layers from 1–3; and the number of neurons per each hidden layer from 5 to 30 on estimating the E_{static}.

Table 1. Statistics of the training dataset.

Statistical Parameter	ρ_b, g/cm^3	ΔT_C, µsec/ft	ΔT_S, µsec/ft	E_{static}, Gpa
Minimum	2.312	44.3	73.2	7.5
Maximum	2.968	77.8	136.1	92.8
Range	0.656	33.4	62.9	85.3
Standard Deviation	0.106	4.69	8.39	13.93
Sample Variance	0.011	22.0	70.3	194.0
Kurtosis	0.569	4.262	1.673	0.167
Skewness	0.011	1.569	0.564	0.186

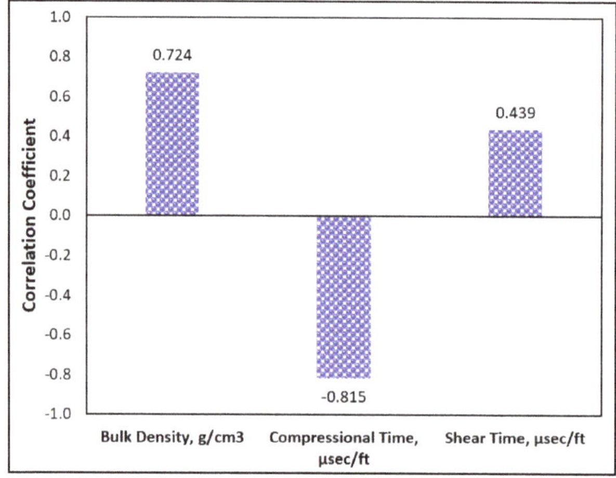

Figure 1. The relative importance of the training dataset input parameter.

The SaDE algorithm was applied using MATLAB software developed by MathWorks (Natick, Massachusetts, U.S.A.) to select the optimum combinations of the ANN design parameters. Based on the optimization process, the combination of the parameters summarized in Table 2 was found to optimize the ANN performance for E_{static} prediction. As listed in Table 2, trainbr is the best training function that optimizes the E_{static} predictability of the ANN model. trainbr is a network training function that updates the weight and bias values of for the ANN model based on Levenberg–Marquardt optimization, and it determines the correct output variable after minimizing a combination of weights and squared errors in a process called Bayesian regularization [33]. logsig is the optimum transfer function. The use of a single hidden (training) function (i.e., a single layer) with 20 neurons also optimized predictability of the ANN model for E_{static}. Figure 2 shows the structure of the suggested ANN model for E_{static} prediction.

Table 2. Combination of the design parameters.

Parameter	Value
Learning function	trainbr
Transfer function	logsig
Number of hidden layers	1
Number of neurons	20

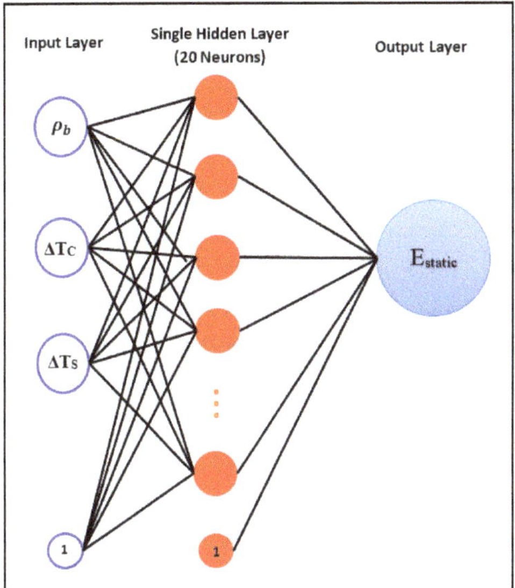

Figure 2. The designed structure for the ANN model with a single hidden layer and 20 neurons.

3.3. Evaluation and Validation of the Developed ANN-Based Empirical Correlation

The remaining 31% of the Well-A dataset, which comprises 183 data points, are considered for evaluating the developed ANN-based empirical correlation. Model validation is important step that is preferably performed on unseen data. The three wells (Well-B: 38 data points, Well-C: six data points, and Well-D: 11 data points) are used in model validation. All testing and validation data are within the range of the training data which used to develop the model to ensure high accuracy in predicting E_{static}. The ability of the developed ANN-based empirical correlation in evaluating the E_{static} for the validation data collected from Well-B will be compared with four of the available correlations, namely, Eissa and Kazi [20], Canady [21], Najibi et al. [22], and Fei et al. [23] correlations are presented earlier by Equations (2)–(5).

3.4. Evaluation Criterion

The predictive power of the developed ANN-based empirical correlation in estimating E_{static} will be evaluated based on the AAPE, R^2, R, and visualization check.

4. Results and Discussion

4.1. Training the ANN Model

The ANN model was trained using 409 randomly selected data points of bulk density, compressional time, and shear time as inputs, and core-derived E_{static} as output. The training data were collected from Well-A. The optimization process was conducted using the SaDE algorithm. The optimized design parameters of ANN model are summarized in Table 2. Figure 3 shows the well log data and their depth corresponding core-derived and predicted E_{static} values for the training dataset. As shown in Figure 3 the ANN model predicted the E_{static} with very high accuracy where the AAPE is only 0.98% and R equals 0.999. A visual check of the plot confirms the excellent matching between the core-derived and the predicted E_{static}.

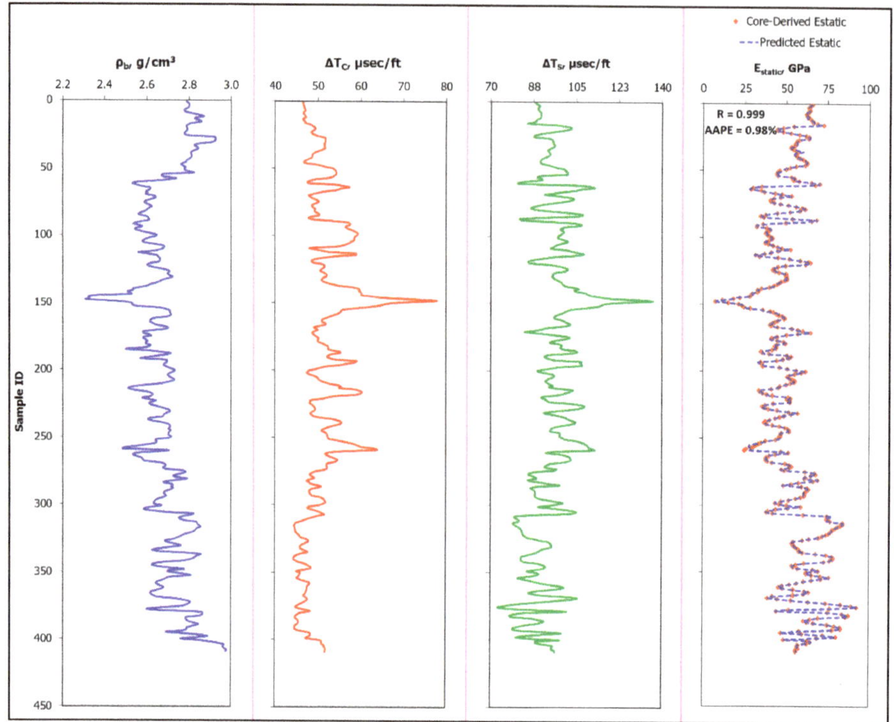

Figure 3. From left to right, bulk density, compressional transit time, shear transit time, and their corresponding predicted and core-derived static Young's modulus values of the training set of Well-A.

Figure 4 presents the cross-plot of the core-derived and the predicted E_{static} of the training dataset. The ANN model is highly accurate in estimating the E_{static} as confirmed by the very high R^2 of 0.9978.

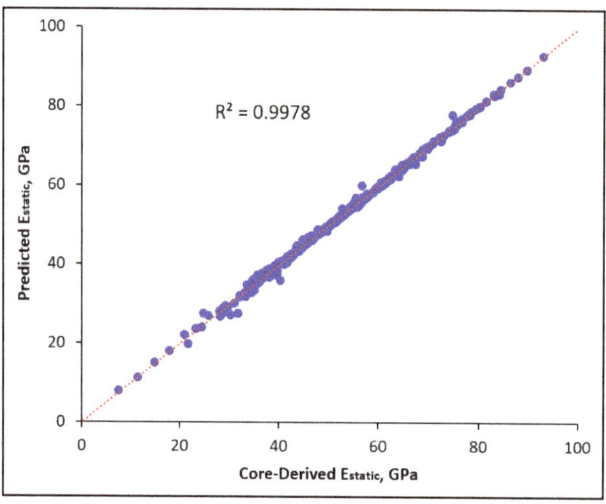

Figure 4. Cross-plot of the core-derived and predicted E_{static} of the training dataset.

4.2. Developing the ANN-Based Empirical Correlation

The proposed ANN-based empirical model is given in Equations (6)–(10):

$$E_{s_n} = \sum_{i=1}^{N} w_{2_i} \frac{1}{1+e^{-(w_{1_{i,1}}\rho_{b_n}+w_{1_{i,2}}\Delta T_{cn}+w_{1_{i,3}}\Delta T_{sn}+b_{1_i})}} + b_2 \quad (6)$$

where E_{sn} is the normalized E_{static}, N represents the number of neurons in the hidden layer ($N = 20$ neurons), i is the index of each neuron in hidden layer, w_{1_i} denotes the weight associated with input and hidden layers for each input parameter, b_{1_i} is the bias associated with hidden and input layers, w_{2_i} represents the weight associated with hidden and output layers, b_2 is the bias associated with hidden and output layers ($b_2 = -1.0767$). All weights and biases associated with the hidden and output layers are summarized in Table 3.

Table 3. The extracted weight and biases of the hidden layer of the optimized ANN model.

i	$w_{1_{i,1}}$	$w_{1_{i,2}}$	$w_{1_{i,3}}$	b_{1_i}	w_{2_i}
1	−4.368	20.303	−14.485	−3.638	5.437
2	−0.216	1.507	2.017	4.178	−2.494
3	1.792	−2.322	−21.160	−3.877	−4.132
4	0.269	0.057	−0.949	1.777	6.325
5	1.498	−19.164	3.620	6.726	9.261
6	13.802	12.907	−0.662	2.170	0.640
7	3.466	8.897	1.549	−3.110	4.174
8	−4.369	−0.142	17.692	1.732	5.813
9	−1.604	21.932	−1.059	−9.030	5.608
10	10.803	−12.301	24.520	16.752	9.748
11	14.298	13.932	0.070	0.895	−5.642
12	−49.173	−25.972	−1.332	15.768	9.929
13	−3.062	15.989	−11.554	−3.115	−4.766
14	−18.124	−17.674	0.206	−1.307	−3.944
15	40.409	23.746	0.690	−14.273	−11.118
16	6.280	3.526	−8.930	2.199	−0.945
17	7.010	3.251	−11.579	1.502	1.544
18	19.888	7.137	0.149	−8.787	−1.596
19	−32.100	−16.426	−1.400	11.426	−23.697
20	−3.053	1.125	19.499	2.853	−9.194

In Equation (6) the values of ρ_{b_n}, ΔT_{c_n}, and ΔT_{s_n} are the normalized input parameters calculated from Equations (7), (8), and (9) respectively:

$$\rho_{b_n} = 2.994(\rho_b - 2.312) - 1 \quad (7)$$

$$\Delta T_{c_n} = 0.0578(\Delta T_c - 44.341) - 1 \quad (8)$$

$$\Delta T_{s_n} = 0.0318(\Delta T_s - 73.187) - 1 \quad (9)$$

The computed value of E_{s_n} in Equation (6) is in the normalized form and should be converted to real value by using Equation (10):

$$E_s = \frac{E_{s_n} + 1}{0.0234} + 7.4987 \quad (10)$$

4.3. Testing the Developed ANN-Based Empirical Correlation

The developed ANN-based empirical correlation, Equations (6)–(10), was tested using the 183 unseen, randomly selected, data points from Well-A. Figure 5 compares the core-derived E_{static} and the estimated E_{static}. As shown in Figure 5, the ANN-based empirical correlation predicted the E_{static} for the unseen data with very high accuracy, where the AAPE is only 1.46% and R is 0.998. A visual check of the plot in Figure 5 confirms the excellent matching between the core-derived and predicted E_{static}.

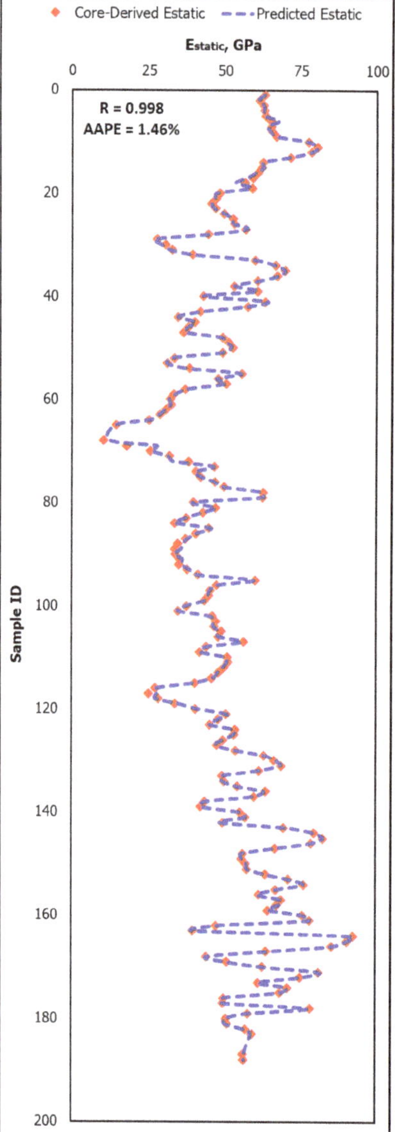

Figure 5. Predicted and core-derived E_{static} of the testing data, Well-A.

Figure 6 presents the cross-plot of the core-derived and the predicted E_{static} of the testing dataset. The ANN-based empirical correlation is highly accurate in estimating the E_{static} as confirmed by the very high R^2 of 0.9951.

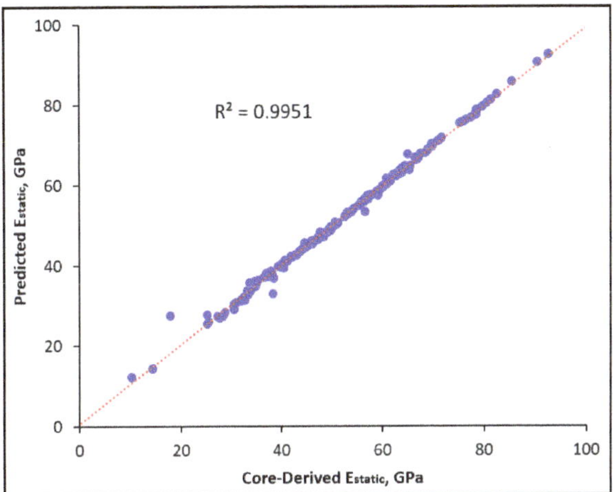

Figure 6. Cross-plot of the core-derived and predicted E_{static} of the testing dataset.

4.4. Validation of the Developed ANN-Based Empirical Correlation

The developed ANN-based empirical correlation (Equations (6)–(10)) was finally verified using 38 data points from Well-B, six data points from Well-C, and 11 data points from Well-D. Figure 7 compares the core-derived E_{static} of Well-B and the predicted E_{static} that developed using ANN-based empirical correlation.

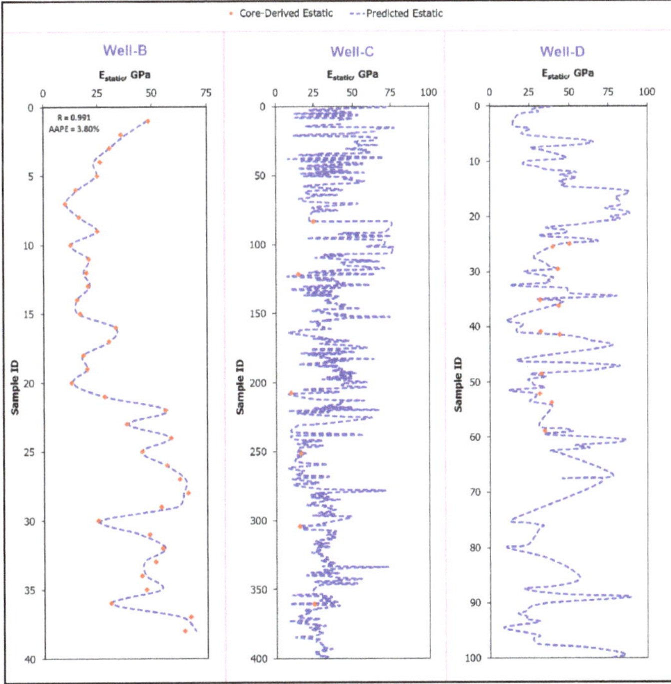

Figure 7. Plot of the predicted and the core-derived E_{static} for the validation datasets collected from Well-B, Well-C, and Well-D.

The plot in Figure 7 confirms the highly accurate predictive accuracy of the ANN-based empirical correlation in estimating the E_{static}; which is validated by the low AAPE of 3.8% and high R of 0.991 in addition to the visual check of the plot. Similarly, the core-derived and the predicted E_{static} values for Well-C and Well-D were plotted in Figure 7. Due to the limited number of data points in Well-C and Well-D, it was enough to check the predictive accuracy of the ANN-based empirical correlation visually. In all three wells used in validation, the ANN-based empirical correlation was able to provide a continuous profile of the predicted E_{static} that conforms to the available core-derived values.

Figure 8 presents the cross-plot of the core-derived and the predicted E_{static} of the 38 data points of Well-B, which are used in validating the ANN-based empirical correlation. This plot affirms the high accuracy of the developed ANN-based empirical correlation in estimating E_{static} as confirmed by the very high R^2 of 0.9816.

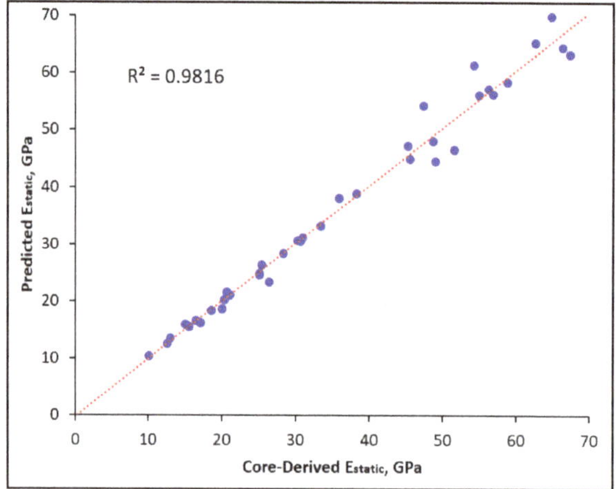

Figure 8. Cross-plot of the core-derived and predicted E_{static} of the validation dataset.

4.5. Comparing the Developed ANN-Based Empirical Correlation to the Available Correlations

Predictive accuracy of the ANN-based empirical correlation was compared with the accuracy of four available developed correlations. Data collected from Well-B was used for the purpose of this comparison. Figure 9 compares predictive accuracy of the ANN-based correlation with the accuracy of Eissa and Kazi [20], Canady [21], Najibi et al. [22], and Fei et al. [23], correlations in estimating E_{static} for the validation dataset of Well-B. Eissa and Kazi [20], Canady [21], Najibi et al. [22], and Fei et al. [23] correlations are presented earlier by Equations (2)–(5). As indicated in Figure 9, the ANN-based correlation overperformed all the four correlations in evaluating = E_{static} with an AAPE of 3.80% compared to an AAPE of 37.2%, 36.3%, 61.7%, and 68.5% for the E_{static} values predicted using Eissa and Kazi [20], Canady [21], Najibi et al. [22], and Fei et al. [23]. These results confirm the high accuracy of the developed ANN-based empirical correlation for E_{static} estimation.

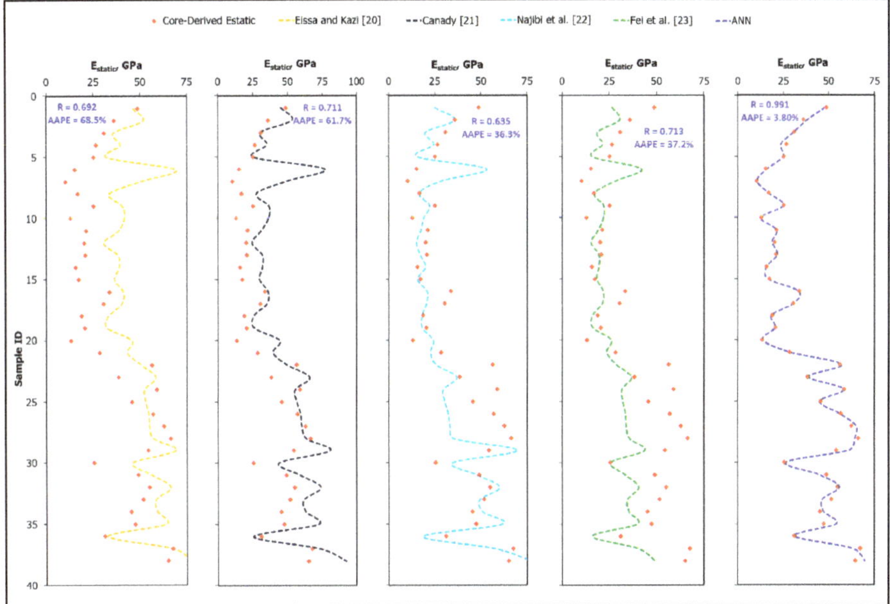

Figure 9. Comparison between predictive accuracy of Eissa and Kazi [20], Canady [21], Najibi et al. [22], Fei et al. [23], and ANN-based correlations in estimating E_{static} for the validation dataset of Well-B.

5. Conclusions

In this study, self-adaptive differential evolution (SaDE) was employed to optimize the ANN design parameters to predict the static Young's modulus (E_{static}) for sandstone formations using well log data of the bulk density, compressional time, and shear time. The ANN model was trained and tested using real field measurements of 592 data points. Based on the results of this study, the following points are concluded:

1. The developed ANN model was capable of estimating E_{static} for the training dataset with very high accuracy, as indicated by the low AAPE of 0.98%, very high R of 0.999, and R^2 of 0.9978. The ANN-based empirical correlation was able to predict E_{static} for the testing dataset (unseen) accurately; the E_{static} values of the testing dataset were estimated with AAPE, R, and R^2 values of 1.46%, 0.998, and 0.9951, respectively.
2. The developed empirical correlation was validated using a dataset composed of unseen 55 data points of three wells. Validating the developed correlation using the dataset of Well-B (38 points) revealed a highly accurate prediction of the developed correlation where AAPE, R and R^2 valueswere 3.8, 0.991 and 0.9816 respectively.
3. The ANN-based empirical correlation is useful in predicting continuous profile of static Young's modulus for sandstone formation using conventional log data, bulk density, shear transit time and compressional transit time when there are no available cores.
4. Comparing the predictive accuracy of the ANN-based correlation with four of the available empirical equations confirmed the high accuracy of the ANN-based correlation which was able to estimate the E_{static} for the validation data of Well-B with an AAPE of 3.8% compared with an AAPE of more than 36.0% for all available correlations.

Author Contributions: Conceptualization, S.E.; Methodology, A.A.M., S.E., and T.M.; Software, T.M.; Validation, A.A.M., S.E., and T.M.; Formal Analysis, A.A.M, S.E., and A.A.; Data Curation, A.A.M. and T.M.; Writing-Original Draft Preparation, A.A.M.; Writing-Review & Editing, S.E. and A.A.; Visualization, A.A.M. and A.A.; Supervision, S.E.

Funding: This research received no external funding.

Conflicts of Interest: The authors declare no conflict of interest.

Nomenclature

AAPE	Average absolute percentage error
ANN	Artificial neural networks
ΔT_C	Compressional transit time
ΔT_S	Shear transit time
E	Young's modulus
E_{static}	Static Young's modulus
$E_{dynamic}$	Dynamic Young's modulus
logsig	Log-sigmoid
R	Correlation coefficient
R^2	Coefficient of determination
ρ_b	Bulk density
SaDE	Self-adaptive differential evolution
trainbr	Bayesian regularization backpropagation

References

1. Fjaer, E.; Horsrud, H.P.; Raaen, A.M.; Risnes, R. *Petroleum Related Rock Mechanics*; Elsevier B. V.: Amsterdam, The Netherlands, 2008.
2. Chang, C.; Zoback, M.D.; Khaksar, A. Empirical relations between rock strength and physical properties in sedimentary rocks. *J. Pet. Sci. Eng.* **2006**, *51*, 223–237. [CrossRef]
3. Gatens, J.M.; Harrison, C.W.; Lancaster, D.E.; Guldry, F.K. In-situ stress tests and acoustic logs determine mechanical properties and stress profiles in the devonian shales. *SPE Form. Eval.* **1990**, *5*, 248–254. [CrossRef]
4. Nes, O.M.; Fjaer, E.; Tronvoll, J.; Kristiansen, T.G.; Horsrud, P. Drilling time reduction through an integrated rock mechanics analysis. *J. Energy Res. Technol.* **2012**, *134*, 2802:1–2802:7. [CrossRef]
5. Meyer, B.R.; Jacot, R.H. Impact of stress-dependent Young's moduli on hydraulic fracture modeling. In Proceedings of the 38th U.S. Symposium on Rock Mechanics, Washington, DC, USA, 7–10 July 2001. ARMA-01-0297.
6. Howard, G.C.; Fast, C.R. Hydraulic Fracturing. In *Doherty Memorial Fund of AIME, Society of Petroleum Engineers of AIME*; Henry L.: New York, NY, USA, 1970; Monograph Volume 2 of SPE.
7. Barree, R.D.; Gilbert, J.V.; Conway, M.W. Stress and rock property profiling for unconventional reservoir stimulation. In Proceedings of the SPE Hydraulic Fracturing Technology Conference, The Woodlands, TX, USA, 19–21 January 2009. SPE-118703-MS.
8. Colin, C.; Potter, S.; Darren, F. Formation elastic parameters by deriving S-wave velocity logs. *CREWES Res.* **1997**, *9*, 1–10.
9. King, M.S. Wave Velocities in Rocks as a Function of Changes in Over burden Pressure and Pore Fluid Saturants. *Geophysics* **1966**, *31*, 50–73. [CrossRef]
10. Rinehart, J.S.; Fortin, J.-P.; Baugin, P. Propagation Velocity of Longitudinal Waves in Rock. Effect of State of Stress, Stress Level of the Wave, Water Content, Porosity, Temperature Stratification and Texture. In Proceedings of the 4th Symposium on Rock Mechanics, University Park, PA, USA, 30 March–1 April 1961. ARMA-61-119.
11. Simmons, G.; Brace, W.I. Comparison of Static and Dynamic Measurements of Compressibility of Rocks. *J. Geophys. Res.* **1965**, *70*, 5649–5656. [CrossRef]
12. Abdulraheem, A.; Ahmed, M.; Vantala, A.; Parvez, T. Prediction of Rock Mechanical Parameters for Hydrocarbon Reservoirs Using Different Artificial Intelligence Techniques. In Proceedings of the Saudi Arabia Section Technical Symposium, Al-Khobar, Saudi Arabia, 9–11 May 2009. SPE-126094-MS.

13. Larsen, I.; Fjær, E.; Renlie, L. Static and Dynamic Poisson's Ratio of Weak Sandstones. In Proceedings of the 4th North American Rock Mechanics Symposium, Seattle, WA, USA, 31 July–3 August 2000. ARMA-2000-0077.
14. Bai, P. Experimental research on rock drillability in the center of junggar basin. *Electron. J. Geotech. Eng.* **2013**, *18*, 5065–5074.
15. Ca, H.; Xi, L.; Guo, L. Rock mechanics study on the safety and efficient extraction for deep moderately inclined medium-thick orebody. *Electron. J. Geotech. Eng.* **2015**, *20*, 11073–11082.
16. Li, P.; Liu, X.; Zhong, Z. Mechanical Property Experiment and Damage Statistical Constitutive Model of Hongze Rock Salt in China. *Electron. J. Geotech. Eng.* **2015**, *20*, 81–94.
17. King, M.S. Static and Dynamic elastic moduli of rocks under pressure. In Proceedings of the 11th U.S. Symposium on Rock Mechanics, Berkeley, CA, USA, 16–19 June 1969. ARMA-69-0329.
18. Wang, Z.; Nur, A.A. Dynamic versus static elastic properties of reservoir rocks. *J. Seism. Acoust. Veloc. Res. Rocks* **2000**, *19*, 531–539.
19. Khaksar, A.; Taylor, P.G.; Fang, Z.; Kayes, T.; Salazar, A.; Rahman, K. Rock strength from core and logs, where we stand and ways to go. In Proceedings of the EUROPEC/EAGE conference and exhibition, Amsterdam, The Netherlands, 8–11 June 2009. SPE-121972-MS.
20. Eissa, E.A.; Kazi, A. Relation between static and dynamic Young's moduli of rocks. *Int. J. Rock Mech. Min. Sci. Geomech. Abstr.* **1988**, *25*, 479–482. [CrossRef]
21. Canady, W.J. A Method for Full-Range Young's Modulus Correction. Presented at the North American Unconventional Gas Conference and Exhibition, The Woodlands, TX, USA, 14–16 June 2011. Paper SPE-143604-MS. [CrossRef]
22. Najibi, A.R.; Ghafoori, M.; Lashkaripour, G.R.; Asef, M.R. Empirical relations between strength and static and dynamic elastic properties of Asmari and Sarvak limestones, two main oil reservoirs in Iran. *J. Pet. Sci. Eng.* **2015**, *126*, 78–82. [CrossRef]
23. Fei, W.; Huiyuan, B.; Jun, Y.; Yonghao, Z. Correlation of Dynamic and Static Elastic Parameters of Rock. *Electron. J. Geotech. Eng.* **2016**, *21*, 1551–1560.
24. Mahmoud, M.A.; Elkatatny, S.A.; Ramadan, E.; Abdulraheem, A. Development of Lithology-Based Static Young's Modulus Correlations from Log Data Based on Data Clustering Technique. *J. Pet. Sci. Eng.* **2016**, *146*, 10–20. [CrossRef]
25. Tariq, A.; Elkatatny, S.A.; Mahmoud, M.A.; Zaki, A.; Abdulraheem, A. A New Approach to Predict Failure Parameters of Carbonate Rocks using Artificial Intelligence Tools. In Proceedings of the SPE Kingdom of Saudi Arabia Annual Technical Symposium and Exhibition, Dammam, Saudi Arabia, 24–27 April 2017. SPE-187974-MS.
26. Elkatatny, S.M.; Tariq, Z.; Mahmoud, M.A.; Abdulraheem, A.; Abdelwahab, A.Z.; Woldeamanuel, M. An Artificial Intelligent Approach to Predict Static Poisson's Ratio. In Proceedings of the 51st US Rock Mechanics/Geomechanics Symposium, San Francisco, CA, USA, 25–28 June 2017. ARMA 17-771.
27. Tariq, Z.; Elkatatny, S.M.; Mahmoud, M.A.; Abdulazeez, A. A Holistic Approach to Develop New Rigorous Empirical Correlation for Static Young's Modulus. In Proceedings of the Abu Dhabi International Petroleum Exhibition & Conference, Abu Dhabi, UAE, 7–10 November 2016. SPE-183545-MS.
28. Tariq, Z.; Elkatatny, S.M.; Mahmoud, M.A.; Abdulazeez, A. A New Artificial Intelligence Based Empirical Correlation to Predict Sonic Travel Time. In Proceedings of the International Petroleum Technology Conference, Bangkok, Thailand, 14–16 November 2016. IPTC-19005-MS.
29. Tariq, Z.; Elkatatny, S.M.; Mahmoud, M.A.; Abdulraheem, A.; Abdelwahab, A.Z.; Woldeamanuel, I.M. Development of New Correlation for Unconfined Compressive Strength for Carbonate Reservoir Using Artificial Intelligence Techniques. In Proceedings of the 51st US Rock Mechanics/Geomechanics Symposium, San Francisco, CA, USA, 25–28 June 2017. ARMA 17-428.
30. Tariq, Z.; Elkatatny, S.M.; Mahmoud, M.A.; Abdulraheem, A.; Abdelwahab, A.Z.; Woldeamanuel, M. Estimation of Rock Mechanical Parameters using Artificial Intelligence Tools. In Proceedings of the 51st US Rock Mechanics/Geomechanics Symposium, San Francisco, CA, USA, 25–28 June 2017. ARMA 17-301.
31. Omran, M.G.H.; Salman, A.; Engelbrecht, A.P. Self-adaptive Differential Evolution. In *Computational Intelligence and Security*; Hao, Y., Liu, J., Wang, Y., Cheung, Y.-m., Yin, H., Jiao, L., Ma, J., Jiao, Y.-C., Eds.; CIS 2005. Lecture Notes in Computer Science; Springer: Berlin/Heidelberg, Germany, 2005; Volume 3801.

32. Al-Anazi, A.F.; Gates, I.D. A support vector machine algorithm to classify lithofacies and model permeability in heterogeneous reservoirs. *Eng. Geol.* **2010**, *114*, 267–277. [CrossRef]
33. MathWorks. Available online: https://www.mathworks.com/help/deeplearning/ref/trainbr.html;jsessionid=7cc70c77fdb3f0bb58ed870c69c7 (accessed on 1 April 2019).

MDPI
St. Alban-Anlage 66
4052 Basel
Switzerland
Tel. +41 61 683 77 34
Fax +41 61 302 89 18
www.mdpi.com

Energies Editorial Office
E-mail: energies@mdpi.com
www.mdpi.com/journal/energies

www.ingramcontent.com/pod-product-compliance
Lightning Source LLC
LaVergne TN
LVHW070126170726
843515LV00007B/1755

* 9 7 8 3 0 3 9 2 8 8 8 9 2 *